Buildings and Semantics

The built environment has been digitizing rapidly and is now transforming into a physical world that is at all times supplemented by a fully web-supported and interconnected digital version, often referred to as Digital Twin. This book shows how diverse data models and web technologies can be created and used for the built environment. Key features of this book are its technical nature and technical detail. The first part of the book highlights a large diversity of IT techniques and their use in the AEC domain, from JSON to XML to EXPRESS to RDF/OWL, for modelling geometry, products, properties, sensor and energy data. The second part of the book focuses on diverse software solutions and approaches, including digital twins, federated data storage on the web, IoT, cloud computing, and smart cities. Key research and strategic development opportunities are comprehensively discussed for distributed web-based building data management, IoT integration and cloud computing. This book aims to serve as a guide and reference for experts and professionals in AEC computing and digital construction including Master's students, PhD researchers, and junior to senior IT-oriented AEC professionals.

Buildings and Semantics

Data Models and Web Technologies for the
Built Environment

Edited by

Pieter Pauwels
Department of the Built Environment, Eindhoven University of Technology,
Eindhoven, The Netherlands

Kris McGlinn
ADAPT Centre, School of Computer Science and Statistics,
Trinity College Dublin, Dublin, Ireland

CRC Press
Taylor & Francis Group
Boca Raton London New York Leiden

CRC Press is an imprint of the
Taylor & Francis Group, an **informa** business

A BALKEMA BOOK

Cover image: Selahattin Dülger

First published 2023
by CRC Press/Balkema
Schipholweg 107C, 2316 XC Leiden, The Netherlands
e-mail: enquiries@taylorandfrancis.com
www.routledge.com – www.taylorandfrancis.com

CRC Press/Balkema is an imprint of the Taylor & Francis Group, an informa business

Library of Congress Cataloging-in-Publication Data
A catalog record has been requested for this book

ISBN: 978-1-032-02312-0 (hbk)
ISBN: 978-1-032-06897-8 (pbk)
ISBN: 978-1-003-20438-1 (ebk)

DOI: 10.1201/9781003204381

Typeset in Times New Roman
by codeMantra

Contents

Figures

Tables

About the authors

Pieter Pauwels works as an Associate Professor at the Eindhoven University of Technology (TUe), the Department of the Built Environment. He previously worked at the Department of Architecture and Urban Planning at Ghent University (2008–2019). His work and interests are in information system support for the building life cycle (architectural design, construction, building operation). With a lot of experience and knowledge in computer science and software development, he is involved in a number of industry-oriented research projects on topics affiliated to AI in construction, design thinking, Building Information Modelling (BIM), Linked Building Data (LBD), Linked Data in Architecture and Construction (LDAC), and Semantic Web technologies.

Kris McGlinn, Ph.D., is a Research Fellow and Computer Scientist in the ADAPT Centre, Trinity College Dublin. His research focus is knowledge engineering, in particular the application of Web of Data technologies for managing data from heterogeneous data sources. He has extensive experience working within the building domain, having worked on several EU and Irish national projects which explored topics ranging from smart building application development and evaluation, energy management in buildings, and the integration of building data with geospatial data. He was Principal Investigator for the H2020 SWIMing project, a Coordination and Support Action which explored the use of Semantic Web technologies for information modelling across EU projects and industry, and he was a Founding Chair of the W3C Linked Building Data community group, with the stated goal of developing ontologies for managing building data.

Contributors

Rubén Alonso
R2M Solution
Pavia, Italy

Jakob Beetz
Faculty of Architecture
    RWTH Aachen University
Aachen, Germany

Calin Boje
Luxembourg Institute of Science and
    Technology (LIST)
Luxembourg

Mathias Bonduel
Department of Civil Engineering
    KU Leuven
Ghent, Belgium

Jan Brouwer
DigiBase
    VolkerWessels
Eindhoven, Netherlands

Ozum Calli
TUBITAK-BIDEB
Ankara, Turkey

Eva Coscia
R2M Solution
Pavia, Italy

Gonçal Costa
Department of Architecture
    La Salle - Ramon Llull University
Barcelona, Spain

Aaron Costin
M.E. Rinker Sr., School of Construction
    Management
    University of Florida
Gainesville, Florida

Mohamed Elagiry
R2M Solution
Pavia, Italy

Tamer El-Diraby
Department of Civil & Mineral
    Engineering
    University of Toronto
Toronto, Canada

Annie Guerriero
Luxembourg Institute of Science and
    Technology (LIST)
Luxembourg

Ralf Klein
Department of Civil Engineering
    KU Leuven
Ghent, Belgium

Sylvain Kubicki
Luxembourg Institute of Science and
    Technology (LIST)
Luxembourg

Tim Pat McGinley
Department of Civil Engineering
    Technical University of Denmark
Lyngby, Denmark

Janise McNair
Department of Electrical & Computer
  Engineering
University of Florida
Gainesville, Florida

Saha Nirvik
School of Architecture
  Rensselaer Polytechnic Institute
Troy, New York

Jeffrey W. Ouellette
Independent BIM consultant
Austin, Texas

Mads Holten Rasmussen
NIRAS
Copenhagen, Denmark

Diego Reforgiato Recupero
University of Cagliari
Cagliari, Italy

Yacine Rezgui
School of Engineering
  Cardiff University
Cardiff, United Kingdom

Madhumitha Senthilvel
Faculty of Architecture
  RWTH Aachen University
Achen, Germany

Dennis Shelden
School of Architecture
  Rensselaer Polytechnic Institute
Troy, New York

Álvaro Sicilia
Department of Architecture
  La Salle - Ramon Llull University
Barcelona, Spain

Soroush Sobhkhiz
Department of Civil & Mineral
  Engineering
  University of Toronto
Toronto, Canada

Devon Sparks
Trimble Consulting
Portland, Oregon

Anna Wagner
PROSTEP AG
Darmstadt, Germany

Jeroen Werbrouck
Department of Architecture and Urban
  Planning
  Ghent University
Ghent, Belgium

Hendro Wicaksono
Department of Industrial Engineering
  and Management
  Jacobs University Bremen
Bremen, Germany

Baris Yuce
College of Engineering, Mathematics and
  Physical Sciences
  University of Exeter
Exeter, United Kingdom

Alain Zarli
R2M Solution
Roquefort les Pins, France

Foreword

The built environment digitizes rapidly in all possible ways: software, data management, hardware, processes, and so forth. The diversity in data and semantics needs good management, and new developments and research show how this can be achieved. This work aims to give an overview of the way in which web technologies are used in the full life cycle of the built environment. It aims to include a full state-of-the-art review of all sorts of web technologies used for a variety of use cases and objectives. As such, this book aims to be at the forefront of research and a reference work for forthcoming research and innovation in the Architecture, Engineering, and Construction (AEC) domain.

At the moment, the AEC industry relies very much on local software solutions and traditional information exchange practices. Stakeholders, including architects, engineers, contractors, and manufacturers, keep resorting to their own silos of information, which are protected by firm protective legal boundaries. Under the influence of Building Information Modelling tools and methods, the industry has gone through a major digital transformation process. Yet, many processes still occur in a very traditional manner. A next step in the digital transformation of the construction industry lies into the full adoption of web technologies and distributed data management. In many ways, a transformation is needed towards the reliance of distributed data management tools, especially in an industry that is heavily networked by default, yet the more standard legal and ownership boundaries need to be preserved. Therefore, the book will primarily focus on technology, aiming to support innovation and progress in that regard.

This book brings together several chapters on the use of web technologies (web services, linked data, semantic web, IoT, cloud-based algorithms) for improving the built environment. The book will not focus on the specific technologies only. It will also focus on key topical areas of relevance in the built environment. Many applications and research initiatives namely use several techniques and data storage mechanisms in combination. Throughout the chapters, a diversity of applications and initiatives is collected to give an indication of what currently can be done with web technologies in the built environment, including construction sites, engineering, architecture, facility management, infrastructure works, geospatial platforms, urban studies, and demolition or circularity. As such, a wide overview is given of potential works and innovation in the life cycle of the built environment.

The key benefit of this book will be to provide a state-of-the-art review of current practices, challenges, barriers, and shortcomings in the management and use of life cycle building data, and how web technologies can be used to address these. As such, this book can be used as a handbook of reference towards future research and innovation, both in academia and industry.

A lot of effort has been put into the publication of this book, by many people. With this book, we aim to advance and further innovate our AEC industry. This book aims to do this by documenting the technical core of our digital technologies. It is not an easy task to write down technical content in sufficient and reliable detail, especially also as it changes over time; and several different viewpoints exist on the topics and on how to move forward. Therefore, this book aims to document a multitude of technologies and perspectives. We hope that it can be an inspiration and source of information and knowledge for the engineers and bright minds of tomorrow.

Pieter Pauwels
Kris McGlinn

Preface

Building data is essential to the planning and design, construction, operation and maintenance, and destruction/recycling of buildings. At its most basic form, building data can simply include the address or geolocation of a home required to deliver mail and goods, or a 2D floor plan of a home used when advertising to buy or sell. When designing a new building, building data can include a 3D solid model of the building, describing the geometry and semantics of each product (walls, windows, doors, columns), Heating, Ventilation and Air Conditioning (HVAC), the materials used, plumbing and electrical wiring in the building, space boundaries, and the relationships between all these. All this data is required to support the wide range of stakeholders involved along the building's life cycle.

Web technologies can not only help streamline existing processes like these, but by making data easier to access and reuse they also open the door to improved integration of data with non-traditional Building Information Modelling (BIM) domains, such as geospatial and energy grid data, making buildings become a living, evolving part of an integrated set of smart environments ranging from scales of rooms, offices and homes, to districts, cities and nations. Today, the construction industry relies only shallowly on web technologies (websites, web platforms, web services). In its digital transformation, it needs to move to the web and link its distributed silos of information.

This book gives the reader a clear understanding of the current status regarding the use of web technologies in the built environment. The importance of web technologies to address some of the key identified challenges in the built environment will be highlighted, with a critical assessment of their limitations and further discussion on the challenges and barriers to their implementation. The reader will be introduced to several chapters which explore in detail how web technologies are used across the built environment, their successes and failures. The book is subdivided in two parts, focusing on (1) semantics and data, and (2) applications and algorithms. The combination of all this material gives readers a full oversight of all potential in the use of web technologies in the construction industry. As such, the book will be open-ended, so that it can be an open cook book and guideline for further research and innovation in this industry. The editor would also like to acknowledge the ADAPT centre, for their continuing support.

Acronyms

6LoWPAN:	IPv6 over Low-power Wireless Personal Area Networks
ABox:	Assertion Box
ACL:	Access Control List
AAT:	Arts and Architecture Thesaurus
AEC:	Architecture, Engineering and Construction
AECO:	Architecture, Engineering, Construction and Operations
AI:	Artificial Intelligence
AJAX:	Asynchronous JavaScript and XML
AMQP:	Advanced Message Queuing Protocol
ANN:	Artificial Neural Network
AP:	Application Profile
API:	Application Programming Interface
AR:	Augmented Reality
ASP:	Application Service Provider
B2B:	Business to Business
B2C:	Business to Consumer
BACnet:	Building Automation and Control Networking Protocol
BACS:	Building Automation and Control System
BAS:	Building Automation System
BCF:	BIM Collaboration Format
BDS:	Building Description System
BDTP:	Building Digital Twin Prototype
BEM:	Building Energy Model
BEMS:	Building Energy Management System
BEO:	Building Element Ontology
BEP:	Building Energy Performance
BIM:	Building Information Modelling
BIMDO:	Building Information Model Design Ontology
BIMSO:	Building Information Model Shared Ontology

BIPV:	Building-Integrated Photovoltaics
BLC:	Building Life Cycle
BMS:	Building Management System
BOT:	Building Topology Ontology
BoQ:	Bills of Quantity
BPMN:	Business Process Model Notation
BPO:	Building Product Ontology
BREEAM:	Building Research Establishment Environmental Assessment Method
BREP:	Boundary Representation
bSDD:	buildingSmart Data Dictionary
CAD:	Computer-Aided Design
CAM:	Computer-Aided Manufacturing
CCTV:	Closed-Circuit Television
CDE:	Common Data Environment
CDT:	Custom Datatypes
CEDR:	Conference of European Directors of Roads
CFD:	Computational Fluid Dynamics
CG:	Community Group
CLI:	Command Line Interface
CNC:	Computer Numerical Control
CoAP:	Constrained Application Protocol
COBie:	Construction Operations Building Information Exchange
CPS:	Cyber-Physical System
CRUD:	Create, Read, Update, Delete
CRS:	Coordinate Reference System
CSG:	Constructive Solid Geometry
CSO:	Central Statistics Office
CSS:	Chirp Spread Spectrum
CSV:	Comma Separated Values
DABGEO:	Domain Analysis-Based Global Energy Ontology
DCAT:	Data Catalog Vocabulary
DCT:	Dublin Core Terms
DDC:	Direct Digital Control
DDL:	Data Definition Language
DDM:	District Data Model
DIM:	District Information Model
DL:	Description Logic
DMS:	Document Management System
DNN:	Deep Neural Network

DNS:	Domain Name System
DNS-SD:	Domain Name System Service Discovery
DOT:	Damage Topology Ontology
DSS:	Decision Support System
DT:	Digital Twin
DTC:	Digital Twin Consortium
DTDL:	Digital Twin Definition Language
EAN:	European Article Number
EMRS:	Enterprise Management and Reporting System
EPBD:	Energy Performance of Buildings Directive
EPC:	Energy Performance Certification
EPC:	Electronic Product Code
EPW:	EnergyPlus Weather
ER:	Exchange Requirement
ERD:	Entity Relationship Diagram
ERP:	Enterprise Resource Planning
ESD:	Environmentally Sustainable Design
EU:	European Union
EUROTL:	European Road Object Type Library
EV:	Electric Vehicle
FD:	Field Device
FEM:	Finite Element Model
FM:	Facility Management
FMIS:	Facility Management Information System
FOG:	File Ontology for Geometry Formats
FOL:	First Order Logic
GA:	Genetic Algorithm
GBC:	Green Building Certification
gbXML:	Green Building XML
GIS:	Geographical Information System
glTF:	Graphics Language Transmission Format
GML:	Geography Markup Language
GOM:	Geometry Metadata Ontology
GPS:	Geographical Positioning System
GUI:	Graphical User Interface
GUID:	Globally Unique Identifier
HCI:	Human Computer Interaction
HMM:	Hidden Markov Model
HTML:	HyperText Markup Language

HTTP:	HyperText Transfer Protocol
HVAC:	Heating, Ventilation and Air Conditioning
I/O:	Input/Output
IaaS:	Infrastructure as a Service
IAI:	International Alliance for Interoperability
IBDT:	Intelligent Buildings Digital Twin
IC:	Industrialized Construction
ICDD:	Information Container for linked Document Delivery
ICT:	Information and Communication Technology
ID:	Identification
IDM:	Information Delivery Manual
IEEE:	Institute of Electrical and Electronics Engineers
IEA:	International Energy Agency
IFC:	Industry Foundation Classes
ifcOWL:	Industry Foundation Classes in the Web Ontology Language
ifcWoD:	Industry Foundation Classes Web of Data
IFD:	International Framework for Dictionaries
IGES:	Initial Graphics Exchange Specification
IMS:	Issue Management System
IoT:	Internet of Things
IP:	Internet Protocol
IPR:	Intellectual Property Rights
IPv#	Internet Protocol version number
ISO:	International Organisation for Standardisation
IT:	Information Technology
IIC:	Industrial IoT Consortium
JSON:	JavaScript Object Notation
JSON-LD:	JavaScript Object Notation - Linked Data
KIF:	Knowledge Interchange Format
KOS:	Knowledge Organization System
KR:	Knowledge Representation
LAN:	Local Area Network
LBD:	Linked Building Data
LCA:	Life Cycle Assessment
LDP:	Linked Data Platform
LEED:	Leadership in Energy and Environmental Design
LLP:	Low-Power and Lossy Network
LoD:	Level of Development
LOD:	Linked Open Data

LOV:	Linked Open Vocabularies
LoRaWAN:	Long Range Wide Area Network
LPWAN:	Low-Power Wide Area Network
LTE:	Long-Term Evolution
M2M:	Machine-to-Machine
MAC:	Media Access Control
mDNS:	Multicast Domain Name System
MAPE:	Monitor-Analyze-Plan-Execute
MEP:	Mechanical, Electrical and Plumbing
MKr:	Multiple Kernel Regression
ML:	Machine Learning
MMC:	Multi-Model Container
MMS:	Model Management System
MQTT:	Message Queuing Telemetry Transport
MRA:	Multiple Regression Analysis
MVD:	Model View Definition
NIST:	National Institute of Standards and Technology
NoSQL:	Not Only SQL
NRA:	National Road Authority
NTA:	Dutch Technical Agreement
NURBS:	Non-uniform Rational Basis Spline
O&M	Operations and Maintenance
OASIS:	Organization for the Advancement of Structured Information Standards
OCC:	OpenCascade
OCCS:	OmniClass Construction Classification System
ODBC:	Open Database Connectivity
OGC:	Open Geospatial Consortium
OGD:	Open Goverment Data
OIDC:	OpenID Connect
OIL:	Ontology Inference Layer
OM:	Ontology of units of Measure
O-MI:	Open Messaging Interface
O-DF:	Open Data Format
OMG:	Ontology for Managing Geometry
OMG:	Object Management Group
ONS:	Object Name System
OPC:	Open Platform Communications
OPC UA:	Open Platform Communications Universal Access

OPM:	Ontology for Property Management
OSC:	OpenSensingCity
OS:	Operating System
OSi:	Ordnance Survey Ireland
OTL:	Object Type Library
OWL:	Web Ontology Language
PaaS:	Platform as a Service
PAS:	Publicly Available Specification
PCA:	Principal Component Analysis
PDF:	Portable Document Format
PHY:	Physical Layer
PIR:	Passive Infrared
PLM:	Product Lifecycle Management
PM:	Process Map
PNG:	Portable Graphics Format
POD:	Personal Online Data storage
PROV:	Provenance
PSD:	Property Set Definition
PSet:	Property Set
PT:	Physical Twin
PV:	Photovoltaic
QoS:	Quality of Service
QUDT:	Quantities, Units, Dimensions, and Types Ontology
RCT:	Randomized Controlled Trial
RDB:	Relational Database
RDF:	Resource Description Framework
RDFS:	Resource Description Framework Schema
REC:	RealEstateCore
REST:	REpresentational State Transfer
RFID:	Radio Frequency Identification
RH:	Relative Humidity
RIF:	Rule Interchange Format
RML:	RDF Mapping Language
RPL:	Routing Protocol for LPNs
SaaS:	Software as a Service
SAREF:	Smart Appliances Reference Ontology
SDCSN:	Software-Defined Clustered Sensor Network
SDK:	Software Development Kit
SDN:	Software Defined Network

SEAS:	Smart Energy Aware Systems
SenML:	Sensor Markup Language
SFA:	Simple Feature Access
SHACL:	Shapes Constraint Language
SimModel:	Simulation Domain Model
SKOS:	Simple Knowledge Organisation System
SME:	Small- or Medium-sized Enterprise
SLA:	Service Level Agreement
SOA:	Service-Oriented Architecture
SOAP:	Simple Object Access Protocol
SOSA:	Sensor Observation Sample and Actuator
SPARQL:	SPARQL Protocol and RDF Query Language
SPF:	STEP Physical File
SPFF:	STEP Physical File Format
SQL:	Structured Query Language
SSN:	Semantic Sensor Network
SSOT:	Single Source of Truth
STEP:	Standard for the Exchange of Product Data
STL:	Standard Triangle Language
SVR:	Support Vector Regression
SWRL:	Semantic Web Rule Language
TBox:	Terminology Box
TCP:	Transmission Control Protocol
TSDB:	Time Series Database
UAV:	Unmanned Aerial Vehicle
UCUM:	Unified Code for Units of Measure
UDP:	User Datagram Protocol
UI:	User Interface
UML:	Unified Modelling Language
UPC:	Universal Product Code
URI:	Uniform Resource Identifier
URL:	Uniform Resource Locator
URN:	Uniform Resource Name
UTF:	Unicode Transformation Format
UX:	User Experience
VMM:	Virtual Machine Monitor
VR:	Virtual Reality
VRML:	Virtual Reality Modelling Language
W3C:	World Wide Web Consortium

WAC:	Web Access Control
WAN:	Wide Area Network
WGS:	World Geodetic System
WKT:	Well-Known Text
WKB:	Well-Known Binary
WLAN:	Wireless Local Area Network
WPAN:	Wireless Personal Area Network
WSDL:	Web Services Description Language
WSN:	Wireless Sensor Network
WWW:	World Wide Web
XML:	Extensible Markup Language
XSD:	XML Schema Definition Language
YAGO:	Yet Another Great Ontology

Part I

Semantics and data

Chapter 1

Building product models, terminologies, and object type libraries

Aaron Costin, Jeffrey W. Ouellette, and Jakob Beetz

CONTENTS

DOI: 10.1201/9781003204381-2

The built environment consists of a wide variety of digital representations of physical products and building elements. These product models have enabled new and efficient digitised methods for the architecture, engineering, construction, and operations (AECO) industry to design, construct, and operate a facility. As a result, product modelling is a key competence for anyone operating with assets in the built environment. In this first chapter of the book, we provide an overview of the way in which product modelling is performed in the AECO industry. Importantly, we hereby distinguish between product modelling in the AECO industry for (1) the design and construction of buildings, which typically needs to support continuous updates between a wide variety of stakeholders; and (2) the operational phase of buildings, in which product data is needed for operations and maintenance (O&M) over long spans of time (facility and asset management). We present the current concepts, methods, and tools required to understand the rest of the chapters of this book.

1.1   INTRODUCTION

The architecture, engineering, construction, and operations (AECO) industry has seen great advancements in technology, both in hardware and software, used in various tasks required during the life cycle of a project, from design to delivery and through its operational life span. Artificial intelligence (AI) and machine learning (ML), virtual and augmented realities (VR & AR), and computational design methodologies are striving to make the delivery faster and safer, as well as more economically and environmentally sustainable. Primary methods and technologies include Computer-Aided Design (CAD) or Computer-Aided Design and Drafting (CADD) and Building Information Modelling (BIM), which enable a wide variety of digital representations of physical products and building elements, sometimes referred to as building product models. *A building product model is a digital information structure of objects making up a building, capturing the form, behavior and relations of the parts and assemblies within the building* [115].

As one of the pioneers in BIM, Charles "Chuck" Eastman envisioned computers being able to capture more than just the details of typical 2D design drawings but also 3D geometry and spatial information analogous to a real, physical construct, being further manipulated to allow a wide range of views. These views would enable a multitude of digitally-based workflows for design, construction, and operational analysis. This early vision has come into reality, and we are currently witnessing further digital advancements with integrating real-time sensors into buildings and connecting them to BIM-based views, effectively creating a dynamic virtual representation of the physical elements of a building, also known as a Digital Twin (DT).

It is important to note that processes and requirements in the design and construction phases of the building life cycle are different from those in the operations phase. Thus, we hereby distinguish between product modelling in the AECO industry for (1) the design and construction of buildings (AEC), which typically needs to support continuous updates between a wide variety of stakeholders; and (2) the operational phase (O) of buildings, in which product data is needed for maintenance operations over long span of time (facility and asset management). This chapter provides an overview of the concepts, methods, and tools needed to create and utilise building product models for the design and construction of a building. The information is presented through the perspective of product modelling, which includes structured vocabularies, object-type libraries (OTLs), and data exchange.

1.1.1   A brief history of CAD/BIM

The idea of using computers to digitise the design and manufacturing of products began in the 1950s during the early days of computing. In 1957, Pronto was the first commercial computer-aided manufacturing (CAM) software, developed only a few years after the first automated robots were used in assembly lines. Soon, this use was extended to computer-aided design (CAD), a term coined by Douglas T. Ross around 1959 at Massachusetts Institute of Technology (MIT), as part of the earliest stages of what ultimately became the MIT Computer-Aided Design Project [80]. This concept for object-based design and parametric manipulation was published in Douglas C. Englebart's 1962 paper *Augmenting Human Intellect* [120]. Cognitive design was also a popular topic of research for architecture, thus the combination of the two seemed destined. In 1963, Ivan Sutherland's SketchPad (a.k.a. Robot Draftsman) was a breakthrough CAD program to show the benefits of using computer graphics for designing.

From his 1975 article, *The Use of Computers Instead of Drawings in Building Design*, Eastman writes:

> A building can be conceived, though, as a collection of three-dimensional elements arranged in space... A detailed building representation might be provided by a computer, if it could store descriptions of a very large number of different elements arranged in space. Designing would consist of interactively defining elements, according to their shape and other properties, and are arranging them, much as one would a balsa-wood model... It should be possible, then, to derive sections, plans, isometrics or perspectives from the same description of elements on an automated plotter... Approached this way, the range of drawings would be infinite. If a consultant or contractor wanted any particular drawing, it could be generated on demand. Any change of arrangement would be made only once for all future drawings to be updated. All drawings derived from the same arrangement of elements would automatically be consistent. The representation would be truly three-dimensional [113].

Eastman's prescient vision of buildings being represented using computers, 3D graphics, and databases would continue to play out over the succeeding decades along multiple paths, leading to our modern BIM platforms. Building Description System (BDS) was the first prototype system that represented custom-designed building systems [114]. BDS mapped out building components into taxonomies of a building's elements, paving the way to product modelling. Specific attention was given to the features distinguishing general-purpose from special-purpose building description systems, especially data structures, access schemes, and the method of interaction between the database and analysis programmes.

With the further advancement of computer graphics and the creation of mass-market personal computers, the 1980s saw an explosion in 2D/3D CAD software options for the AECO industry, including Autodesk AutoCAD, GRAPHISOFT Archicad, Bentley Systems MicroStation, Diehl Graphsoft MiniCAD (now Vectorworks), GIMEOR Inc. Architrion, MICROTECTURE DataCAD, Nemetschek Allplan, and Sigma ARRIS CAD to name a few. Since those early days, some platforms have further evolved their functionality into modern BIM authoring applications still available to the market or they have quietly faded into obscure history. New BIM-centric platforms, such as ACCA Edificius, Autodesk Revit, BricsCAD BIM, CYPE Architecture, and Vertex BD, have been launched into the continually growing market. Concurrently, general computing has evolved from server-(thin) client models to networked or standalone desktop workstations back to a familiar server-client model of web-based software as a service (SaaS). This has resulted in a new class of web browser-based BIM authoring applications with various degrees of functional complexity, such as TestFit.io, Hypar, Cedreo, and Space Designer 3D. Today there are thousands of applications in various forms (web, desktop, and mobile) that attempt to serve the AECO market, in smaller local or regional markets as well as internationally, and its desire to create and leverage the information associated to semantic 3D models of the built environment.

1.1.2 Tackling CAD/BIM data exchange

Exchanging data between these systems has been evolving over the same period as well. With Autodesk's AutoCAD establishing itself early as a dominant global market solution, much of the initial data exchange work was finding ways to import and export DWG files, native to AutoCAD, or DXF(TM) (Drawing eXchange Format) files, Autodesk's data exchange protocol for AutoCAD and related products. Other early neutral, open 2D/3D CAD formats, such as CGM, IGES, STEP (ISO 10303), and VRML, were supported by many early CAD systems, but most never really offered a sufficient alternative to DWG or DXF for building designs and their documentation. As the data grew in complexity and size, especially with the maturity of BIM and the need for building-centric semantic data models and formats, it became apparent a new kind of interoperability was needed.

In 1994, the first iteration of today's buildingSMART International was established, the International Alliance for Interoperability (IAI), to tackle the issue of exchanging data between disparate applications. Started as a proposal by Autodesk to develop an open C++ class specification for this purpose, it quickly evolved into an open, object-oriented, EXPRESS-based schema and file format specifications (closely tied to ISO 10303, STEP), known as Industry Foundation Classes (IFC). First demonstrated in 1995 at the A/E/C Systems '95 conference in Atlanta, GA, the IFC semantic data model schema would quickly mature to version 1.0, published in 1996, to standardise the representation of buildings and constituent parts, processes, contexts, and relationships for data exchange among industry CAD/BIM software. By 2003, support for IFC2x2 (v2.2.0.0) was showing up in a relatively small number of AECO software platforms. But the publishing of IFC2x3 TC1 (v2.3.0.1) in 2007 as an ISO PAS (Publicly Available Specification) was really the beginning of the modern era of open BIM data interoperability. Since then, buildingSMART International and its community of industry members, including software companies, building designers, contractors/builders, and facility owners, have continued to develop, adopt, and implement the schema (IFC4.3 as of October 2021) to keep up with industry needs and technology improvements.

1.1.3 Seamless data exchange: the endemic problem

Even in the early days of CAD, when there were only a few programmes and simple computations, an efficient and effective method to exchange data between the systems was still a challenge. As Douglas T. Ross, the early CAD pioneer, and Jorge E. Rodriguez eloquently write:

> There is a great many existing specialized languages and programming systems of the individual areas which must be covered by a Computer-Aided Design System,

but each of these languages and systems has its own restrictions and interwoven computational complexities so that it would be completely impractical to attempt to integrate such systems in a straight-forward manner. Furthermore, such a brute force approach would not satisfy the Computer-Aided Design Requirements in the first place, since there would be little or no cross fertilization between the various systems, even if the mechanics of translating data and control information from one system to another could be solved in a moderately satisfactory manner [329].

Herein lies the endemic problem: there will never be a single unified data language, model, or schema that will accommodate all software, program languages, or user data requirements to enable seamless, efficient, and error-proof data exchange. Data exchange, namely, the manipulation of data from one source into another, will forever be a requirement for today's computer systems. That is not to say, however, that we will never be close enough to a unified system or method to achieve the efficient data exchanges at a practical level. Ross and Rodriguez continue:

> The first step in this direction is to recognize once and for all that it is completely impossible to construct a system which will satisfy the requirements immediately and without modification..., In fact to postulate the existence of a closed system for Computer-Aided Design as we mean it is completely and absolutely contradictory to the very sense of the concept... the very nature of the system must be such that its area of application is continuously extended by its users to provide new capabilities as the need arises... If, in fact, the system can be so organized that it can naturally be moulded to suit the needs and interests of individual users, then the concept of a general Computer-Aided-Design System not only begins to seem possible, but practicable as well [329].

Instead of making all the internal functions freely available, computer systems utilise data exchanges for a multitude of reasons, including keeping proprietary information secret, different programming languages, non-compatible schemas, and different end user data needs. Most end users only require a subset of the data from one system, and there are a few low-level methods to achieve this:

1. Writing a translator, in which data from the first is exported in the native file, the file is converted (also known as mapping) into a textual format, the text is conversely mapped to the native file of the second system, and the file is imported in the second system.
2. Adding internal plugins or scripts directly into system environment that manipulate and export the data for reading into another application using method 1 above.
3. Writing a standalone application that uses the system's available application programming interfaces (APIs) to access the software's programming function to produce, manipulate, and export the data using method 1 above.

These types of data exchange methods only function at the lowest level of data, since there is much brute force and single encoding of data, which will be infeasible to conduct on a large scale. Solutions on larger scales, such as complete BIM systems, IoT-based applications, and semantic web approaches, require more methodical and extendable methods. End users of the systems typically do not rely on a single software solution, so today's software and computing hardware vendors understand this, and are enabling various ways to achieve this. To achieve seamless data exchange today, we need to have a paradigm shift from the mindset that developing a single, one-fits-all unified data model is the only approach. Instead, we need to shift our mindset to understand that data models do not need to be fully complete to take advantage of the benefits of achieving seamless data exchange. Understanding the basic methods, functions, and features is important to leverage the current approaches of achieving seamless data exchanges today. Herein lies the objective of this chapter and the rest of the book: to provide the state-of-the-art review of all sorts of web technologies used for a variety of use cases to enable future research and innovation in the Architecture, Engineering, Construction, and Operations (AECO) domain.

1.2 CONCEPTS AND DEFINITIONS

With the invention of the internet and the increasing power of computing, visions that Englebart, Eastman, Ross, and others have had, in fact, came to fruition. Even with these technologies, many of the concepts used back then are still relevant in today's modern systems, including among applications involving the Internet of Things (IoT), smart cities, and linked data. This section reviews some of these concepts and definitions (Table 1.1).

Table 1.1 Terms and acronyms

Acronym	Term
AEC	Architecture, Engineering, and Construction
AI	Artificial Intelligence
BIM	Building Information Modelling
BOT	Building Topology Ontology
BSD	Building Description System
bSDD	buildingSmart Data Dictionary
DL	Description Logic
FOL	First Order Logic
HTTP	Hypertext Transfer Protocol
IDM	Information Delivery Manual
IFC	Industry Foundation Classes
JSON	JavaScript Object Notation
KIF	Knowledge Interchange Format
KR	Knowledge Representation
LBD	Linked Building Data
OIL	Ontology Inference Layer
OWL	Web Ontology Language
RDF	Resource Description Framework
RDFS	Resource Description Framework Schema
RIF	Rule Interchange Format
SPARQL	SPARQL Protocol And RDF Query Language
SQL	Structured Query Language
SWRL	semantic web Rule Language
UML	Unified Modelling Language
URI	Uniform Resource Identifier
URL	Uniform Resource Locator
W3C	World Wide Web Consortium
XML	Extensible Markup Language

1.2.1   Chapter definitions

Business Process Model Notation (BPMN) A standardised graphical specification language for modelling business processes and workflows.

Class (type, family) An object-oriented data concept for typing and describing the structure and behaviour of similar objects.

Common Data Environment (CDE) A single source of information, used to collect, organise, manage, and disseminate all relevant project documentation among multidisciplinary project team members.

Computer-Aided Design (CAD) Computer-aided design of products based on 2D and 3D geometric models. In the past also known as Computer-Aided Drafting, which primarily applies to digital design based on 2D plans.

Computer-Aided Manufacturing (CAM) Computer-aided manufacturing of products, traditionally using numerically controlled milling machines, more recently using 3D printing techniques.

Data exchange Process of data export from a software program and subsequent data import into another software program using a data exchange format.

Data exchange format Specification describing how data exchanged between programmes should be saved, loaded, and edited, for example, IFC.

Entity A graphical or physical element within a CAD system, data model, or the physical world. It may be either a class or instance.

Entity class The abstract grouping of elements having similar structure and properties. An entity class may be a generic concept or a specific type of product.

Entity instance A unique, individual entity of any entity class.

Exchange Requirements (ER) Part of the IDM method. Tabular catalog of requirements for data or information exchange.

EXPRESS A declarative data modelling language specified in the STEP standard part 11, with which object-oriented data models can be defined. EXPRESS is used to specify the IFC data model.

Geometric modelling A methodology for describing the geometry and topology of products.

Industry Foundation Classes (IFC) A vendor-independent open, standardised and object-oriented data format for exchanging Building Information Models (BIM).

Information Delivery Manual (IDM) Methods for capturing and specifying data exchange processes and flows of information in the lifecycle of a building. Composed of roles and tasks, process maps, exchange requirements, and model view definitions.

Interoperability (Software Interoperability) Compatibility of software systems with respect to the lossless exchange of data.

Linked Data Structured information that can be shared, interconnected and queried over networks using open standards. The reuse of common data models, vocabularies, and semantics enhances interoperability between heterogeneous information systems.

Model View Definition (MVD) Part of the IDM method. Specification of a subset of a model or schema (e.g., the IFC) is needed to satisfy the ERs for a particular task (e.g., determining energy demand).

Object-Oriented Modelling method for the structured description of data or information on the basis of objects and their inter-relationships.

Ontology In the context of computer systems, an ordering system. A formally organised collection of concepts/categories in digital form, typically formulated verbally or graphically.

Process Maps (PM) Part of the IDM method. Standardised process diagrams for selected sub-processes of the planning, construction, and use of buildings/building constructions.

Process Modelling Methods and concepts for describing processes and workflows (planning, communication, data exchange, business, controlling, construction, and operating processes).

Semantics In the context of computer systems, the meaning of a sequence of characters, symbols, data or information, usually related to non-geometrical information.

Solid Modelling A consistent set of principles for mathematical, and computer modelling of three-dimensional solids.

Syntactics In the context of computer systems, the syntax or arrangement of a sequence of characters, symbols, data, or information, usually related to non-geometrical information.

I.3   STRUCTURED VOCABULARIES

Structured vocabularies are an important means of defining and structuring the meaning of concepts and terms used in the building industry to ensure their consistent use by all stakeholders over the life cycle of a construction. In their traditional form as text documents and tables, they are designed for use by domain experts to facilitate the creation and use of unambiguous specifications, requirement documents and mutual agreements. In their digital, machine-readable form, they can be used in a BIM and web-based context for the semantic annotation of model objects to further enhance exchange and interoperability in data exchange scenarios.

Structured vocabularies offer an efficient way to organise knowledge for subsequent retrieval, such as querying the semantic web. This chapter introduces the fundamental concepts, application areas, and technical implementations of such terminologies and structured vocabularies.

1.3.1   Structured vocabulary types

1.3.1.1   Classification systems

A classification system is a method of organising "things" based on a fundamental concept or predefined classes. There are many ways and degrees of specification of classifying things, but the major aim is to provide a logical order to differentiate certain things from others.

A classification system is important to have because it can supply the core definitions used throughout a specific domain, enabling clarification and consistency. The OmniClass Construction Classification System (OCCS) is the leading classification system for the AECO in the United States. It is structured in 15 tables (e.g., elements, phases, disciplines, materials) that provide the structured data for many applications. Likewise, Uniclass is a unified classification system for all sectors of the UK construction industry. It contains consistent tables classifying items of all scales.

1.3.1.2   Taxonomy

A taxonomy is a hierarchical structure of terms that represent the relationships and attributes among those terms. Essentially, a taxonomy defines how terms are related by organising and displaying them as nodes in a tree. For example, a bridge can be decomposed into a superstructure and substructure, which then each can be further decomposed into other parts (see Figure 1.1). Many computer schemas, e.g. the Industry Foundation Classes (IFC), and classification systems include such hierarchical formalisations.

1.3.1.3   Ontology

In computer and information science, *an ontology is an explicit specification of a conceptualisation,* in which conceptualisation refers to the *objects, concepts, and other entities that are presumed to exist in some area of interest and the relationships that hold among them* [157]. An ontology is the formal classification of entities in a particular domain that includes the types, properties, relationships, and other significant attributes

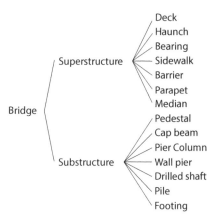

Figure 1.1 Tree node representation of a simple bridge taxonomy.

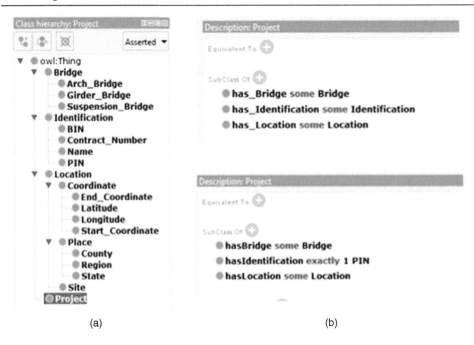

(a)   (b)

Figure 1.2 Example bridge ontology: (a) Hierarchy in taxonomic form. (b) Property restrictions.

about the entities within the domain. Ontologies are used to define the logic and semantics needed for computer systems and software applications. In other words, an ontology can be viewed as the foundation, or sublevel, needed to support computer systems. A taxonomy and ontology are very similar (see Figure 1.2a), and, in a non-technical sense, can be difficult to distinguish. The major difference is that the taxonomy is the classification structure, and the ontology contains the information and properties about those terms (see Figure 1.2b). In essence, a taxonomy is often a subpart of an ontology.

Ontologies are powerful; in that, they can contain assignment relations, formal assertions, and constraints by using axioms. Axioms supply the rules for how the terms, properties, and attributes are related and are used. Axioms are an important part of developing an ontology because they provide truths and assumptions that give meaning to the terminology. Axioms can be defined both abstractly (logically) and operationally (structured). An ontology without axioms would essentially be a hierarchy of terms with meaning limited to the intrinsic meaning of the terms used. Moreover, definitions alone in the form of classification systems and product data models lack effective modelling of concept semantics, which is a fundamental requirement for human-based exchange of knowledge. A classification system is important to have because it can supply the core relations of a specific domain.

All standardised terminology needs to be semantically consistent across organisation boundaries, since the communication aspects of information require that communicating parties have the same understanding of the meaning of the exchanged information.

1.3.1.4 Data dictionary

A data dictionary is a centralised repository of information about data such as meaning, relationships to other data, origin, usage, and format [248]. A data dictionary is used to

Domain Metadata

Field	DataType	Required	Translatable	Description
OrganizationName	Text	Yes	No	buildingSMART International
DomainName	Text	Yes	No	IFC
DomainVersion	Text	Yes	No	2.2
VersionDate	Date	No	No	2017-10-01
LanguageCode	Text	Yes	No	de-DE
LanguageOnly	Boolean	Yes	No	yes
License	Text	No	No	No license

Property Metadata

Field	DataType	Required	Translatable	Description
Id	Text	Yes	No	ifc-99088-01
PropertyName	Text	Yes	Yes	IsExternal
ClassificationId	Text	Yes	No	ifc-00123-01
Dimension	Text	No	No	10-20000
MethodOfMeasurement	Text	No	Yes	Thermal transmittance
DataType	Text	No	No	integer

Classifications Metadata

Field	DataType	Required	Translatable	Description
Id	Text	Yes	No	ifc-00123-01
ClassificationName	Text	Yes	Yes	IfcBridge
Definition	Text	No	Yes	A Bridge is civil ..
Status	Text	No	No	Active
DocumentReference	Text	No	No	ISO 6707 1 2014
CountryOfOrigin	Text	No	No	DE
CountriesOfUse	Text	No	No	EN;NL;DE
SubdivisionsOfUse	Text	No	Yes	US-MT
ClassificationType	Text	No	No	ComposedProperty

IfcClassifications

Attribute	Type	Cardinality	Description
Source	IfcLabel	?	Source (or publisher) for this classification.
Edition	IfcLabel	?	The edition or version of the classification system from which the classification notation is derived.
EditionDate	IfcDate	?	The date on which the edition of the classification used became valid.
Name	IfcLabel		The name or label by which the classification used is normally known.

Figure 1.3 Example data dictionary with multiple tables of metadata.

catalog and communicate the structure and content of data by providing meaningful descriptions for individually named data objects. Essentially, a data dictionary gives context to the data being stored. Data dictionaries can be created using a number of tables to define any information about a data point. Figure 1.3 displays a data dictionary with four tables of metadata, in this case describing the domain, element properties, element classifications, and the associated IFC classification properties.

1.3.1.5   Object-type library

An OTL is a collection of standardised object types, names, and properties or specifications. This is a semantic model that includes predefined objects, such as shapes and building elements, and how relevant data should be delivered to ensure compliance. An object is described with its object-type data, geometry data, and metadata, which are important since each object type has its own properties. OTLs utilise the organisational features of an ontology. Likewise, an OTL can be linked to a data dictionary, with the definitions of object types. By using an OTL, BIM elements can be described with a standard language, syntax, and semantics.

1.3.2   Functionality and features

1.3.2.1   Object-oriented functionality

Object-oriented is a methodology that enables a system to be modelled as a set of "objects" which can be controlled and manipulated in a modular manner. Each object can have unique values to the properties defined in the class and code. Four major features of object-oriented systems include abstraction, encapsulation, inheritance, and polymorphism.

Abstraction is the creation of abstract concept-objects from subsuming features or attributes of various non-abstract objects or systems of study. This enables system/software developers to omit the "abstract" class, but still has access to the properties and methods.

Encapsulation is the protection of data by packaging and providing restricted access to the user or programmer. A user is able to gain access to the data by calling an instance of an object, but cannot add or modify the properties about that object.

Inheritance is the access of metadata from a class higher in the hierarchy by objects lower in the hierarchy. For instance, since a girder bridge is a type of bridge, anything defined in the bridge term is inherited by the girder bridge. Hierarchy, such as the case of a taxonomy, is used to avoid the duplication of properties.

Polymorphism is the ability for data to change functionality or form based on a stringent set of defined rules. Polymorphism enabled objects to take on more than one form depending on the context. For example, the functionality of a "door" is to restrict access in and out of a room. However, what happens if a door is flipped horizontal and used as a table... is it a door or a table? If a table is modelled using a door (i.e. a door is selected and rotated), it would still appear and function as a table, but the properties of a door would not be consistent with that of a table (e.g. strength properties, materials, manufacturers, etc.). This need can be met by defining the instances in a taxonomy which would allow for polymorphism. Once the case has been defined, the door flipped into a table would be polymorphic, and therefore the door would assume the properties of the table if it is being used as a table.

1.3.2.2   *Semantics and logic*

Structured vocabularies need to capture two important elements: semantics and logic. Semantics are the meanings and interpretations of a word or phrase in a specific context. At the core level of computer science, data are essentially bits and bytes that the computer uses in processes and which are essentially useless to both human and computer function without any context. Therefore, it is important that the data be given the semantic information needed to represent what human function represents. For example, when data are exchanged between BIM software, it is insufficient to solely rely on 3D visual properties of the objects. Although the geometries of the objects in the 3D model are important, they alone are not sufficient to describe the needed meaning of the modelled objects. At the exchange level (i.e. passing information), semantics may cause issues for humans and computers that are interpreting the context. The goal is to have semantic consistency in an information exchange in which the human-based knowledge and computer-based interpretation of the information are equivalent, i.e. the computer understands what the user intends.

Logic is the reasoning behind the development of the structured vocabulary. Logic is represented by Description Logic (DL), which is the basis of most ontology languages. DL is the formal knowledge representation (KR) used to express the conceptualisation of domains in an organised and formally well-understood manner. The logic structure contains the additional axioms (logic assertions provided by the ontology language in a common form or structure). As part of mathematical logic, these types of rules are associated with type theory. Table 1.2 gives the description logic and definitions for an example bridge ontology, and Figure 1.4 displays the corresponding visualisations.

Building a structured vocabulary with the full semantic information and well-organised structure is challenging because of the complexity of the natural languages and the broad scope and large amount of scientific knowledge accumulated. Therefore, it is important to understand the usage and the vernacular of the terminology based on the domain.

Table 1.2 Examples of defining axioms

Relation	Description logic	Definition	Example
ComposedOf	Composed-of(A,B) $\leftrightarrow (B \subseteq A) \wedge (A \nsubseteq B)$	B is composed of A, if A is a subset of B, and B is not a subset of A	Bridge is composed of smaller parts, e.g. columns, beams, etc.
InverseTo	$\forall$ A,B**f** (A,B)$\leftrightarrow$ **g** (B,A)	For all A and B, relation **g** is the inverse of relation **f** if A maps to B and B maps to A.	If beam is *partOf* bridge, then bridge *hasPart* beam.

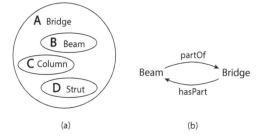

(a) (b)

Figure 1.4 Example relations: (a) *ComposedOf*. (b) *InverseTo* relation.

1.4 DIGITAL BUILDING EXCHANGE FORMATS AND SCHEMAS

There have been many data exchange formats and schemas developed throughout the years, from the early geometric focused formats, such as SIF, DXF, IGES, and SET, to the more knowledge-based formats, such as IFC, BRICK, and BOT. This section provides a short documentation of current relevant formats and schemas.

1.4.1 Semantic web and linked data

A fundamental problem in structuring knowledge and information for automated processing is the heterogeneity of technical representations. In the past, different vocabularies, classification systems, conceptual models, and ontologies have been created and presented using different modelling languages, data formats, and interfaces. Up to now, considerable effort has been invested by both software vendors and users to access and harness relevant classification systems. These efforts impose severe obstacles to facilitating the semantically unambiguous exchange of information in the building industry.

To address such interoperability problems, methods, and technologies for the distributed modelling of and access to information resources were developed that are referred to as the semantic web initiative [35]. The core idea is to standardise generic means of modelling and representing knowledge and information that enables their uniform, decentralised creation, and the publication and linking of resources in a global network. At the core of this standardisation effort under the umbrella of the World Wide Web Consortium (W3C), the Resource Description Framework (RDF) is the ability to capture atomic statements in any model in the form of a triple consisting of a subject, predicate, and object. Each of these components is identified by a Uniform Resource Identifier (URI), the most common form of which is the network address URL (Uniform Resource Locator), which has the inherent ability to distribute and link information across network structures. This allows the reuse of concepts, properties,

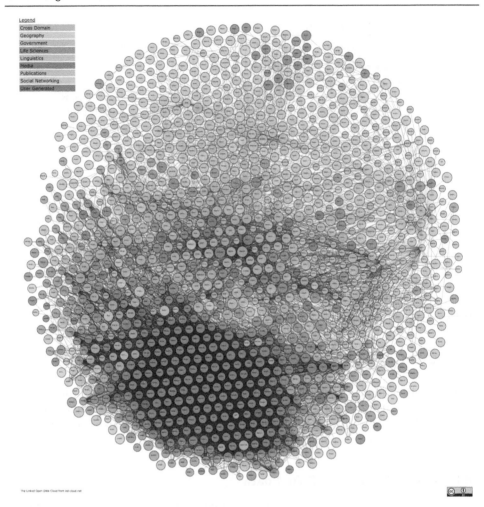

Figure 1.5 Linked open data (LOD) cloud from lod-cloud.net.

models, and instance data even across domain boundaries. Basic concepts of information and knowledge modelling such as "Class", "Property", and "Data Value" are provided by standardised vocabularies such as the RDF Schema (RDFS), the Web Ontology Language (OWL), or the Simple Knowledge Organisation System (SKOS).

The Web of Data has exploded over the last decade. The Linked Open Data (LOD) Cloud is a graphical representation of all of the databases that have been published using linked data format.[1] Figure 1.5 is the LOD Cloud Figure, in which each shaded circle represents a linked dataset in the respective sub-domain based on shaded. The lines represent the dataset links, in which the dark shaded shows a multitude of links. As of May 2021, it contains 1301 datasets with 16283 links and billions of triples.

1.4.1.1 Resource description framework (RDF)

The Resource Description Framework (RDF) is a framework for conceptual description and the modelling of information that is implemented in web resources. The RDF Schema (RDFS) provides a data modelling vocabulary for RDF data. RDF is composed of three components, known as RDF triples: subject, predicate, and object. RDF triples state a single fact about a resource, in which the subject is the subject being described, the predicate is the relationship of the subject, and the object represents what is related to the subject by the predicate. RDF and RDFS are managed by the World Wide Web Consortium (W3C), the main international standards organisation for the World Wide Web (WWW). RDF is a good basic language for building many other languages, although it has limitations with being expressive and in describing resources.

1.4.1.2 Web ontology language (OWL)

One such language that incorporates RDF is the Web Ontology Language (OWL). There are three levels of OWL: OWL Lite, OWL DL (Description Logic), and OWL Full. The simplest level, OWL Lite, supports only a subset of the OWL language constructs, and provides a classification hierarchy and simple constraints. OWL Lite is used by users who want to support OWL full, but want to start at a basic level. In addition to rules and requirements of OWL Lite, OWL DL adds the tools and features of Description Logic to represent the relations between objects and their properties. Description Logic, the basis of any ontology language, is the formal knowledge representation used to express the conceptualisation of domains in an organised and formally well-understood manner. OWL Full provides the highest freedom of using the OWL language and RDF constructs, but takes considerably more computing power to run the inference engines. The current release, OWL 2 (Figure 1.6), can be found in the W3C standards pages.[2] Additionally, the Rule Interchange Format (RIF) defines a standard for exchanging rules among systems on the Web that specifies how RDF, RDFS, and OWL interrelate.

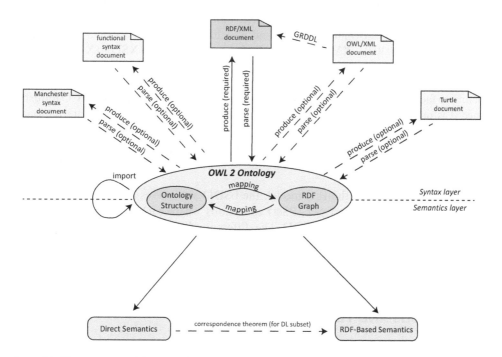

Figure 1.6 The structure of OWL 2. (Copyright ©2012 W3C ®(MIT, ERCIM, Keio), All Rights Reserved.)

1.4.1.3   Simple knowledge organisation system (SKOS)

Simple Knowledge Organisation System[3] (SKOS) is a W3C recommendation designed for representation of thesauri, classification schemes, taxonomies, subject heading systems, or any other type of structured controlled vocabulary. SKOS is an area of work-developing specifications and standards to support the use of knowledge organisation systems (KOS) such as thesauri, classification schemes, subject heading systems, and taxonomies within the framework of the semantic web. SKOS provides a standard way to represent knowledge organisation systems using RDF. Encoding this information in RDF allows it to be passed between computer applications in an interoperable way. Using RDF also allows KOSs to be used in distributed, decentralised metadata applications. Decentralised metadata is becoming a typical scenario, where service providers want to add value to metadata harvested from multiple sources.

1.4.2   ISOs for building classifications

ISO 12006, "Building construction - Organisation of information about construction works", defines a framework for the development of built environment classification systems. The two parts of ISO 12006 include a framework for classification and a framework for object-oriented information. Derived from the Swedish SfB system, it is referred to as the International Framework for Dictionaries (IFD) and is an official standard of the buildingSMART organisation. In part 2 of this standard, a central conceptual framework is provided for concepts such as "construction result", "process", and "resource". This framework, however, only provides a general recommendation for possible classifications that it can describe.

ISO 16739 establishes the Industry Foundation Classes (IFC) for data sharing in the construction and facility management industries. IFC specifies a data schema and an exchange file format structure, and is defined in EXPRESS data specification language (ISO 10303), XML Schema Definition language (XSD) (XML Schema W3C Recommendation), Web Ontology Language (OWL), and recently into the buildingSMART Data Dictionary (bSDD). IFC has been the most common neutral file format to facilitate efficient workflows and data exchanges for BIM in the AECO industry since it is public, non-proprietary, and is facilitated by buildingSMART International. Attributes along with the constraints and structure of IFC are defined in a schema. In addition, IFC provides methods to define entities and properties needed for design, production, and maintenance of buildings and other major civil structures (roads, bridges, rails, etc.).

Since version 2x4 of the IFC model, the bSDD also serves as a central repository for the standardised PropertySet (PSet) extensions where each individual property is represented by a concept in the bSDD. The different relationship types between concept nodes (specialisation, part-whole-relationships, etc.) together with the ability to link these concepts to other normative documents, building codes, etc. make the bSDD a valuable body of knowledge that will gain increasing importance in the future.

1.5   METHODS AND TECHNIQUES

In addition to the data exchange formats and standards previously described, this section provides other methods and techniques required for web technologies.

1.5.1   Product and solid modelling techniques

In computer information systems, product modelling is the process of describing the steps and levels of information required to produce or service. In terms of BIM, a product model is the actual 3D visual data model that represents the physical and functional characteristics of the real-life object. Product models are the primary source and storage of data and information throughout different stages of that object's life cycle. Product models incorporate knowledge representation, sharing, and exchanging. Most often, when people think of a BIM model, they are referring to the product model (e.g., the CAD model) since they can see and interact with the actual 3D visualisation.

Product models are developed using the method of solid modelling. Solid modelling is a method of using mathematical equations to computer generate three-dimensional solids. Solid modelling is distinguished from other areas in geometric modelling and computing by its emphasis on informational completeness, physical fidelity, and universality [351]. Two main concepts for doing so include geometric modelling and parametric modelling. Geometric modelling, which is the use of solid representation schemes (e.g., constructive solid geometry (CSG), boundary representation (B-rep)), is the foundation for most product representations because geometry is essential for design, analysis, and production. Parametric modelling, on the other hand, is an advanced technique that uses parameters and dependencies to define a model. Parameters are essential for large models, such as building and bridges, because the user only needs to change the parameters instead of manually changing each dimension or definition with each change in design. The notions of object-based modelling and parametric modelling of buildings have their roots in the concepts of parametric solid modelling for generic product modelling.

1.5.2   Information collection mechanisms

To consistently describe BIM processes and the accompanying information exchange requirements, buildingSMART developed a standardised method called the Information Delivery Manual (IDM), "which is now defined in ISO 29481." This predefined uniform structure and method for presenting process models enables users to develop, agree on and accurately document their BIM processes. The corresponding technical counterpart to the individual IDM specifications is so-called Model View Definitions (MVD) that define the specific sub-elements of the overall IFC data model that can support the specific exchange requirements of the IDMs. MVDs can, therefore, serve as technical specifications for software vendors that wish to support IFC. In the user interface of the IFC import/export facility of BIM software, the user should have a choice of relevant MVDs. However, because users are usually unaware of MVDs, the user interface needs to use terms that describe the underlying MVDs in more user-friendly terms, for example that describe the purpose of the data exchange.

1.5.3   Development and management

Developing a structured vocabulary requires thoughtful planning for it to be useful and practical. It is important to develop a strategy by asking competency questions such as: *what do you want the structured vocabulary to do?*; *whom will the structured vocabulary serve?*; and *what is the scope of the structured vocabulary?*

Although specifically focused around ontology development, [277] present three fundamental rules that can be applied to any structured vocabulary:

- Rule 1: There is no one correct way to model a domain; there are always viable alternatives. The best solution almost always depends on the application that you have in mind and the extensions that you anticipate.
- Rule 2: The development is necessarily an iterative process.
- Rule 3: Concepts should be close to objects (physical or logical) and relationships in your domain of interest. These are most likely to be nouns (objects) or verbs (relationships) in sentences that describe your domain.

1.6   PRACTICAL EXAMPLES IN THE AECO INDUSTRY

In building and construction, a number of vocabularies, classifications, and ontologies exist that have been developed using the technologies described in this chapter. This section provides practical examples that can be utilised.

1.6.1   Core vocabularies and linked datasets

Schema.org,[4] a collaborative, community activity with a mission to create, maintain, and promote schemas for structured data, is a reference website that publishes documentations and guidelines to using structured data. Schema.org provides a core, basic vocabulary for describing the kind of entities the most common web applications need that is used for more defined extensions. The vocabulary currently consists of 792 Types, 1447 Properties, 15 Data types, 83 Enumerations, and 445 Enumeration members. One extension of Schema.org is GoodRelations, which is a powerful Web vocabulary specifically designed for e-commerce scenarios.[5]

The free classification structure FreeClass provides approximately 2,800 concepts pertaining to building materials that are available in eight languages.[6] This vocabulary has been used to create product catalogues of approximately 70,000 building products by 90 manufacturers in Austria that can be accessed, searched, and indexed by search engines in a uniform way. The global networking of such datasets is summarised by the notion of linked data, which has been previously discussed. Furthermore, a number of general-purpose datasets exist, such as the semantically annotated form of the Wikipedia corpus, DBPedia,[7] and YAGO (Yet Another Great Ontology).[8] Since the emergence of the LOD Cloud in 2007, DBpedia constitutes the main resource of linked open data on the Web containing more than 228 million entities to date. YAGO is another large knowledge base with more than 10 million entities and contains more than 120 million facts about these entities.

Building relevant vocabularies are also available. The Getty Arts and Architecture Thesaurus (AAT)[9] provides structured data for generic concepts related to art, architecture, and cultural heritage. Many smart city and energy efficiency-based datasets can be found from the READY4SmartCities FP7 project.[10]

1.6.2   Existing AECO ontologies

By design, ontologies can be extended from the core ontologies. Thus, there are a variety of ontologies[11] that exist to be leveraged by the AECO domain [85,301]. One important group of such ontologies is available from the W3C Linked Building Data (LBD) Community Group.[12] LBD work has been aiming at several critical elements or modules of use for representing building data, and it has resulted in a combination of ontologies that is welcomed to be further extended with additional domain ontologies (product data, properties, and geometry). The group focuses on information exchanges, and therefore, creates smaller and modular ontologies rather than a single large one. A key ontology in this regard, is the Building Topology Ontology (BOT), which aims to be a central AECO ontology that allows for domain-specific extensions and reuse. BOT takes the core building concepts defined by IFC, ultimately enabling practitioners the ability to extend the ontology for their own interests.

The W3C LBD CG has discussed the representation of product data and product properties at length, in which a number of different approaches can be found in this regard. Two key references in this regard are the Building Product Ontology (BPO) [406] and the Ontology for Property Management (OPM) [316]. Although not explicitly or formally part of the LBD CG, these ontologies represent the mindset behind the W3C LBD CG in the sense that these OWL ontologies provide the means to describe product and properties for building elements, yet do not encode on their own the more specific building data classes. Classes such as "wall", "door", are not supported. Instead, the aim is for national standardisation bodies and individual organisations to define such content while relying on ontologies like the BPO and OPM.

Finally, the W3C LBD CG also worked on the representation of 3D geometry, although this has never been a core topic in this group because 3D geometry is not easily represented using OWL ontologies and RDF graphs. The most notable results to represent such 3D geometric data can be found in the Ontology for Managing Geometry (OMG) [405] and the File Ontology for Geometry formats (FOG) [43].

1.6.3   Existing OTLs and data dictionaries

Due to the recent advancement of the technology, there are only a few widely accessible AECO OTLs and data dictionaries. One such OTL of transportation assets was developed by the INTERLINK project.[13] The project was initiated by the Conference of European Directors of Roads (CEDR) in 2015 into the use of BIM for information management during the delivery and operation of civil infrastructure. The research aimed at improving interoperability within European National Road Authorities (NRAs) and their stakeholders by digitalisation of the sector. A major outcome is the European Road Object-Type Library (EUROTL). The published CEDR-INTERLINK Approach, along with the EUROTL, can enable NRAs to improve their Asset Information Management.

One of the premier examples of a data dictionary in the AECO industry is the buildingSMART Data Dictionary (bSDD).[14] In addition to being a data dictionary, the bSDD is an online service that hosts classifications and their properties, allowed values, units and translations. The bSDD allows linking between all the content inside the database by providing a standardised workflow to guarantee data quality and information consistency. Figure 1.7 shows a screenshot of a web-based interface for browsing and searching the contents of the dictionary. To reduce complexity and filter information relevant to a particular use case, the notion of "contexts" makes it possible to scope associated codes and concept hierarchies to a particular local standard. The bSDD has undergone new development to become more interactive, accessible, and expressive.

Figure 1.7 Searching the term "Bridge" in the bSDD.

For example, the publishing of the data as linked data, compliance to ISO 12006-3 and ISO 23386 are built-in functionalities now.

1.7   OPEN RESEARCH CHALLENGES

As noted previously in this chapter, there is an endemic problem relating to efficient and seamless data exchange. This chapter outlines solutions, tools, and methods to help alleviate such a prevalent problem; however, these in and of themselves also present some challenges. This remaining section highlights some of these open research challenges currently faced.

1.7.1   System limitations

Early CAD systems were limited by: (1) display technology, (2) processor capability, and (3) software capability [115] which have been alleviated with modern day computing systems. On the flip side, earlier CAD systems were able to exchange data efficiently since they only exchanged geometries and there were only a handful of software to translate. Nowadays, there are dozens of exchange formats and thousands of software that make efficient data exchange a major challenge.

1.7.2   Open standard limitations

The establishment of an open standard to enable efficient data exchange, and to theoretically solve interoperability problems, is a noble method; however, some challenges still exist. Since there are too many open challenges, we listed them in four main groups: (1) development, maintenance, and extension of the neutral exchange,

(2) functionality of the CAD systems, (3) market availability, readiness, and vendor cooperation, and (4) governance.

The first group of challenges is grouped around the development, maintenance, and extension of the neutral exchange. The scope of what the neutral file will capture is essential because representing all possible knowledge and instances is (nearly) impossible. Identifying all of the entities and properties is a balance of scope and usability: in which the higher the scope, the less defined the entities are; and the less defined the entities are, the more difficult to pass detailed properties. For example, a building is a simple domain, and developing IFC has taken nearly a decade to produce a useful product. Even yet, there are still challenges in adopting the IFC model to other non-building types, such as a bridge or roadway.

The second group of challenges pertains to the functionality of the CAD systems. Often the functional differences such as intended uses, for example, BIM geometric design versus structural analysis, will have varying functionalities and rely on different representations that may be ambiguous or incompatible with each other. Such differences may result in data mismatches or loss when there are no adequate representations in the sending or receiving CAD systems. In addition, each system may only require only a subset of data from the original data model, which serves as the source of truth. Will the exchange be one-way, bi-directional, or multi-directional through various software? If so, how do we maintain consistency to ensure data alignment and assurance?

The third group of challenges relates to market availability, readiness, and vendor cooperation. The first problem is the adoption and integration of the neutral file. Many software vendors are hesitant, if not refusing, to share their proprietary information. Incorporating a neutral exchange takes quite a substantial amount of effort of the software vendor to map out the neutral schema into their own software. Even though the purpose of the neutral file is to have one "shared" open standard, all existing code needs to be mapped to it.

The fourth group of challenges pertains to governance, or the management and oversight of a certain standard or framework. For example, who is responsible for creating a standard? Who dictates what governs and implements the standard? How are changes and updates handled? What if there are multiple domains represented?

1.8 CONCLUSION

The structured vocabularies, application areas, and technical implementations introduced in this chapter are gaining increasing importance in many use case scenarios of the building industry and are already an indispensable part of building information modelling technology. Their main purpose is to serve the unambiguous exchange of information by all stakeholders in the building process by providing clear definitions of concepts and terms. Augmenting static models such as IFC, a prime application area is the annotation of spatial structures, their components, and elements. The digitisation and machine-readability of building codes, norms, and specifications are progressing rapidly and is used in many business practices already today. New developments in the areas of the semantic web and linked data will further accelerate the integration and dynamic composition of semantically rich information resources in digital building models.

The challenges and issues with interoperability will always be prevalent since the essence of CAD and BIM systems are intended to be ever evolving. This is the endemic problem. The idea that there can be one unified model that will solve all the challenges to interoperability and enable seamless data sharing is not only infeasible but also detrimental to the current methods that can achieve efficient data transfer. If researchers and practitioners are too focused on achieving this non-existent holy grail, then they will never be able to harness the current methods, many of which are discussed in this book, that can enable data sharing right now. A paradigm shift of the current mindset of what a BIM is and how data is created, stored, and shared is required for the advancement of the next generation of BIM and computing technologies. Only when the industry can recognize that a perfect one-fits-all solution is not the answer, but rather linking subsets of data models together, can the industry realize the full benefits of seamless data exchange.

NOTES

1 https://lod-cloud.net/.
2 https://www.w3.org/TR/owl2-overview/.
3 https://www.w3.org/2004/02/skos/intro.
4 https://schema.org/.
5 http://www.heppnetz.de/projects/goodrelations/.
6 http://www.freeclass.eu/freeclass_v1.html.
7 https://www.dbpedia.org/.
8 https://yago-knowledge.org/.
9 https://www.getty.edu/research/tools/vocabularies/aat/.
10 http://smartcity.linkeddata.es/datasets/index.html.
11 http://smartcity.linkeddata.es/datasets/index.html.
12 https://www.w3.org/community/lbd/.
13 https://roadotl.eu/static/ireport/index.html.
14 https://search.bsdd.buildingsmart.org/.

Chapter 2

Property modelling in the **AECO** industry

Mathias Bonduel, Pieter Pauwels, and Ralf Klein

CONTENTS

DOI: 10.1201/9781003204381-3

In the semantic representation and documentation of our built environment, properties of building elements, spaces, and construction-related concepts in general, in general, play a tremendously important role. Such properties – e.g. thermal transmittance, door height, and fire resistance classification according to specifications – capture the state of the built environment at a given moment and place. In a global world with very diverse stakeholders and contexts, a multitude of property definitions are made, leading to different ways in which properties can be defined, exchanged, and handled. In order to avoid misunderstandings, confusion, and continuous redefinition of properties, there is a need to distinguish between those methods for describing properties and using them in a connected World Wide Web environment. In this chapter, we therefore look into property modelling approaches on the web and their associated functionality and limitations. We focus particularly on a linked data and semantic web context, yet parallels are made with general property modelling approaches as applied in relational databases, custom JSON and XML formats, and IFC-SPFF (Industry Foundation Classes in STEP Physical File Format). This includes explicit parallels with international standardisation initiatives in CEN, ISO, W3C, and buildingSMART. These property modelling approaches are compared, diverse implementations are documented, and guidelines are presented on when to use which property modelling approach – applied to the AECO (Architecture, Engineering, Construction, and Operations) industry. The resulting framework can be used to easily distinguish between property modelling approaches and make corresponding agreements for an information exchange.

2.1   INTRODUCTION

In a web-based world where plenty of stakeholders collaborate in many different ways, properties of objects are tremendously important, as those properties provide the actual data about the objects modelled and exchanged. Within Building Information Modelling (BIM) datasets, the attributes of objects are used to digitally describe construction project information, construction components, construction materials, and spatial zones in constructions such as a building site, a building, a storey, or a space. In these BIM datasets and many related models (e.g. Facility Management Information Systems (FMIS), product manufacturing databases and configurators), properties help to create a rich digital representation of a designed or existing construction, together with geometry descriptions as well as (construction) topological relations between and classification of construction-related information objects.

Properties can be very diverse and are needed for a large variety of use cases ranging from quantity surveys (e.g. dimension and cost properties), driving parametric models, handling building data for building simulation purposes (structural analysis, thermal analysis, etc.), exchange of data in a circular building economy, declarations of performance for manufactured building elements, tracing components on-site, specifying and controlling requirements over the assertion of measurement results coming from sensors, and so forth.

In this very broad domain of modelling and exchanging properties related to construction objects, several data modelling approaches exist. A reoccurring topic related to the modelling of properties in datasets, whether they rely on relational databases, JSON data, XML, graph databases, STEP or RDF data, is the *degree of detail and standardisation* with which those properties are described.

2.1.1 Simple property names and values

On the one hand, the AECO industry is well-versed in a direct, ad-hoc exchange of properties that is often unstandardised and underdefined. The exchange of properties, for example, maintained in an organisation's BIM environment, can be arranged by sharing entire BIM datasets in IFC format, exporting spreadsheets from Autodesk Revit, importing and/or exporting lists through custom scripts, uploading of COBie spreadsheets to online FMIS platforms, retrieving the JSON output from one of several web services, and endless more examples. Most commonly, such properties only have a label and a value, and do not allow or support the exchange or storage of any other metadata (e.g. unit, date, author). In other words, the definition of properties is very light.

An example can be seen in Listing 2.1 for an I-Shape Profile Definition, which can be used to describe a structural I-beam component. This example shows properties (e.g. overallWidth) as simple discrete values assigned to property names using a simple key-value construct: property names are represented as simple 'keys' and property values are listed as simple 'values'. There is no inclusion of metadata for the property values or names.

Listing 2.1   Simple property modelling focusing on direct exchange of values, shown in JSON syntax

```
 1  "sweptArea" : {
 2      "type" : "IfcIShapeProfileDef",
 3      "profileType" : "AREA",
 4      "profileName" : "UC305x305x97",
 5      "position" : [[ 0.0, 0.0 ], [ 0.0, 1.0 ]],
 6      "overallWidth" : 305.3,
 7      "overallDepth" : 307.9,
 8      "webThickness" : 9.9,
 9      "flangeThickness" : 15.4,
10      "filletRadius" : 15.2,
11      "flangeEdgeRadius" : 0.0,
12      "flangeSlope" : 0.0
13  }
```

2.1.2 More complex property names and values with metadata included

On the other hand, a number of initiatives exist that support and strongly recommend well-defined vocabularies of standardised property definitions, thereby typically supporting full definition, representation, and exchange of associated metadata. In particular, systems with a long lifespan, in which logs need to be kept of changes and a sequence of property values (e.g. FMIS environments, the buildingSMART Data Dictionary, national OTLs), tend to prefer the latter property modelling practice). It is insufficient to store property names and attributes as simple key-value combinations in such cases.

Listing 2.2 shows a conservative example of the extent to which the data in Listing 2.1 can be defined more strictly and elaborately. For example, the property overallWidth of the I-Shape Profile Definition in JSON is not a simple value ('305.3') any more, but it becomes an object in itself, which has several properties itself. This is often referred to as 'objectified properties', 'qualified relations', or 'data wrapping' in the AECO industry. In this case, the value for the property `overallWidth` is objectified, and additional data for this property are included: `id`, `type`, `value`, `unit`, `definedBy`, `date`.

This objectification pattern can be further extended and possibilities in adding further properties' (meta)data are endless. For example, the `unit` of the property `overallWidth` can also be objectified in itself, having more data definitions and metadata in itself again (e.g. translations, conversion factors, etc.).

Listing 2.2   Complex property modelling focusing on extensive descriptions of property data, shown in JSON syntax. Example for illustration purposes only, based on Listing 2.l, and limiting to the `overallWidth` property of a `sweptArea`.

```
1   "sweptArea" : {
2       "type" : "IfcIShapeProfileDef",
3       "profileType" : {
4           "type" : "IfcProfileTypeEnum",
5           "label" : "AREA",
6           "description" : "The resulting geometric item is of type
                  surface. The resulting geometry after applying a
                  sweeping operation is a swept solid with defined
                  volume."
7       },
8       "profileName" : "UC305x305x97",
9       "position" : {
10          "type" : "IfcAxis2Placement2D",
11          "location" : {
12              "type" : "IfcCartesianPoint",
13              "coordinates" : [ 0.0, 0.0 ]
14          },
15          "refDirection" : {
16              "type" : "IfcDirection",
17              "directionRatios" : [ 0.0, 1.0 ]
18          }
19      },
20      "property" : {
21          "id" : "b7432442-a7be-4474-b52a-b76ad51774dc",
22          "type" : "overallWidth",
23          "value" : 305.3,
24          "unit" : {
25              "id" : "e0687323-b2ef-454c-829f-852ee8a9173f",
26              "type" : "CENUnit" ,
27              "label" : "mm",
28              "symbol" : "mm",
29              "unitType" : "LengthUnit",
30              "SIUnitType" : "SI derived unit",
31              "fullName" : "millimeter"
32          },
33          "definedBy" : "Mathias Bonduel",
34          "date" : "2020-09-15"
35      }
36  }
```

For this chapter on property modelling approaches used in a web context, a separation is made between the *definition of properties* (meaning) on the one side and the *application of such defined properties* (use) in digital representations of the built environment on the other side. Although the examples in this chapter document this topic from a semantic web perspective (graphs), parallels can easily be made with other modelling approaches on the Web (e.g. web services with a relational database backend, JSON-based web services) and traditional file-based collaboration formats such as IFC-SPFF.

Researchers and industry actors actively examine the opportunities offered by web technologies for application in the AECO domain. Especially the uptake of semantic web technologies is more and more considered for describing constructions in a collaborative environment. semantic web technologies have demonstrated several advantages over conventional BIM technologies that are mainly file-based [301]. semantic web technologies are built around the Resource Description Framework (RDF) graph data model and are in essence domain-agnostic. Consequently, the technology has already been applied over its +20 years of existence by a large variety of researchers and software developers and this in a plethora of knowledge domains. Semantic web aspects that are particularly useful for the construction domain are fine-grained data exchange, standardised querying, generic reasoning engines to infer additional statements, the interlinking of previously disparate datasets, and the flexibility in defining and extending schemas for interoperability [301]. semantic web technologies allow us to rethink the decision process for and structure of property definitions, as well as modelling patterns for asserting properties and property values in digital models representing constructions.

First, an overview is given of best practices for defining and exchanging property data in Section 2.2. Continuing with Section 2.3, crucial aspects related to diverse property modelling approaches are extracted from the most recent international standards as a guideline for the remainder of this chapter. This includes short examples of the implementation of properties using the IFC standard (STEP version), yet similar conclusions can be made for other data schemas based on JSON, XML, and web ontologies in RDF.

A thorough analysis of approaches for asserting properties with semantic web technologies in digital construction datasets is contained in Section 2.4. This leads to a number of graph patterns for property modelling (L1, L2, and L3) documented in Section 2.5. Section 2.6 subsequently describes compatible methods for creating reliable property definitions using semantic web technologies that can be referenced from digital construction datasets. In a short Section 2.7, initial steps are made towards a recommendable property modelling method. Section 2.8 combines final conclusions together with indications of future work.

2.2 GUIDELINES AND STATE OF PRACTICE FOR MODELLING AND EXCHANGING PROPERTIES

In this section, we document how properties can be modelled and defined, after which we indicate how these defined properties can be used and exchanged.

2.2.1 Definition of properties

Modelling and defining data schemas, including properties and attributes of objects, is a very well-studied domain and has a long state of practice. Guidelines and state of practice are, therefore, widely available. The best places to start with, in terms of these guidelines and the state of practice, are likely the Unified Modelling Language (UML) [138] and Entity Relationship Diagrams (ERD) [65].

Before setting up a relational database or developing an application, best practice dictates that a number of actions need to be performed first:

1. Define the scenario and context for which the database/application is built (incl. use of competency questions)
2. Clearly define the scope of the database/application
3. Prepare a draft ERD (relational database) or UML diagram (software application) that lists all concepts, relationships, and properties needed to comply with the provided scope and use

This section will briefly indicate how properties are modelled using those notations.

2.2.1.1   Entity relationship diagrams (ERD)

ERD are of particular use in the creation and definition of relational databases that can be queried using the Structured Query Language (SQL). An example ERD can be found in Figure 2.1. This diagram follows the notation that was originally specified by [65], a notation that is still widely used today, although no agreed standard exists.

Only when the ERD is finished and relatively stable, one can proceed with the definition of the relational database using the available statements in a relevant Data Definition Language (DDL), e.g. SQL CREATE TABLE. This chapter will of course not deal in detail with how to create an ERD diagram. Yet, of particular relevance for this chapter is the distinction between attributes and concepts. Figure 2.1 shows concepts in rectangles and attributes in ellipses. Relationships are represented by diamond shapes, including their cardinality restrictions. Identifying attributes are often underlined.

When creating a relational database for this ERD, all entities (rectangles) will result in single tables, and the associated attributes (ellipses) are included as columns in this table. The identifying attributes define which columns are used as primary keys. Relationships (diamonds) find their way into the database definition either as foreign keys in the existing tables (one-to-one and one-to-many relationships) or as extra

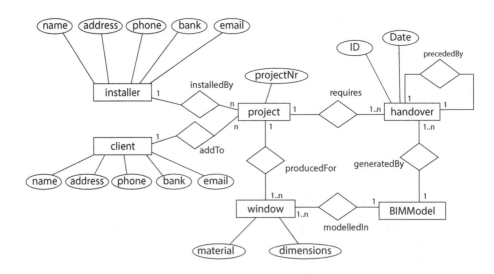

Figure 2.1 Example ERD with different types of properties included.

'intermediate' tables with foreign keys pointing to the associated tables (many-to-many relationships).

A crucial challenge when defining such a database is the question whether certain data needs to be included as either an attribute (ellipse) or as an entity with attributes on its own (rectangle). A very simple and often encountered example in this regard is the definition of an 'address', but this can easily be applied as well to the cases of 'overallWidth' and 'unit' in Listings 2.1 and 2.2. In some databases, the address data is absolutely critical (e.g. governmental database with population data), in which case it is necessary to define addresses in full detail. In other cases, the address is a simple additional fact and can simply be included in the form of a single `string` value. In the first case, separate tables (entities) are needed to represent street names, city names, postal codes, and so on, thus increasing the complexity level of the database in this particular content (similar to Listing 2.2). In the second case, the database is kept intentionally simple, in order to keep it manageable and focused on what is in scope for the database design (similar to Listing 2.1).

2.2.1.2 UML class diagrams

Where ERD diagrams typically result in relational databases, UML Class Diagrams typically result in object-oriented code written in modern programming languages such as Java and C#. Other than that, its key features are very similar, and it is often intelligent to align a UML Class Diagram with an ERD diagram in case one models software that communicates with an underlying (relational) database. In a UML Class Diagram, it is possible to define key classes in a domain, including its relations, attributes, properties, functions, and so forth. An example diagram is provided in Figure 2.2.

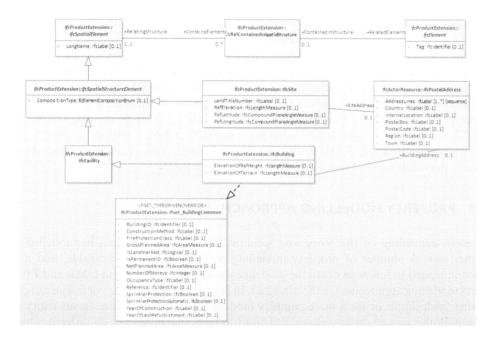

Figure 2.2 Example UML class diagram.

Although the notation of the UML Class Diagram is quite different from the ERD notation, the idea is very similar. Instead of entities, the diagram consists of Classes, named at the top of each rectangle/table. Attributes are not marked with ellipses, but they are rather listed underneath the class name, including the datatypes (integer, boolean, etc.) of the property value. Relationships exist between rectangles/tables, marked through the lines and arrows between the classes. These relationships rely on attribute names for their definition. For example, in the case of Figure 2.2, the `ReplacementOrder` class relates to the `Warehouse` class through the `warehouseLocation` attribute.

Similar to the ERD method, it is critical to reflect about whether a property should be modelled as a class of its own or rather as an attribute of another class. In the example in Figure 2.2, it has been considered important enough to define `Counterparty` as a separate class, although it has only one `name` attribute. On the other hand, the `location` of a `Warehouse` is modelled as an attribute, and is therefore not important enough in the scope of this class diagram to justify the overhead of a full class definition.

The outlined data modelling paradigms in UML and ERD have their effects on the availability of properties in code (e.g. Application Programming Interfaces or APIs) and databases, as either attribute values or fully defined property classes or entities. This of course has its effect on the retrieval and use of these data. Representation of properties in one way makes it not possible to retrieve them in the other way, and vice versa. Hence, these are very important data modelling decisions that need to be made, which depend almost fully on the context and scenario of use (cfr. competency questions).

2.2.2   Application and use of defined properties

In the application, exchange, and use of properties, one commonly relies on data serialisation and exchange mechanisms. Applications seldom make their databases openly available. Data and functions are typically only accessible using the custom API of the tool. In particular, in the context of the Web, users and applications rely heavily on the communication of data through JSON (e.g. through REST or Representational State Transfer APIs) and XML (e.g. through SOAP or Simple Object Access Protocol APIs) to enable its wide use of a service-oriented architecture (SOA) and inherent use of web services and micro-services. The data in this format often reflects the underlying data structure, implying that attributes in a database or in code are often exported as XML attributes or simple key-value pairs in JSON, whereas entities or objects are typically serialised into XML Entities or JSON objects. This effect has already been briefly demonstrated in Listings 2.1 and 2.2 in case of JSON.

2.3   PROPERTY MODELLING APPROACHES

Clearly, depending on the scope and scenario of use, properties can be modelled in either way. A number of property modelling patterns are hereby available, and it is recommended to follow in detail the guidelines and state of practice in UML and ERD data modelling approaches (see Section 2.2). In general, based on this review, one can opt either for a simple or for a more complex modelling method (attribute versus entity or class). Both have certain advantages and disadvantages that need to be considered, which will be explained in more detail in this section. We will hereby also refer to Listing 2.3 for a small example in the SPFF version of IFC.

Listing 2.3   Excerpt of an IFC-SPFF file: IFC attributes and IFC properties

```
1  /* construction component with IFC attributes */
2  #217=IFCWALLSTANDARDCASE('0jf0rYHfX3RAB3bSIRjmmy',#1,'Basic Wall:
       Exterior - Brick on Block:184944',$,'Basic Wall:Exterior -
       Brick on Block:130892',#17866,#18858,'184944');
3  /* ... */
4  /* IFC properties related to IfcWallStandardCase of line #217 */
5  #25664=IFCRELDEFINESBYPROPERTIES('1QzuEumazE4veTqU_HR5$b',#1,$,$
       ,(#217),#23904);
6  #23904=IFCPROPERTYSET('0hUJB39M5EYxXWY$tZw5Av',#1,'Pset_WallCommon
       ',$,(#322,#39,#35,#41));
7  #39=IFCPROPERTYSINGLEVALUE('LoadBearing',$,IFCBOOLEAN(.F.),$);
```

In any case, these different ways of defining properties have led the construction industry to a large variety of property modelling patterns in use, each supporting and leading to a different kind of functionality. This of course makes sense when considering that data models are typically defined in support of a particular scenario or use (cfr. competency questions in Section 2.2). However, this also leads to a diversity of data standards with regard to property modelling. The need for a uniform approach towards the modelling of both qualitative and quantitative properties thus becomes apparent. To a large extent, this has been at stake of the EN ISO 23386:2020 standard[1] that focuses on a methodology for establishing, linking, and sharing property definitions and their metadata: 'Methodology to describe, author and maintain properties in interconnected data dictionaries'.

2.3.1   Simplified property modelling

In a simplified property modelling approach, properties are modelled predominantly as attributes with property values of particular datatypes (Figure 2.3).

2.3.1.1   *Advantages*

The very clear advantage of this simple modelling method is typically recognised in the human-friendliness and high level of readability. Indeed, when looking at the example in Listing 2.1, it is relatively easy to recognise what data and property values are transmitted in one glimpse. This is a big advantage for tool developers that want to have as easy and straightforward access to data as possible.

The simple representation of property values and names, as well as the exclusion of large sets of metadata also minimise the size of data represented and exchanged using this

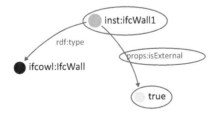

Figure 2.3   Simple property modelling (attributes).

approach. This is a second advantage that can be recognised here: it is straightforward to exchange small data snippets with plenty of data values, e.g. in a service-oriented architecture.

2.3.1.2 Disadvantages

Besides the mentioned advantages, several disadvantages can be named for this simple property modelling approach. For example, data is very loosely defined and it is necessary in the majority of cases to define property values in more detail. The lack of metadata makes it difficult to come to standardisation and agreement, as data is underdefined.

Furthermore, systems and databases typically define data in much more detail. They hereby serve as the single source of truth (SSOT) for the data that they contain. When data is transmitted to and from such native systems, using a simplified modelling approach, the data detaches from its source. Data in this simple property modelling approach therefore tends to be less trustworthy as it is near to always derived from a more complex source.

2.3.1.3 Primary scenarios of use

These types of property representations and definitions are behind the example in Listing 2.1. Obviously, this approach in property modelling is unfit for modelling properties in full detail, which is of strong importance in any data management platform. For example, FMIS, BIM modelling tools, CDE (Common Data Environment) platforms, and any framework that is required to keep track of the development of property values over time, incl. relevant details and metadata, cannot use this data modelling approach at the core.

They can use this property modelling approach at their edges, i.e. where they import and export data. Indeed, when handing over data, an overload of metadata is often unneeded and holding back the data transfer. Because of its typical small size and good readability, this property modelling approach is of use in quick and direct exchanges of property data. This includes small responses by web services and query engines, which simply deliver small responses to specific queries. Commonly used visual programming environments such as Grasshopper and Dynamo, but also many programming libraries are quite well able to consume such data fast and flexibly.

The very important downside in this approach, however, is that this data is fully detached from its source: once downloaded or retrieved from a server, there is no direct connection to the original dataset anymore (no single source of truth).

2.3.1.4 Requirements

When aiming at this particular simplified modelling approach and thus representing properties in simple data exchanges, the following practical requirements need to be met:

1. Readability of the data, both from human and machine perspective
2. Straightforward filter and query mechanisms supported
3. Support for out of the box parsing and (de)serialisation toolboxes
4. Uniform and well-defined property naming strategy, in order to make up as good as possible with a missing property identification mechanism
5. Alignment with available tools, frameworks, and methods used in practice (e.g. Speckle, bSDD API, Dynamo, web services)

2.3.1.5   IFC-SPFF example

The vendor-neutral IFC data schema, documented in open buildingSMART specifications[2] as well as in EN ISO 16739-1:2020, allows us to analyse how properties are modelled and defined. Two types of properties are implemented, i.e. 'IFC attributes' (directly connected to the object) and 'IFC properties' (grouped in an IFC property set). Note that 'IFC relations' between objects, e.g. a construction component that is contained in a certain space (IfcRelContainedInSpace), are implemented in IFC using other mechanisms which are out of scope here.

The IFC attributes are excellent examples of properties defined in a simple modelling pattern. All IFC attributes are predefined in the IFC data schema and are directly associated to construction components, spatial concepts (IfcSite, IfcBuilding, IfcBuildingStorey, IfcSpace, or IfcZone) as well as IFC properties and property sets. IFC attributes are either required or optional and tend to describe more generic information managed by applications. An example of an IFC attribute is the Globally Unique Identifier (GUID) of the wall object, e.g. `0jf0rYHfX3RAB3bSIRjmmy` in Listing 2.3. These attributes are directly available together with the object and they tend to be easily found and usable.

2.3.2   Complex property modelling

In a complex property modelling approach, properties are modelled predominantly with entities (ERD) or classes (UML) or similar (objectification and data wrapping) in order to allow an as complete as possible semantic representation of relevant data and metadata. An example is shown in Figure 2.4.

2.3.2.1   Advantages

In this more complex approach, very strong advantages lie in the full and complete definition of all related data. Modelling properties using distinct classes and entities also provides a large amount of flexibility, because it is trivial to add an additional property with unit, name, value, and metadata as preferred, because the schema is often domain-agnostic (e.g. class 'Property' can be filled in with any new property as needed).

Furthermore, every property can be defined with a unique identifier, which can be used to unambiguously reference it. This identification system is used in the buildingSMART Data Dictionary[3], in the form of GUIDs. In a semantic web context, the mechanism of URIs (Uniform Resource Identifiers) in RDF graphs allows to define both property name *and* location for retrieval with a well-proven URI referencing mechanism. This property identification mechanism is not available in a simplified modelling approach as identified earlier, in which there is only a key and a label available. When using a URI, which is inherently available in any RDF serialisation (JSON-LD, RDF/XML, Turtle, etc.), but not in custom JSON, XML, and other formats, it is possible to look up the definition of a property and find more relevant information (definitions, translations, connections to other standards, etc.).

2.3.2.2   Disadvantages

As for the downsides of this approach, this data tends to be cumbersome and complex. Especially when the data from this single source of truth property management system is made available outside of its system, it is often unclear to an end user why the data

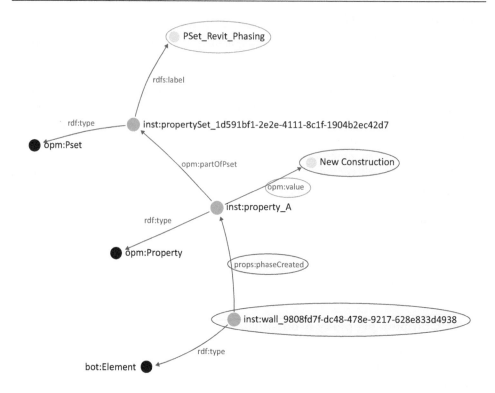

Figure 2.4 Complex property modelling, in this case including a propertyset definition.

needs to be defined in such a complex manner (need for simplified views). Furthermore, it is not self-evident to apply the data in the more agile and simpler data workflows often used in the AECO domain.

2.3.2.3 Primary scenarios of use

This data modelling method for properties is typically used in systems that store and log all relevant data, including CDEs, FMISs, BIM modelling systems, and so forth. It is key that all data is logged, and a software developer can query for any of the available data in this single source of truth (SSOT), in order to display it in a user-friendly manner to an end user. This data modelling method for properties is less fit for simple data exchanges.

2.3.2.4 Requirements

When aiming at this particular modelling approach and building a web-compliant information management system, it should be able to fulfil at least the following practical requirements:

1. The value of a property can either be another object or a literal (Boolean, integer, double, string, etc.).
2. Units of measurement can be represented for quantitative properties such as 'length', in order to make sense of a property value.

3. Metadata can be represented for asserted properties including accuracy, tolerances, grouping, measurement metadata and dependencies between properties.

4. Complex numeric property values can be represented (min, max, function, list, table, etc.).

5. Links between property types and potential domains of the property can be represented (classes of instances described by the property).

6. Defining links between property types and construction standards that define them is possible.

7. Versioning of data can be applied, in order to enable keeping track of changes supported by change metadata (time, actor, etc.).

8. It is possible to define different values for the same property, e.g. measured value, designed value, and targeted value.

2.3.2.5 IFC-SPFF example

The large majority of properties in IFC is modelled using a complex property modelling method. An example of how this is done, is the LoadBearing property in line number #39 of Listing 2.3 (see also Figure 2.5). Similar to this one example, most properties in IFC datasets are included using IfcPropertySingleValue, IfcPropertySet, and IfcRelDefinesByProperties. Namely, single value IFC properties are grouped in IfcPropertySet (Pset) instances that are either predefined in the IFC standard or are defined by (groups) of users (e.g. exported as custom Revit property sets) or other standardisation bodies. The definition of these property

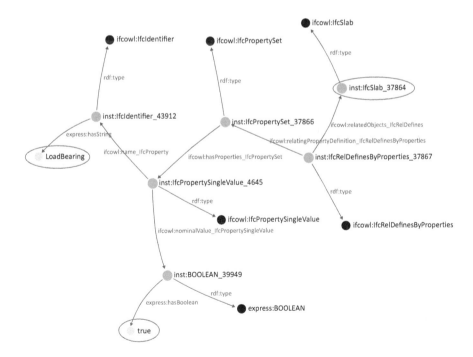

Figure 2.5 Complex property modelling for the IFC sample.

sets happens with PSD XML files[4]. These properties are related to the component to which they belong through the many-to-many (objectified) relationship identified by the `IfcRelDefinesByProperties` entity.

In contrast to simple IFC attributes (e.g. GUID, Name, etc.), these complexly defined IFC properties require a more elaborate data structure according to the IFC schema, making it less intuitive to retrieve such properties. A limited amount of additional metadata can be added to these individual properties such as units, text descriptions and property value types. The Pset instance containing properties can be connected to a slightly broader range of metadata including external documents or specifications. For example, the property `LoadBearing` in Listing 2.3 is part of `Pset_WallCommon`, which has more information attached about this propertyset (metadata).

Summarising the above, IFC properties can be categorised as closer to the complex property modelling approach from Section 2.3.2. In Listing 2.3, the same wall has a connected Pset named `Pset_WallCommon` corresponding with a predefined Pset Definition[5] (PSD) from the IFC standard. The Pset documentation can also be retrieved through the buildingSMART Data Dictionary[6] (bSDD).

2.4 ASSERTING PROPERTIES IN A SEMANTIC WEB CONTEXT

In the following sections, we will show in more detail how the above property modelling approaches can be applied in the context of the semantic web, thereby focusing mostly on the more complex approaches. This section particularly discusses available approaches for asserting properties and related information such as units in datasets. To abbreviate long URIs, the prefixes given in Table 2.1 are used throughout the remainder of this chapter.

Table 2.1 Prefixes and namespaces used in this chapter

Prefix	Namespace
rdf	http://www.w3.org/1999/02/22-rdf-syntax-ns#
rdfs	http://www.w3.org/2000/01/rdf-schema#
owl	http://www.w3.org/2002/07/owl#
xsd	http://www.w3.org/2001/XMLSchema#
ifcowl	Multiple schema versions, e.g. v4.1:
	https://standards.buildingsmart.org/IFC/DEV/IFC4_1/OWL#
bot	https://w3id.org/bot#
beo	https://pi.pauwel.be/voc/buildingelement#
dot	https://w3id.org/dot#
qudt	http://qudt.org/2.0/schema/qudt
unit	http://qudt.org/vocab/unit/
prov	http://www.w3.org/ns/prov#
dct	http://purl.org/dc/terms/
schema	http://schema.org/
cdt	http://w3id.org/lindt/custom_datatypes#
skos	http://www.w3.org/2004/02/skos/core#
opm	https://w3id.org/opm#
contax	https://w3id.org/con-tax#
ex	A fictive namespace for example terminology and data:
	http://ex.org/

2.4.1   Methods to attach properties

The modelling of properties using semantic web technologies can be done by attaching properties to class instances and/or classes. The most frequently used method to define properties using semantic web technologies is to connect properties to class instances (see Figure 2.6). Clearly, the property name (predicate `ex:propName2`) is then ideally defined in an OWL ontology for semantic disambiguation purposes. This approach of regular instance properties is most commonly used, since no reasoning engine is needed and the modelling of properties is more straightforward. This approach resembles the concept of attributes in a database or in object-oriented code.

Alternatively, properties can be attached directly to class definitions, as shown in Figure 2.7. When the class is to be logically interpreted as an individual, this technique implies 'OWL punning'[7]. For example, one may want to define that certain walls are all defined by a person called Bob, in which that wall type is defined using a class, and Bob is defined as an instance. In that case, the triple beo:Wall ex:constructedBy ex:Bob would mean that *all* instances of that (predefined) Building Element Ontology (BEO) class are installed on-site by someone identified by the URI ex:Bob. This approach is also used in several BIM modelling tools, e.g. Revit, to be able to define new types (families) with predefined sets of type properties.

A third alternative is to infer instance properties through so-called OWL class restrictions as demonstrated in [318]. This example is shown in Figure 2.8. An advantage of this approach is that it is more formal and thus more stable than the first approach (Figure 2.6). This approach also requires an inference engine to infer also the simpler properties in the data, which is an additional but not insurmountable overhead.

2.4.2   Units for quantitative properties

A standardised modelling approach for attaching unit information to quantitative properties in a semantic web context does not exist. Three approaches explored in literature and practice are discussed hereafter.

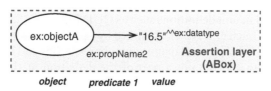

Figure 2.6 Instance property.

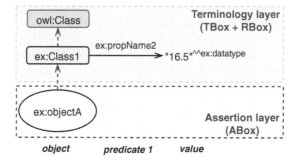

Figure 2.7 Type property.

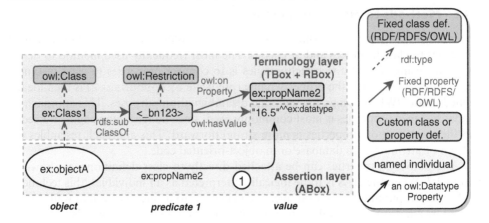

Figure 2.8 Inferred instance property, marked by I, via an OWL class restriction.

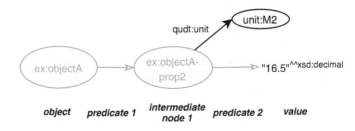

Figure 2.9 Units with QUDT, using a qualified relation.

A first approach is to informally mention a fixed unit in the text description as part of the property definition. This approach is often taken in the simple property modelling methods. Each time the property is used in a dataset, the same unit has to be assumed.

An alternative and more stable and scalable method is to apply URIs from recognised ontologies specialised in unit definitions such as QUDT [313] or the Ontology of units of Measure (OM) [324]. Each unit is then identified by a predefined URI and is connected to additional metadata such as textual definitions, abbreviations and links to specifications. As an example, Figure 2.9 contains a part of the definition of the 'square metre' unit in the QUDT ontology. The same URI representing the unit can be attached to an example property definition ex:doorWidth in its terminology box (TBox) using a (still to be defined) owl:AnnotationProperty. If the unit is not fixed for each occurrence of the property, a qualified relation should be installed between the construction object and the property value using the approach illustrated in Figure 2.9. The intermediate node can now be used to indicate the used unit by linking to the correct URI using qudt:unit.

The third method is based on Custom Datatypes (CDT) and extends the content of an RDF literal with a unit from the Unified Code for Units of Measure (UCUM) specification [223] as demonstrated in Figure 2.10. This takes the approach of Figure 2.6, only with a specialised CDT datatype (cdt:ucum) and an embedded UCUM string code to declare the unit.

An extended SPARQL engine[8] is available to demonstrate that values of numeric properties with different CDT units can be compared while querying. Listing 2.4 contains such a SPARQL query over the triple depicted in Figure 2.10. Unit conversions are managed by the query engine.

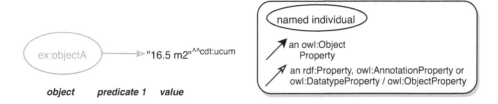

object   predicate 1   value

Figure 2.10 Units with CDT.

Listing 2.4 Query to retrieve all properties containing areas with the units defined using CDT and smaller than 1500 dm²

```
1  SELECT ?object ?propName ?propValue
2  WHERE {
3    ?object ?propName ?propValue .
4    FILTER ( ?propValue < "1500 dm2"^^cdt:ucum )
5  } ORDER BY ?propValue
```

2.5 GRAPH PATTERNS FOR PROPERTY MODELLING

An initial overview on property modelling approaches was already given in Section 2.3. Further to this overview, a more elaborate comparison between several property modelling approaches in existing ontological models is available in the *Towards a PROPS ontology*[9] presentation of the author to the W3C Linked Building Data Community Group (LBD-CG). References are made in this presentation to ifcOWL [299], ifcWoD [260], simpleBIM [298], and BIMSO/BIMDO [273] as well as other ontologies such as the SEAS (Smart Energy Aware Systems) ontologies [222] for measured and simulated properties. From all this material, it can be concluded that a considerable variety of property modelling patterns is implemented in ontologies, each with a different amount of functionality.

In general, the amount of functionality supported by an approach is proportional with the complexity of the modelling pattern. Based on all the previous material and research, three uniform modelling patterns or properties 'levels of complexity' (L1, L2, and L3) can be distinguished, with rising complexity. The number (1, 2, or 3) hereby represents the number of 'steps' or predicates between object and the value of its property, which will become clear in the below examples. Depending on the use case, it is considered beneficiary to choose one specific property modelling level.

2.5.1 Level 1

L1 uses a single triple to link an object (building element, spatial zone, material, etc.) to a property value (Figure 2.11). Properties versioning, complex numeric property values, and metadata connected through regular Assertion Box (ABox) triples cannot be defined following this pattern, unless the triples themselves are annotated using alternative techniques such as RDF reification, named graphs or RDF* [178].

In level 1, the units can be defined in the definition of the property if they have to be assumed for each occurrence (see Section 2.4.2). If, however, a property can have

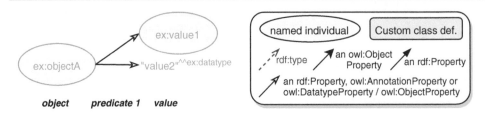

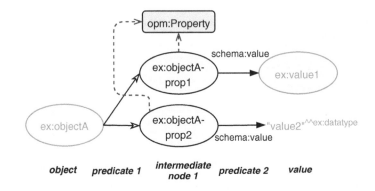

Figure 2.11 Level I.

Figure 2.12 Level 2.

values of different units, one can optionally rely on the CDT approach [223] to make this modelling approach more standard and explicit on data level. Other popular ontologies for units such as QUDT and OM cannot be used in an L1 property modelling approach since they rely on qualified relations (objectification).

2.5.2   Level 2

L2 properties include two predicates between an object and the value of its properties. In other words, they are based on single qualified relations (Figure 2.12). The first predicate links the object to an intermediate node (objectified property), while the second predicate connects the intermediate node to the property value.

This second predicate can be expressed using a generic `hasValue` property. The schema:value property from schema.org can be used for that purpose. Other properties can be used to specify more complex numeric property values (e.g. schema:minValue, schema:maxValue) in parallel. In addition, terminology from the Building Product Ontology (BPO) [406] can be used to define a flat list, interval steps and 2D lists for numeric values. Besides CDT, it becomes possible to use ontologies such as QUDT or OM to assign units to numeric properties and stay within the boundaries of existing W3C standards for RDF and SPARQL.

Another advantage of the intermediate node is the possibility to attach metadata using ABox triples, e.g. accuracy, provenance, derived properties or grouping for a specific occurrence of property type. For provenance metadata, terminology from the standardised PROV-O ontology can be used (prov:generatedAtTime, prov:wasGeneratedBy, prov:wasAttributedTo). Any other metadata can be added as well, similar to any other complex property modelling approach (cfr. example in Section 2.1.2).

2.5.3   Level 3

Finally, L3 is based on double qualified properties resulting in three predicates between the object and the property value (Figure 2.12). Extending the functionality of L2, Level 3 properties allow to conveniently version individual property values inside an RDF graph. In other words, this explicitly allows versioning and property states, while also still allowing the elaborate representation of metadata. As a result, this graph pattern for property modelling is by far the most powerful representation method.

The metadata and unit information can be linked to the property node (first intermediate node) if it is valid for every property state, or to the property state node (second intermediate node) if it is specific to a certain property state. An example of the second is the connection of property change and creation metadata (prov:generatedAtTime, prov:wasGeneratedBy, prov:wasAttributedTo). Terminology from the Ontology for Property Management (OPM) [316,317] can be used to link the first intermediate node (an opm:Property node) to the second intermediate node (an opm:PropertyState node) using opm:hasPropertyState. The opm:CurrentPropertyState and opm:Deleted classes, both subclasses of opm:PropertyState, can be asserted to respectively classify the property state as 'most recent' or as currently being 'inactive'. The generic schema:value, schema:minValue, and schema:maxValue terminology from schema.org or alternative terminology from BPO can be used on the position of the third predicate.

2.5.4   Summary

A summary of the findings discussed per property level is available in Table 2.2. This table shows what property modelling method may be useful for what purposes and what their characteristics are. It relies on the diverse requirements identified in Section 2.3 and thus includes the potential both for use cases needing simple and more complex modelling approaches.

From this overview, Level 3 is clearly most relevant when the project history needs to be preserved, for example during an iterative design and engineering process that is internally managed within a company [317]. For the majority of use cases related to the construction industry, L2 is likely sufficient. This property modelling pattern aligns most closely to how it is done in the IFC standard, as well as other property modelling standards in the AECO industry. L1 cannot be used for complex property values, metadata and units defined using ontologies, yet excels for user-friendly representation of property values in small and purposed data exchanges.

It is possible to convert all the properties in a dataset to a specific property level using SPARQL Update queries. When downgrading, some information is of course lost such as the version history or property metadata. Hence, it is mainly recommended to choose your property modelling approach for the use case situation in which you are in.

2.6   PROPERTY DEFINITIONS FOR USAGE IN A SEMANTIC WEB CONTEXT

The property name can be located at different positions in the three mentioned property levels. Two main approaches can be considered. First, one option would be to define the property at the first predicate in each of the three levels of property modelling (Figures 2.11–2.13). RDF/RDFS and OWL define several fixed classes to type the property, i.e. respectively rdf:Property and owl:DatatypeProperty, owl:ObjectProperty or

Table 2.2 Evaluation of the functionality of the property levels

Practical requirements for properties	*Level 1*	*Level 2*	*Level 3*
Req 1a: Readability of the data	YES	NO	NO
Req 2a: Straightforward filter and query mechanisms	YES	YES	NO
Req 3a: Support for out of the box parsing and (de)serialisation toolboxes	YES	YES	NO
Req 4a: Need for uniform and defined property naming strategy	YES	NO	NO
Req 5a: Alignment with available tools	YES	YES	YES
Req 1b: value can be literal or URI	YES	YES	YES
Req 2b: units for quantitative properties	YES[a]	YES[b]	YES[b]
Req 3b: metadata for asserted properties	NO[c]	YES	YES
Req 4b: complex quantitative properties	NO	YES	YES
Req 5b: link between property type and potential domains of the property	YES	YES	YES
Req 6b: link between property type and standard(s) that defines it	YES	YES	YES
Req 7b: versioning of properties	NO	NO	YES
Req 8b: alternative states of properties (e.g. measured vs. designed)	NO	NO	YES

[a] using CDT for units
[b] using CDT or QUDT/OM/... for units
[c] unless alternative approaches for metadata of triples are used
(RDF reification, named graphs, RDF*)

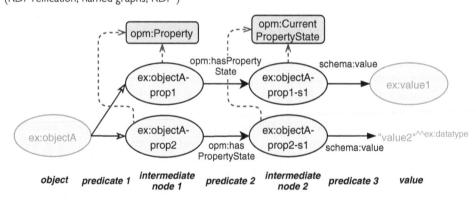

Figure 2.13 Level 3.

owl:AnnotationProperty. An alternative approach is to use a class (owl:Class or rdfs:Class) or an individual (skos:Concept) linked to the first intermediate node in Figure 2.12 or 2.13. Since this position is only available for L2 and L3, these alternative approaches cannot be used with L1. Five possible approaches for the modelling of the property name are identified and analysed hereafter.

2.6.1 Approach 1: hierarchy of rdf:Property

The generic rdf:Property class can be used to classify property names, while the hierarchy between these property names can be established through rdfs:subPropertyOf.

Using rdf:Property without additional owl:DatatypeProperty, owl:ObjectProperty or owl:AnnotationProperty would result in an underspecified property definition according to a strict interpretation of the OWL standard [177]. Consequently, some OWL reasoners might behave in an unexpected way, returning no, wrong or incomplete answers. Users wishing to infer such properties through OWL class restrictions might also notice strange behaviour of reasoners as that construct only supports owl:DatatypeProperty and owl:ObjectProperty properties. This approach is hence not recommended.

2.6.2   Approach 2: hierarchy of owl:AnnotationProperty

Instead of rdf:Property, the owl:AnnotationProperty class can be used to define properties in a way that is accepted by OWL inference engines. Again, a hierarchy of properties is made using rdfs:subPropertyOf relations. Again, OWL class restrictions are not possible since that construct only supports owl:DatatypeProperty and owl:ObjectProperty properties. This approach is hence not recommended.

2.6.3   Approach 3: hierarchy of owl:DatatypeProperty and owl:ObjectProperty

The following approach can be recognised as the most 'normal' way of modelling properties with OWL, i.e. defining properties as instance of owl:ObjectProperty (always for L2 and L3; L1 with URIs as value), or as an instance of owl:DatatypeProperty (L1 with literals as value). Figure 2.14 shows how this is modelled in the case of an object property with L2, while the earlier Figure 2.6 shows this in the case of a datatype property with L1.

Property definitions of properties that have a literal value (e.g. a 'height' property) will need certain URIs when they are used as an owl:ObjectProperty and other URIs when they are as a owl:DatatypeProperty. It is not possible to use the same URI, simply because a property definition cannot simultaneously be an owl:DatatypeProperty and owl:ObjectProperty. If one wants to enable L1, L2, and L3 properties with standard names and URIs that point to each other, it is proposed to give the owl:ObjectProperty all metadata and to create a reference link between both property URIs using the informal rdfs:seeAlso relation. This is not the most formal method, yet it does bind property

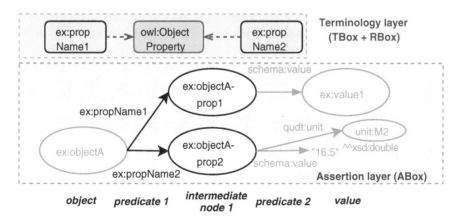

Figure 2.14 Approach 3.

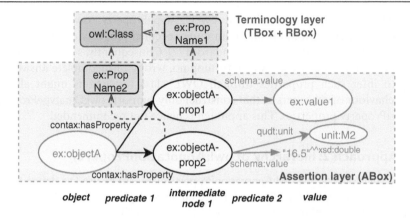

Figure 2.15 Approach 4.

definitions of very different sort. Furthermore, it reduces the amount of duplication of metadata (multilingual labels, definitions, potentially related objects, etc.). A hierarchy of properties can be defined using rdfs:subPropertyOf relations, both in the case of an owl:DatatypeProperty and an owl:ObjectProperty.

2.6.4   Approach 4: hierarchy of owl:Class

Alternative to the first three approaches, the first intermediate node between the object and the property value in L2 and L3 patterns can get a class assigned that indicates the actual name of the property. In this case, a single taxonomy of classes suffices as there is no need to differentiate between owl:ObjectProperty and owl:DatatypeProperty taxonomies. A hierarchy of classes representing property definitions is made using rdfs:subClassOf relations. The object and the first intermediate node can be connected using for example the contax:hasProperty from the ConTax ontology. Of course, this approach is not available for simple L1 properties. An example overview diagram of this approach is given in Figure 2.15.

2.6.5   Approach 5: hierarchy of skos:Concept

Similar to approach 4, the first intermediate node links to another node that indicates the name of the property. In this approach however, the node is a metaclass as it is an instance of skos:Concept connected through contax:propertyName, a specific property defined in the ConTax ontology. Any other ontology may be used as well. A single SKOS-based hierarchy of concepts indicating property names is made using the skos:broader relation. As mentioned for approach 4, the object and the first intermediate node can be connected using for example the contax:hasProperty from the ConTax ontology. An example overview diagram of this approach is given in Figure 2.16.

2.7   TOWARDS A RECOMMENDED MODELLING OF PROPERTIES

Each approach has its advantages and drawbacks, which are summarised in Table 2.3, making it difficult to move one uniform modelling pattern forward. This particular issue is currently under discussion within different organisations, such as the W3C LBD-CG,

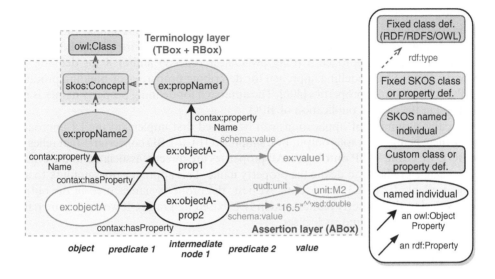

Figure 2.16 Approach 5.

Table 2.3 A summary of the five approaches for modelling and defining the actual
name of the property.

Evaluated approach	Prop. level	OWL2 profile	owl:Class restrictions for properties	Taxonomy size
I) rdf:Property	LI/2/3	None	NO	$P_{lit} + P_{URI}$
2) owl:AnnotationProperty	LI/2/3	All	NO	$P_{lit} + P_{URI}$
3) owl:DatatypeProperty and owl:ObjectProperty	LI/2/3	All	YES	$2 * P_{lit} + P_{URI}$
4) owl:Class	L2/3	All	YES	$P_{lit} + P_{URI}$
5) skos:Concept	L2/3	All	YES	$P_{lit} + P_{URI}$

P_{lit} refers to the properties that have an RDF literal as property value while P_{URI} denotes
the properties that have another URI as value.

buildingSMART International, and the CEN/TC 442 WG4 under WI00442021[10]. Also
in the AECO industry, several approaches are commonly supported and advertised, and
standardisation across approaches is considered to be difficult.

2.7.1   Available implementations

For example, the Dutch standardisation institute NEN published with the Dutch
Technical Agreement (NTA) 8035:2020 [275] a set of general recommendations for
ontological modelling related to construction projects. Section 7.11 of the NTA describes
the modelling of properties which resembles approach 3 but limited to L2 patterns for
properties that have an RDF literal as value and limited to L1 for properties that have
a URI as value. As a result of this limitation, only a taxonomy of owl:ObjectProperty
instances is necessary. Units are only to be defined using QUDT.

Alternatively, OPM is an ontology implementation that provides terminology
for L3 property patterns [316,317]. It does not enforce a certain approach for the

definition of the property name, however. OPM extends concepts from the Smart Energy Aware Systems (SEAS) Evaluation ontology [222] which has some similarities with the modelling patterns described above. The SEAS ontologies support L2 and L3 patterns and are oriented to approach 3 (only owl:ObjectProperty) and/or 4 for defining the property name. The modelling approach for describing properties of building products in BPO is based on L2 properties [406]. The approach for defining property names is not discussed in the current publication of BPO.

Finally, a number of applications can be named that implement the L1 approach, one of them being the graphs output by a simplified IFCtoLBD converter[11] that relies on the BOT, BEO and MEP ontologies. All properties are directly associated to the object, using a single predicate to define the property name. This approach does not link to any of the available L2 or L3 approaches. At best, the property names used in the output are identical to the property names used in IFC-based L2 graphs. It would require name matching to be able to find such cases.

2.7.2   Recommendations

According to the summary in Table 2.3, approach 3 appears to have support for all evaluated features. To allow support for all three property levels, a slightly larger taxonomy of property definitions will be needed compared to the other approaches. Approach 3 is also closest to the approaches in related literature (OPM and SEAS) and the recent NTA 8035 specification. To continue, van Berlo [395, p. 651] indicates that this approach for defining construction-related properties is *well-known and informally very well accepted*.

Even in the case of approach 3, it is possible to include datatype properties and/or object properties. Although it is possible to relate object properties and datatype properties that have the same meaning, practice indicates that it is recommended to use only a single property modelling approach in a dataset. This also helps reducing the complexity of queries. For example, when we use only instance properties modelled with L2 patterns, QUDT units and schema.org to refer to values, the SPARQL query to obtain properties of an object can be simplified to the query presented in Listing 2.5.

Listing 2.5   Query to retrieve all L2 instance properties (without reasoning) connected to a certain object

```
1  SELECT ?object ?propName ?propValue ?unit
2  WHERE {
3     ?object ?propName ?propNode1 .
4     ?propNode1 schema:value ?propValue .
5     OPTIONAL { ?propNode1 qudt:unit ?unit }
6  } ORDER BY ?object ?propName ?propValue
```

2.8   CONCLUSION

Different approaches for the modelling of properties are in use throughout the AECO industry. In the first part, a distinction between simple and more complex property modelling was made independently of the used data technology. Datasets with simple properties are easy to grasp and have a smaller size, suggesting their aptness for direct data exchange. In contrast to an approach with more complex properties often used internally in data management systems, no metadata (units, versioning, sources, accuracy, etc.) can be attached to a simple property. In addition to the benefits and consequences of simple and complex properties, general requirements were listed for data modelling patterns.

In the second part, specific data modelling patterns related to properties, based on semantic web technologies and linked data graphs, were analysed and compared against the general requirements for simple and more complex properties.

First, three different methods to attach properties to objects such as construction components, spatial building zones, damages, etc. were identified. The discussed methods include regular instance properties (directly connected to the object), type properties (connected to class of the object), and inferred properties through OWL restrictions (inferred to the object, through an OWL restriction on the class of the object).

Second, modelling patterns for units of measurement of quantitative properties were illustrated, including the embedding of units in property values (literals) and the application of specialised ontologies for units such as QUDT.

Third, in line with the general distinction between simple and more complex properties, three property levels were defined. Level 1 represents a directly connected property, while levels 2 and 3 correspond with respectively a single qualified (one intermediate node between object and property value) and double objectified relation (two intermediate nodes between object and property value). The additional functionality of level 3 properties comprises the versioning of properties inside a dataset using so-called property states.

Following, five approaches for defining properties, influencing the position of the property name in a graph pattern, were discussed. In the first three approaches, the first predicate is used for respectively an RDF property, an OWL annotation property and an OWL object/datatype property. The fourth and fifth approaches require objectified properties as the property name is respectively represented by a class or a SKOS instance node on the first intermediate node between the object and the property value. For each of the approaches, the compatibility with the three methods to attach properties and the three property levels is analysed.

Finally, recommendations for the modelling of properties in a semantic web context are formulated, by referring to existing implementations, literature, and specifications. Approach 3, using OWL object and/or datatype properties, is most referenced and can be used with each of the three property levels and the three methods to attach properties to objects (instance, type, and inferred through restrictions). While different property patterns can be mixed in a single dataset, it is recommended for fluent data management (e.g. query writing) to select a single modelling pattern for a dataset which functionality complies with the needs of the use case.

ACKNOWLEDGEMENTS

This research is funded by the Research Foundation Flanders (FWO) in the form of a personal Strategic Basic research grant (grant agreement 1S65917N).

NOTES

1 EN ISO 23386:2020 - https://www.iso.org/standard/75401.html.
2 https://technical.buildingsmart.org/standards/ifc/ifc-schema-specifications/.

3 http://bsdd.buildingsmart.org/.

4 For example, https://standards.buildingsmart.org/IFC/RELEASE/IFC2x3/TC1/ HTML/psd/psd_index.htm.

5 https://standards.buildingsmart.org/IFC/RELEASE/IFC4/ADD2_TC1/HTML/link/ listing-ifc4_add2.htm.

6 http://bsdd.buildingsmart.org/api/4.0/IfdConcept/2VWFE0qXKHuO00025QrE$V.

7 https://www.w3.org/TR/owl2-new-features/#F12:_Punning.

8 https://ci.mines-stetienne.fr/lindt/playground.

9 https://github.com/w3c-lbd-cg/lbd/blob/gh-pages/presentations/props/presentation_ LBDcall_20180312_final.pdf.

10 https://standards.cen.eu/dyn/www/f?p=204:110:0::::FSP_PROJECT,FSP_LANG_ ID:67839,25&cs=1A09AFFEC4DF9D3A023109C35665E4C83 (Accessed 2020-08-13).

11 https://github.com/pipauwel/IFCtoLBD.

Chapter 3

Web technologies for sensor and energy data models

Gonçal Costa and Álvaro Sicilia

CONTENTS

Today, many efforts are devoted to maximising the energy efficiency of buildings throughout the design, operation and maintenance stages of the building's life cycle. While in the design stage, energy simulations are useful to predict energy behaviour, in the operation and maintenance stage, sensor data are crucial to provide informed decisions for an optimal use of building energy systems. To meet the data requirements of different applications used by the experts (e.g., architects, engineers and facility managers) involved in the various stages, a combination of the data from multiple domains (e.g., architecture, energy and economy) is necessary. In this chapter, the authors conduct a comprehensive review of the research works on the use of web-based technologies, sensor data and energy data represented through semantic data models to reduce energy consumption in buildings. The study reveals the convergence of works in various research lines, the differences between methodological approaches, the links to ongoing investigations and their future evolution.

DOI: 10.1201/9781003204381-4

3.1   INTRODUCTION

Reducing energy consumption in buildings has been a priority for many governments and organisations around the world when tackling climate change and meeting environmental goals. Furthermore, reducing the energy footprint and CO_2 emissions has also been part of the Europe 2020 strategy [88]. The need to reduce energy consumption has prompted the development of different kinds of techniques and technologies. These range from those aimed at collecting data on the behaviour of the building, to their processing, analysis and interpretation through web- and cloud-based services to aid decision-making.

The data from monitoring combined with prediction models can serve to forecast the energy performance of a building at multiple levels (areas, floors, rooms, etc.). This way, owners and building energy managers can determine a better fit between energy consumption, comfort and use of their buildings. Another strategy to achieve greater energy efficiency in buildings is by providing simulations that are more accurate at the design stage, whether in new construction or in renovation projects. Energy simulations enable energy performance assessment of a building from its design, which is usually created from scratch using specialised tools (e.g., EcoDesigner). Designs to perform energy simulations can be generated more easily when the Building Information Modelling (BIM) methodology is adopted in the development of projects. With BIM, accurate virtual models of a building can be digitally constructed and their data analysed in multiple ways [116]. Architects, for example, can create architectural designs that incorporate the information required by the energy simulation programmes, so that the building model they have created can be leveraged for that purpose. For some years now, BIM models that include energy information have been called Building Energy Models (BEM). Therefore, in addition to including geometry (see Chapter 4), these models also include data on the properties of materials, HVAC systems, etc. (see Chapter 2) and can play an important role in the wider management of smart cities (see Chapter 11).

Web- and cloud-based technologies, along with the use of standardised data representations (see Chapters 1 and 5), are playing a crucial role in developing more streamlined and efficient building energy simulation processes. A key issue is the exchange and reuse of the energy model data between different systems and platforms. An option to overcome interoperability issues between the energy data (e.g., monitoring data, building models and energy models) and tools (e.g., BIM authoring tool, energy simulation tool and building energy management systems) is to rely on a centralised data model (i.e., based on IFC schema). These models are usually the result of an agreement between members of the user community in their quest to provide a model that encompasses all the data needed for the simulation tools.

However, the lack of flexibility of centralised data models to adapt to changes and the loss of information (see also Chapter 5), after exporting and importing data through applications to/from these models, leads us to consider a decentralised approach as a more plausible alternative. Each tool has its own data schema and models generated according to this schema. These models can be shared on the web therefore enabling other tools to extract data from them. For example, BIM data is shared on the web using IFC servers and BEMs facilitate monitoring data through REST APIs. Within this scope, sharing data models using standard web languages such as OWL and RDF is attracting increasing attention in recent years since it reduces ambiguity by providing a common vocabulary along with formal semantics.

In this chapter, we review the state and the role of the web technologies in sensor data collection and the combination of such with other energy information to facilitate and automate energy simulation processes and decision support in the design, operation and maintenance stages of the building's life cycle. First, Section 3.2 reviews model-based approaches to assessing the energy performance of buildings. Then, in Section 3.3, we review the achievements reached with the development of the energy data models. Finally, Section 3.4 reviews the advances in the application and combination of different web technologies for collecting, integrating, storing and accessing sensor data for energy simulation and decision-making purposes.

3.2 MODEL-BASED APPROACHES TO ASSESSING THE ENERGY PERFORMANCE OF BUILDINGS

This section provides an overview about the benefits of applying a data model-based approach in the analysis and assessment of the energy performance of buildings. The examination focuses on project development and energy certification of buildings.

3.2.1 Analysis and prediction of the energy performance of buildings

Energy performance in buildings can be described as the set of interrelationships that exist between the building envelope, its surroundings, the number of occupants, the building components used, the lighting and mechanical systems, among others. Achieving the highest energy efficiency possible using the fewest resources requires understanding these interrelationships [320].

In the design stage of a building, the main purpose to use energy performance simulations is to compare different alternatives based on certain sustainability criteria to choose the most appropriate design. However, currently there is still a gap between simulated and real performance [396]. Some authors indicate that these differences are due to a lack of integration of energy analysis in the design process. Others claim that part of the problem is due to energy analysis being performed too late in the design process, leaving little room for readjustment [287]. Data model-based approaches can help to mitigate this situation. The growing interest in optimising energy consumption in buildings is leading to the generation of a large volume of monitored data, which in turn has paved the way for the development of new algorithms and predictive models (e.g., based on genetic algorithms and decision trees).

Energy performance can be evaluated and predicted. The evaluation can be done through different types of analysis that can range from basic considerations (e.g., building typology and orientation) into more complex issues (e.g., possible effects when including an additional insulation layer, use of thermal bridges). Methods of evaluating the energy performance can be classified into: (1) statistical analysis, which involves the use of statistical data that are publicly available to evaluate energy use (e.g., energy indicators published by the International Energy Agency (IEA)); (2) input-output analysis, which determine the use of energy in relation to the monetary flow within an economy (e.g., to estimate the carbon footprint); and (3) process analysis, which examines the direct and indirect energy of a process (e.g., energy used in the production of a building component) [338]. With respect to the prediction of energy performance, Shabani and Zavalani [349] classify the methods into: (1) model-based methods (engineering methods), which use mathematical models that include information on the physical properties of the building and its functionality and occupancy; and (2) data-based methods, which use statistical and machine learning methods (e.g., support vector machines, regression and artificial neural networks).

3.2.2   Monitoring and sensor data

The generation and collection of sensor data is subject to different usage scenarios with their own characteristics and limitations. For example, smart meters typically collect data on customers' electricity consumption every few minutes, leading to a massive accumulation of data. This type of collection falls within the domain of big data and, as such, it poses certain challenges with regard to the infrastructure required for the collection and storage of energy big data, the subsequent use of these data to support decision-making and others. Further opportunities and challenges in the use of energy big data are described in [438], such as operational efficiency, cost control, stability and reliability of the system, management of renewable energy, energy efficiency and environmental issues or the participation of the final consumer, among others.

Sensors are key elements in accurately determining the energy performance of a building. The data produced by sensors installed in buildings are a valuable source that can help to minimise the possible discrepancies between the estimated (by simulation) and actual performance, for example in terms of actual environmental conditions and energy consumption. The sensor data can be combined with those from digital building and district models – BIM and District Information Models (DIM) – in order to generate more detailed and rich energy models. Web technologies and the Internet of Things (IoT) approach (see also Chapters 10 and 11) have proven useful in facilitating this combination and keeping model data up-to-date in real time.

IoT protocols, standards and communication technologies make it easy for devices to be operated through platforms and cloud services. IoT enables things that are not strictly speaking computers, to hear, see, calculate and act to communicate and coordinate with each other in decision-making [336]. However, there is still a challenge in making the generated data compatible when it has to be extracted and combined with other kinds of data (see Chapter 6). Here, standards play a relevant role in facilitating the understanding and communication of the data from IoT devices – such as those generated by sensors – when they have to be exchanged with other systems. To accomplish this, it is essential to opt for the use of standardised data formats such as XML or JSON.

For example, in the OPTIMUS project[1], a web-based Decision Support System (DSS) was developed exploiting multi-sourced data (e.g., weather conditions, monitoring data, energy prices) to enable building managers the creation of energy management action plans [245] for smart energy cities. Figure 3.1 shows the architecture of the DSS which is fed by five different sources (weather forecasting, de-centralised sensor-based, social media, energy prices and renewable energy production).

In that line, Bottaccioli et al. [51] developed a software platform designed to manage and simulate energy behaviours in buildings by integrating heterogeneous IoT devices and meteorological services with BIM, Geographical Information System (GIS) and simulation data. In turn, these data are interrelated and enriched with historical data from IoT devices. The Simulation Engine block uses the EnergyPlus engine to perform the simulations of buildings.

3.2.3   Access and use of energy data

Building energy simulation tools may require data that can be used to improve energy efficiency, both in building renovation and in new construction projects. In both cases,

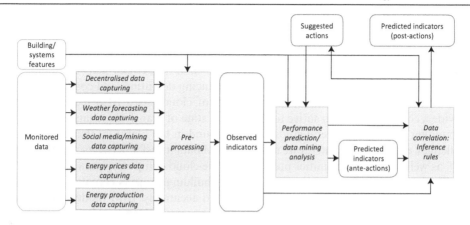

Figure 3.1 OPTIMUS DSS overall architecture scheme [439].

building designers and energy experts need accurate and reliable data about buildings and their performance to take well-informed decisions. These data can come from multiple sources and domains (e.g., sensor data, energy production, user behaviour and building geometry).

In the design stage, the use of energy data can be useful to determine the degree of energy efficiency and savings with greater precision. For example, energy simulation tools can make use of these data in combination with the geometry of the building – either generated manually within the energy simulation environment or obtained from BIM architectural models – to determine its performance. BIM models can also include a significant number of energy parameters such as those representing the climatic conditions in the building location, thermal properties of the building materials and so on. However, energy data can also come from other sources (e.g., occupant schedules or the shading effect from nearby buildings). Access to reliable data from multiple domains and applications is essential for conducting complete, accurate and useful energy simulations.

Having access to the energy data is also necessary for the development of decision support tools to assist building managers and owners in the operation and maintenance stages. For example, tools aimed at supporting the optimised use of HVAC systems can use monitored data from different kinds of building sensors (e.g., indoor/outdoor temperature, humidity, ambient light, energy consumption, bills). Therefore, providing access to these data can require the appropriate combination of different technologies (e.g., wireless transmission, network communication, cloud computing, big data storage and analysis). For example, in [310], the authors present a four-layer architecture (sensing, networking, cloud computing and applications) based on Heterogeneous Internet of Things (HetIoT) with independent function and scalability.

3.2.4   Energy analysis in BIM-based projects

Solutions to improve the energy performance of a building – both in renovation and in new construction projects – can be approached differently depending on the methodology adopted to carry out their development. Here, a distinction can be made between projects carried out through traditional practice – in which the project information is transferred through 2D CAD formats [240] – or when BIM is adopted. Focusing on the latter, BIM methodology facilitates the integration of the fragmented fields of architecture, engineering and construction with the aim of optimising the performance of the processes involved in the building's life cycle [131], through data-rich, object-oriented, intelligent and parametric digital representation of a building [17,116]. The design of a building through an integrated data model can help to enhance the productivity and quality in the workflows by reducing downtime and costs. In turn, the adoption of BIM is also accompanied by a set of cloud-based BIM technologies that provide a cost-effective alternative to the current state of transfer, exchange and storage of building data [435]. Several studies can be found in the literature that deal with the challenges, requirements and future trends of BIM data transfer between cloud platforms [3,7], as well as the opportunities presented by the cloud computing paradigm [293].

The adoption of the BIM methodology in building projects can also be beneficial to estimate energy efficiency more effectively and accurately. BIM models can serve as the backbone for the green building evaluation process [121]. Since developing models for energy simulation is time-consuming and cumbersome [168], the reuse of data from existing BIM models is presented as an alternative. This way, a BIM-based energy simulation facilitates a more efficient and accurate Building Energy Performance (BEP) estimation process, by reusing data from BIM architectural models. It is more efficient because it avoids recreating the geometry of the building in some energy simulation environment and it provides the potential to automate the generation of the energy model from the architectural BIM. Similarly, it is more accurate because it reuses the BIM data to generate automatically, or semi-automatically, an energy model thus reducing the misinterpretation of the data as may occur with a manually generated energy model [392].

In turn, this increase in efficiency makes it easier to simulate optimisation scenarios and determine the best one, as well as rethinking the architectural design solutions if necessary, closing the loop in a shorter time. However, the building energy model generated with BIM data (model-based) must be complemented with other data (e.g., the actual data collected from monitoring activities and energy bills) [51].

The application of BIM in the field of building energy performance and simulation has given rise to a number of associated terms. For example, 'Green BIM' was coined to describe the convergence between BIM and green buildings. According to [141, 212], Green BIM is based on three premises: *(1) environmentally sustainable design (ESD) principles, (2) optimisation of green building certification (GBC) credits and (3) integrated construction systems and design processes supported by tools of object-based modelling and analysis.* Another term that frequently appears in the literature is Building Energy Modelling (BEM), a computerised method to assess the energy performance of buildings through the analysis of various design alternatives in order to obtain an optimised design of the building [142].

3.2.5   Energy performance certification

An evaluation of the energy performance of buildings is also a requirement in the certification processes. Here, the energy performance certification (EPC) has been recognised as an effective assessment methodology to systematically analyse and improve the energy performance of buildings. Some governments use this methodology to reduce energy consumption in buildings, contributing to the improvement of environmental and economic sustainability in the construction sector.

In Europe, and because of the introduction of the mandatory EPC (Energy Performance of Buildings Directive 2010/31/EU; EPBD recast, amended by Directive

2018/844), many Member States have defined building certification guidelines to support data collection. Since then, the energy performance certification has become commonplace, and a large number of data have been collected and stored in many countries to be used for various purposes.

On an EPC, a dwelling or building (residential, commercial, public, etc.) is rated based on the relation between its efficiency and the amount of energy required to meet the needs of the occupants in terms of comfort and functionality. The range is defined along the first seven letters of the alphabet, establishing different degrees of efficiency ranging from 'A' (very efficient) to 'G' (very inefficient). Along with the performance rate, EPCs include recommendations for cost-effective building improvements.

Carrying out an energy certification is a complex process that requires in-depth knowledge about the features of the building, which in turn must be considered as an integrated system rather than as the sum of its parts [4]. While certifications can be issued at the design stage of new buildings, it is during the interior fitting, operation and maintenance stages that it is often most difficult to obtain the necessary data.

In recent years, the European Union (EU) has shown interest in funding various research projects aimed at improving the implementation of EPC processes. An example is ENERFUND[2] [143], which developed a web tool to rate and score deep renovation opportunities based on EPCs, number of certified installers, governmental schemes and other parameters. However, there is still room to explore new solutions to improve EPC processes and to increase the quality and reliability of EPC schemas.

The interlinking of EPC databases with other data sources (cadastre, mobility and energy production) can be useful to obtain a holistic understanding of the building performance. The EPC data enriched with other data sources can be processed with advanced analysis tools in order to facilitate decision-making on a large number of buildings. For example, ENERHAT[3] and ENERPAT[4] are two online applications that allow users (owners, tenants and real estate agencies) and experts (planners and policymakers) to estimate the costs of the building refurbishment measures at the building and city level [238]. Both applications use the same dataset in which data from EPCs (Oracle database), cadastre (Web API), building technical inspection (spreadsheets), geographic boundaries (SHP files) and census data (spreadsheets) has been integrated using semantic web technologies such as:

- R2RML: A mapping language between relational databases and ontologies recommended by W3C [93].
- Map-On: A visual web environment for the manual editing of mappings hiding R2RML complexity [358].
- Morph-RDB: An RDB2RDF engine to generate RDF instances from data in a relational database based on R2RML specifications [307].

3.3 ENERGY DATA MODELS

The concept of energy data modelling first appeared in the early 1980s [176,195]. The concept gave rise to the creation of different energy models that were used to perform simulations for specific purposes (e.g., calculating transient heat flow through multi-layered slabs). This section provides an overview of energy data models including what they represent (energy systems, spaces, etc.), their role as standards and the possible benefits when their schemas are provided through ontological definitions. This last part includes a brief review of the ontologies created for modelling energy data. The section concludes in Section 3.3.5 with a review of some of the research projects developed in recent years in which energy data models have been used to create more efficient solutions for energy simulation and decision support.

3.3.1   System approach definition

Addressing the problem of energy efficiency in buildings in a comprehensive manner, covering all its dimensions and scales, requires a system approach. Therefore, buildings can be seen as systems made of subsystems that produce and consume energy in various forms. One way to understand how the different systems of a building (e.g., HVAC systems and lighting systems) work is by modelling them, that is, by creating a digital representation of the systems from the available data. The combination of these data models with other tools can help stakeholders in their decision-making. For example, an energy model that combines multiple data (energy consumption, energy production, building envelope, etc.) can help energy experts to better understand how a building performs. Then, from this understanding, measures can be taken to reduce energy consumption or improve comfort using simulation tools.

3.3.2   Energy modelling of buildings and cities

The energy modelling of buildings can be carried out on an urban scale. Urban energy models can contribute to a better understanding of the energy consumption patterns – past, present and future – of a building stock. To comply with directives aimed at reducing CO_2 emissions due to building energy consumption, policy makers, municipalities and other public authorities need tools to develop, deploy and monitor climate action plans. Urban energy models can be useful to obtain information, for example, about the amount of energy consumed in an urban area and how it can be reduced, or about the relation between urban density and energy demand. They can also provide information on the energy use profiles of existing buildings in a district or city, help predict future energy supply and demand, and compare alternative scenarios of urban development regarding possible future use of energy [1]. However, the literature review suggests that the widespread use of these models in urban energy estimation tools still focuses mainly on the analysis of the result, ignoring the causes that lead to it.

Building energy modelling has become an interdisciplinary area of study encompassing architecture but also electrical, mechanical and civil engineering [169]. Therefore, the development of energy models of building energy systems requires the collaboration of experts in different domains. When energy improvements need to be applied to a building stock rather than to a single building, simpler representations of the buildings can be used in combination with representations of other elements of the city (e.g., energy networks). These simple models may be sufficient to understand the interaction between the different elements. Detailed energy models for city-scale analysis and simulations require a large amount of memory and computational processing time, which is presented as a limitation for its execution on local computers. However, the proliferation of alternatives based on cloud services in the last decade appears to be breaking this barrier.

3.3.3   Standards

The purpose of standardised data models is to facilitate the exchange, sharing, and reuse the data from one domain to another. In the case of energy data, standardisation can be useful to effectively reuse the data from architectural BIM tools (e.g., Revit, ArchiCAD and AECOsim) in energy simulation tools (e.g., EnergyPlus). It can also be useful for creating decision support tools for energy planning and performance assessment of buildings.

In the same way that in the area of BIM, there are standards such as IFC (IFC4 - ISO 16739) for the detailed definition of buildings, and GML (Geography Markup Language) [110] for the geospatial modelling domain, in the area of energy simulation there are various standards based on open schemas. For example, the Green Building XML[5] (gbXML) is a data model schema and the underlying architecture of Green Building Studio software used to transfer building information between BIM and energy simulation tools. As a limitation, it only accepts geometry with rectangular shape [151]. Another alternative is the Simulation Domain Model (SimModel) specification [278] which includes a terminology semantically aligned with the IFC. Both models enable to leverage BIM data such as building geometry in the energy simulation of entire buildings. These models can also be used for computational fluid dynamics (CFD) and fire and safety simulation.

While some methods for the data transfer between BIM and energy simulation tools are based on energy domain standards (e.g., gbXML, SimModel), others are based on IFC given its universal base and sufficiently generic approach [20]. For example, according to the proposed workflow for energy performance simulation tools proposed by [242], in [142] the authors classify BIM-based BEM methods in six steps in terms of building information (geometry, building materials, building or spaces type, thermal zones, space load and HVAC systems). Methods based on IFC for each of these steps are reviewed and contrasted with those based on gbXML. However, the authors conclude that obtaining reliable energy models in both cases requires manual revision and modification. Furthermore, pre-processing work in BIM and BEM tools is still slow and error-prone.

3.3.4   Ontologies

Despite the existence of standards for interoperability, it is necessary to emphasise that it seems difficult to achieve full interoperability between BIM and energy simulation tools [54] (see also Chapter 9). An alternative to trying to make the building's energy data more interoperable can be using ontologies – a conceptual representation of a domain in a formal manner [157]. Ontologies provide a *right interpretive philosophical underpinning that is more likely to address the information and knowledge sharing requirements of the construction user community* [322]. Ontology-based interoperability is presented as an alternative to move from a model-centric towards a more decentralised semantic approach [357]. The use of ontologies can facilitate the integration of building energy data from multiple sources, even from standards such as IFC. This way, ontologies serve as a bridge to transfer the data from one domain to another. This should facilitate the sharing of data from different software applications under a common representation, known and accessible to all and easily understood through the semantic relations defined in the scheme.

3.3.4.1   Ontologies in the construction sector

In the last two decades, different ontologies have been proposed for the construction industry [29,128,316,382], in part as a consequence of the rise of semantic web technologies. Also, the Linked Building Data (LBD) Community Group, within the

W3C (W3C LBD CG), has been active proposing a set of ontologies following a methodology based on use cases and requirements for applications based on linked data throughout the life cycle of buildings.

Most ontologies are intended to structure information from a specific domain of the industry. Some of them have been generated from versions of existing standards, incorporating semantic meaning in the concepts of their schemes. Another distinction can be made between ontologies that aim to cover a very broad domain (e.g., ifcOWL for BIM data), and ontologies that only include core concepts so that it is easy to extend them as needed to specific domain ontologies (e.g., BOT) [318]. Simple ontologies also have the advantage of being easier to maintain than complex ones.

3.3.4.2 Ontologies in the energy domain

Within the domain of energy modelling and simulation of buildings, an example is the SEMANCO ontology [237], which contains concepts and properties for the domain of urban energy models at different scales based on international energy standards such as ISO/IEC CD 13273, ISO/DTR 16344 and ISO/CD 16346, among others. The SEMANCO ontology was built based on standard tables in which the terminology, descriptions, units of measures and relationships between concepts are described (Figure 3.2).

There are also ontologies that are based on standards, such as SimModel ontology [294], which was generated from the Simulation Domain Model (SimModel) specification, mentioned in the previous section. Considering the use of ontologies to facilitate the integration of data from different sources, in the case of SimModel, this integration includes not only BIM data but also GIS and those that can be necessary for

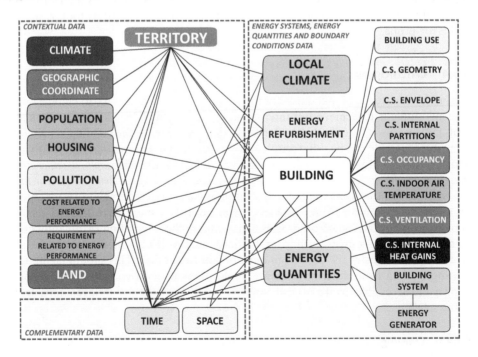

Figure 3.2 SEMANCO standard tables covering different domains (i.e., data categories) and encompassing about 1,000 concepts [81].

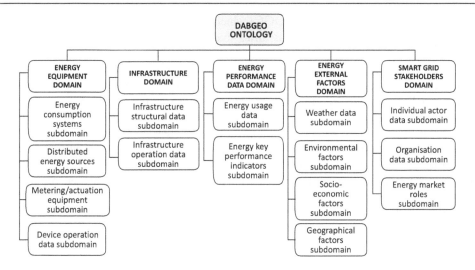

Figure 3.3 DABGEO ontology high-level structure [90].

energy simulation in different tools. Another ontology for the energy domain is Domain Analysis-Based Global Energy Ontology (DABGEO)[6] [90] (Figure 3.3). This ontology has been designed to address the lack of interoperability between ontology-based energy management applications for large-scale energy management scenarios. To deal with the heterogeneity of energy ontologies, DABGEO provides common vocabularies to represent energy subdomains. In turn, and to moderate the effort required to reuse it in different applications, the ontology follows a layered structure with a balance between reusability and usability.

The use of ontologies can also be useful to improve interoperability between the sensor data and energy information systems. The development and expansion of the use of IoT technologies is leading to a greater volume of data that must be processed (see also Chapter 10). These data are often incomplete, follow different representation formats, or have even been developed under different semantic approaches [332]. During the last decade, various ontologies and ontology-based approaches have been developed to deal with this heterogeneity. For instance, the Smart Appliances Reference (SAREF) ontology [92] focuses on the concept of device to represent a reference model for smart appliances (e.g., light switches, temperature sensors, smart metres and washing machines). Another example is the RealEstateCore (REC) Ontology [167], which supports the modelling of: (1) devices (configurations, capabilities, values that they report or receive), (2) buildings (components, room types and locations) and (3) contractual situations of buildings.

Although the use of ontologies in the information systems of the construction industry may represent a benefit for data interoperability, the current reality indicates that from the perspective of standardisation, and unlike other sectors, ontologies have not yet been considered as a standard method for the exchange of information in this sector [263] (see also Chapter 5).

3.3.4.3 *Ontologies and sensors*

Ontologies are being used to represent sensor data in different domains (e.g., earth observations, maritime and IoT). SensorML standardised model has been proposed by Open Geospatial Consortium (OGC) which is part of a suite to represent sensors, observations and sensor data repositories making them discoverable, accessible and usable on the web [52]. Later, the Spatial Data on the Web working group (made up of W3C and OGC) developed the SOSA and SSN ontologies (i.e., Sensor Observation Sample and Actuator ontology and Semantic Sensor Network, respectively), which became a W3C recommendation in 2017 [166]. SOSA is the core ontology that can be extended with modules such as SSN to add expressivity. SOSA has been designed as a lightweight ontology in which common axioms such as rdf:domain, rdf:range and rdf:subClassOf have not been used. The reason behind this design decision is to make SOSA simple enough to be added to schema.org vocabularies [165]. SOSA/SSN has been used for representing meteorological datasets [331], for modelling smart sensors such as smart accelerometers used in sports and personal monitoring [335], as part of a Building Automation and Control System (BACS) [344], and to propose ontology design patterns for energy efficiency in buildings [122].

3.3.5   Research projects

In the last two decades, different energy data models have been proposed to evaluate the performance of energy systems in different fields (building, transportation, energy, etc.), or through a combination of them, with the participation of experts in multiple domains. Part of this research has focused on investigating how to include the possible factors that intervene in energy analysis from a data-based approach perspective, and how the use of models facilitates achieving a semi-automation of the simulation process.

Efforts to address the energy modelling of buildings on the scale of buildings and cities have been made in different research projects. One example is the SEMANCO[7] project (2011–2014), where semantic web technologies were applied to support the creation of urban energy systems and models [237]. The result was a semantic platform that integrates the data from multiple sources and scales (e.g., micro, meso, macro) (Figure 3.4). In this platform, users can create urban energy models that represent the combined knowledge of the various experts involved in the evaluation and planning of energy-efficient cities.

Another example is STREAMER[8] (2013–2017), a project dedicated to improving building energy efficiency in mixed-use health districts. The aim of the project was to optimise semantics-based design methodologies with interoperable tools for GIS and BIM to validate energy performance at the design stage [101].

Two more recent research projects in this area are OPTEEMAL[9] (2015–2019) and BIM4REN[10] (2018–2022). In OPTEEMAL, an energy efficiency design platform optimised for rehabilitation at the district level was developed that integrates and transforms the data from different sources (e.g., BIM and GIS) to the domain of energy simulation tools [83]. Figure 3.5 shows the data integration and transformation process devised, consisting of three steps: (1) input data models are transformed into semantic data models in RDF (e.g., BIM models represented according to the IFC standard in the STEP format are transformed into RDF according to the ifcOWL ontology), (2) semantic data models are transformed into simulation data models and (3) simulation files are generated from simulation data models according to the input requirements of a specific simulation tool (e.g., EnergyPlus tool requires input data models in IDF format).

The core of the platform is a District Data Model (DDM) which contains representations of CityGML and IFC. These schemes are semantically enriched and related to existing ontologies in the main fields for urban sustainable regeneration

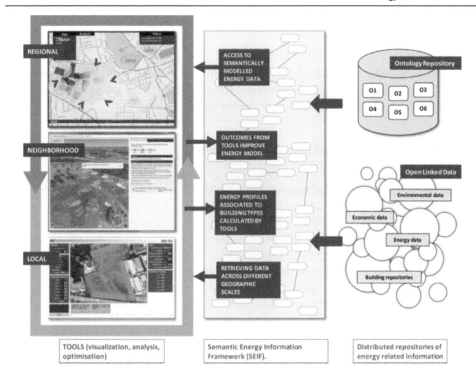

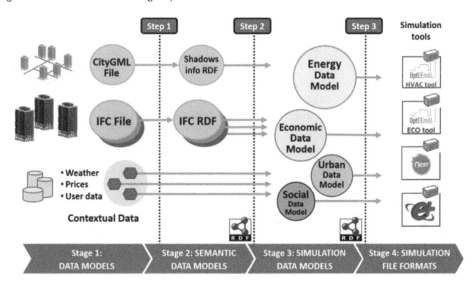

Figure 3.4 SEMANCO technological platform.

Figure 3.5 Data integration and transformation process.

(energy, social, environment, comfort, urban morphology and economic) [82]. Similar to OPTEEMAL, the BIM4REN project proposes a systematic workflow to manage the BIM to BEPS transformation. Within this project, a review of the current tools and approaches to generate BIM models, the verification of their geometry, as well as their enrichment with the data necessary for the BEP tools, was developed [118].

3.4   ENABLING TECHNOLOGIES FOR SENSOR DATA-BASED APPLICATIONS

This section includes a review of the advances in the application and combination of different web technologies to collecting, integrating, storing and accessing sensor data for energy simulation and decision-making purposes.

3.4.1   Building sensor data and technologies

Building energy monitoring technologies rely on devices that can sense the reality: sensors. In a building, not only energy is monitored (e.g., electricity consumption and renewable energy production), but related features are measured such as indoor/outdoor conditions (e.g., temperature, humidity), occupancy (is a room empty?) and status (e.g., whether the windows are open or not), among others. Therefore, the use of sensors in a building can range from monitoring the parameters and conditions of the built environment in real time to the behaviour and occupancy of its users.

In the last decades, smart sensors have become an important asset, as they are devices that can sense reality and can process the results to increase the reliability and integrity of the measured data produced. This is possible, thanks to features such as self-calibration, which will depend on each type of sensor.

- Temperature sensors: Measure the level of heat of space. These sensors are used by HVAC systems to decide to raise or lower the temperature. The impact of temperature monitoring is significant for the thermal comfort of occupants and energy consumption [426].

- Humidity sensors: Monitors the amount of water vapour in the air. These kinds of sensors are used by ventilation systems to maintain proper comfort conditions for occupants of the space. For example, there are machines in hospitals, factories, museums that are sensitive to particular levels of humidity, so an active monitoring is needed. Humidity sensors enable so-called smart ventilation systems that try to reduce energy consumption while indoor air quality is not affected [161].

- Occupancy sensors: Detect the presence of people and objects in a space. These sensors are usually used to understand which spaces get the most use and therefore may need more energy. There are different sensor technologies for occupancy detection – from motion sensors to air pressure change sensors – which have been designed to applications such as lighting/appliance management and energy consumption prediction based on occupancy levels [269].

- Contact sensors: Indicate if something is open or closed (e.g., doors or windows). This kind of sensors are employed in building automation systems and in HVAC systems to detect the status of façade openings (i.e., windows) since having a window opened has a significant impact on the indoor conditions such as temperature and air quality [71].

- Air-quality sensors: Measure the number of gases in the air such as ozone, carbon monoxide, carbon dioxide, sulphur dioxide and nitrogen dioxide. These kinds of sensors are related to ventilation systems, since when the concentration of these gases is high enough, the air in the spaces should be refreshed. Therefore, the use of these sensors makes it possible to understand the impact of poor air quality on human health [126].

- Electrical sensors: Measure parameters such as voltage, current and power factor. They are used in electricity/smart metres to measure electricity consumption. Data from electrical sensors are mainly used for billing purposes and by analytical modules of building energy management systems to characterise the energy load and shift the load when energy prices are lower [172].

Building management systems (BMSs) traditionally access sensor data through a hard-wired network, although it is expensive (e.g., wires have to be hidden inside walls) and it is not flexible in terms of adding new sensors or relocating existing ones. From the last decade, wireless sensor networks (WSN) have gained attention and popularity since they overcome hard-wired network limitations (see also Chapter 10). A WSN is composed of several sensors that have computation and wireless communication capabilities making it possible to transmit data generated by the sensor to a central server and/or to nearby sensors. Sensors used in WSN are fed by batteries, so they have been designed for low resource conditions in terms of bandwidth and sensor energy consumption.

Measurements data generated by sensors is moved to central databases by means of machine-to-machine (M2M) protocols such as Open Platform Communications (OPC) and its successor OPC Universal Access (OPC UA), which have been developed for the industrial domain [241] (see also Chapter 7). OPC UA has been designed to provide high scalability and vertical communication over different methods such as SOAP (Simple Object Access Protocol) and HTTP (Hypertext Transfer Protocol).

When the sensors are in low resource conditions, transferring data using such heavy weight protocols like SOAP may not be the best choice, however, turning to a lightweight protocol can make the WSN robust and sustainable. Examples are protocols based on the REpresentational State Transfer (REST) approach, which have a flexible design where the communications are made through Hypertext Transfer Protocol (HTTP) with the standard HTTP commands (GET, POST, PUT and DELETE). REST or RESTful services have gained popularity in the last decade since it is much simpler than other solutions based on SOAP and WSDL (Web Services Description Language). REST is stateless and generic in terms that can be applied in any domain. It has been designed for performance, reliability and to be able to work with information based on XML and JSON objects. In REST, everything is a resource that is identified by a Uniform Resource Identifier (URI) or Uniform Resource Locator (URL). The resources are handled through a HTTP method such as POST (insert), GET (select), PUT (update) and DELETE (remove). REST supports very simple and effective load-balancing and flexible URI-partitioning mechanisms.

When the communications between client (i.e., sensors) and server (i.e., central database) are so constrained such as WSN scenarios, the Constrained Application Protocol (CoAP) can be an alternative to a REST approach since it has been designed to keep the message overhead as small as possible. Indeed, CoAP is a compact version of REST with extra features for M2M applications. WSN for BMSs have been studied, in particular how RESTful approach can address the challenges of interoperability, integration and low bandwidth problems of those networks [107].

3.4.2   Storing and accessing building sensor data

The fact that more and more sensors are used to monitor buildings in real time is leading to an increase in traffic on the web with millions of entries per second. Sensor data storage generally involves massive data storage technologies, as sensors generally produce data continuously. It is for this reason that the sensor data typically falls into the realm of big data, which involves the corresponding technological infrastructure for their storage, processing, etc.

The database systems to be used in a sensor data management context are relational databases (RDB, also known as SQL databases), NoSQL databases and time series databases (TSDB). Of these three options, the latter seems more appropriate considering the high frequency in which new events are inserted into the database, which can reach a large number of entries per second. In addition, it is assumed that inserts are much more frequent compared to queries, which do not usually address individual values but rather a set in a time interval [125]. Another drawback in the use of relational databases is that they require greater processing and storage capacity. They are also considered too rigid and less scalable [397], as well as expensive in computational cost due to having to maintain the indices on time [308,323]. Although the TSDB is indicated in the literature as the best choice for sensor data, some solutions based on NoSQL like IOTMDB [226] can be found in the literature.

With the need to connect devices such as sensors through the IoT infrastructure, there is also the need for database services adapted to this new interconnectivity context, provided from the cloud environment itself so that greater scalability can be achieved. This way, in recent years, the concept of cloud database has appeared to refer to database services that run on cloud computing platforms (e.g., Amazon Web Services, Microsoft Azure) [9]. Among the advantages of opting for a cloud database service, there is the ease of combining it with other services provided from the same platform in the cloud (e.g., IoT and analytics components). Likewise, in recent years, some time series databases have appeared provided through web platforms. An example is InfluxDB[11], an open source TSDB database with built-in time-centric functions for querying measurement, series and point data. Another example is TimescaleDB[12], a time-series SQL database designed to provide fast analytics and scalability with automated data management.

Regarding the accessibility of building sensor data, the difficulty for their integration, openness and interoperability with other energy data that could be used to produce new value-added services is pushing the search for more practical interoperability approaches. Kubler et al. [213] refer to this current problem through the term 'vertical silos' and propose the use of APIs as a solution to promote the development of open ecosystems (see also Chapter 6). In [213], and as part of the bIoTope H2020 project, these authors introduce building blocks which make use of two Open API standards: Open Messaging Interface (O-MI) and Open Data Format (O-DF), to meet the requirements for backend data-sharing communications. This API interoperability approach is explored in [10] to manage energy data in real time in the context of residential buildings and electric vehicles (EV). The authors propose a layered architecture that uses APIs for district energy management to provide energy information intelligence and support decision-making on energy sustainability. The HTTP protocols (via REST API) and MQTT protocols are integrated into this layered architecture.

Further challenges in using sensor data in the energy efficiency domain include their combination with the data from other sources such as energy simulation tools and occupancy schedules. To provide the required interoperability, this combination of heterogeneous data requires a comprehensive approach. In Section 3.3.4, we have indicated through the literature review how the use of semantic web technologies and ontologies can be an alternative to facilitate this combination of the data in a plausible way through energy data models. However, there are other research works that have tried to address this problem through engineering solutions based on the appropriate combination of different technologies. For example, in [57], a platform developed for urban district data management and energy flow simulations is presented, which enables the correlation of building energy profiles in (near) real-time with environmental data from sensors and BIM and GIS models. The architecture of this platform is based on two communication paradigms: (1) request/response in non-real time (based on REST) and (2) publication/subscription (near) in real time (based on MQTT). The platform includes middleware-based software components that use a common abstraction model to describe sensor devices. The Sensor Markup Language (SenML) data format is used to transmit sensor measurements [57].

Finally, systems that typically require combining different real-time data from building physical sensors with other types of energy information are Building Energy Management Systems (BEMS). The term refers to those systems designed to manage information on systems that involve the use and demand of energy to improve the energy efficiency of buildings and generate savings. An example in this area is [257], which presents a solution for the implementation of energy-saving strategies with low costs for the installation of new equipment like sensors. The BEMS is implemented through a set of technologies, including semantic web technologies, and components to model and integrate building information with that of sensors and actions. One of the central components of the system is a knowledge base that stores the different heterogeneous building data relevant to BEMS. The data are stored in a semantic database (Fuseki Server). The information is provided to the BEMS users through an interface implemented using a combination of AJAX (asynchronous JavaScript and XML) and SPARQL [257].

3.5 CONCLUSIONS

In this chapter, we have reviewed the progress made so far in research work on the use of web- and cloud-based technologies, sensor data and energy data represented through energy data models to improve energy efficiency and reduce energy consumption in buildings. A significant number of articles and research works have been reviewed covering these technologies and data modelling approaches for better collection, combination and management of energy data in the design, operation and maintenance stages of buildings.

The review indicates that solutions based on data generated by sensors and other smart building technologies still require further development. For example, better integration of sensor data into energy data models to achieve greater reuse of the data in energy simulation tools for buildings and cities, as well as in the support of informed decisions for a more optimal management of the energy systems of the buildings. Here, the role of web- and cloud-based technologies can be useful for standardising the application layer to obtain more interoperable, reliable and detailed energy data models. These then can be fed with data collected from disparate sources (e.g. BIM models, sensor devices) further opening up new realms for analysis and decision-making to improve energy efficiency. With the incorporation of larger dynamic data streams, richer energy data models are possible that will enable reliable energy simulations of buildings, districts and cities.

The analysis of the literature also suggests that it will remain a considerable challenge to obtain an energy data model that satisfies the information needs of all energy simulation tools. Several articles suggest that more flexibility can be achieved with a decentralised approach (see also Chapters 5 and 6) through shared models on the web, allowing other tools to extract required data as needed.

Regarding monitoring data for the energy performance of buildings, web and cloud technologies play a key role in solutions aimed at increasing use of wireless devices as they simplify installation in any type of building. In Section 3.4, we reviewed how these technologies are used in building sensor data-based applications. This is made possible by communication protocols such as REST/CoAP in which the data sent by the devices have a reduced overhead making the consumption of the devices very low.

Other challenges in using sensor data in the energy efficiency domain include combining it with data from other sources such as energy simulation tools and occupancy schedules through a comprehensive approach. We have indicated through the literature review how the use of semantic web technologies and ontologies can be an alternative to facilitate this combination of data in a plausible way through energy data models. In Section 3.4.2, we have pointed out that this can also be achieved through the appropriate combination of different technologies, for example, through communication paradigms based on REST and MQTT.

NOTES

1 https://cordis.europa.eu/project/id/608703/es.
2 http://enerfund.eu/.
3 http://www.enersi.es/en/enerhat/.
4 http://www.enersi.es/en/enerpat/.
5 https://www.gbxml.org/.
6 https://innoweb.mondragon.edu/ontologies/dabgeo/index-en.html.
7 http://www.semanco-project.eu/.
8 http://www.streamer-project.eu/.
9 https://www.opteemal-project.eu/.
10 https://bim4ren.eu/.
11 https://www.influxdata.com/.
12 https://www.timescale.com/.

Chapter 4

Geometry and geospatial data on the web

Anna Wagner, Mathias Bonduel, Jeroen Werbrouck, and Kris McGlinn

CONTENTS

DOI: 10.1201/9781003203438-5

Geometry and geospatial data play an important role in digital building and infrastructure models, and its inclusion into web-based models, such as Linked Building Data (LBD), is of importance to obtain complete descriptions. Yet, the description of geometry and geospatial data in a linked data context is neither straightforward nor widely implemented. This chapter aims to give an overview of currently existing approaches to combine web-based building models and (non-web-based) geometry descriptions or geospatial data and discuss their issues. Moreover, existing implementations for integrating geometry descriptions in a semantic web context through the Ontology for Managing Geometry (OMG), the File Ontology for Geometry formats (FOG) and the Geometry Metadata Ontology (GOM) are demonstrated. Finally, available tools for spatial querying or handling and processing geometry in a Web context from ongoing research projects are revised.

4.1 INTRODUCTION

Geometry is a core perspective in the construction industry that is needed across domains and use cases. In fact, the influence of geometry is so prevalent that 2D drawings with minor non-geometric enrichment via hatching have been the basis for file-based information exchange in the AEC industry for decades. With the introduction of BIM and model-based engineering, the focus moved away from plain 2D drawings. Instead, 3D models were propagated that can be extended with non-geometric information. In traditional BIM, geometry is still a key functionality and serves as the core of the building model.

However, with the steadily changing and developing software landscape, and the increase in specialised stakeholders, geometry models that are semantically enriched are stretched to their limits: multiple competing software applications expect different geometry descriptions; stakeholders of varying backgrounds and perspectives require only certain parts of the geometry; and several stakeholders wish to work on the same part of a building model simultaneously. Hence, recently, geometry has been discussed as being a special kind of property instead of the core of a building model [254,404]. With that understanding, this chapter aims to give an overview of the state-of-the-art and ongoing research for handling geometry descriptions in a (semantic) web context.

As a basis for the upcoming topics, Section 4.2 reflects on terminology and context used within this chapter. Thereafter, Section 4.3 presents possibilities to integrate geometry descriptions into graph data models, as proposed in literature and practice. Exploring existing integration efforts, Section 4.4 reflects on ontologies that relate to or annotate geometry descriptions, before Section 4.5 covers existing implementations for integration. Finally, Section 4.6 closes this chapter with a conclusion.

4.2   GEOMETRY AND GEOSPATIAL DATA

Since this chapter covers geometry and geospatial data, a differentiation has to be made. Both the geospatial and construction domain rely on geometry for their tasks. Yet, the considered geometry varies in some aspects, namely the observed subjects, i.e. terrain, vegetation, buildings, infrastructure or landmarks, and their scale, i.e. metres vs. kilometres. These differences have an impact on the type of coordinate system that is used to spatially relate described objects to each other, and the applied detail of the described geometry.

While building models commonly use custom local coordinate systems georeferenced to a publicly known coordinate reference system (CRS), geospatial data attaches coordinates that are often directly expressed in such a CRS. Apart from the general usage of custom local coordinates versus publicly-known coordinate reference systems, the type of coordinate system applied, i.e. ellipsoidal or Cartesian, varies in dependency of the observed objects and their scale.

Moreover, the detail of the geometry is different for larger and smaller objects. Buildings and spatially limited objects can and often need to be described in a high level of detail with all spatial dimensions. If geospatial objects are to be described in the same detail, the amount of data storage needed to do so, would be immense. Hence, geospatial data is often restricted to 2D geometry and – if more detail is needed – enhanced with excerpts in 3D geometry.

4.2.1   Terminology

To ensure a common understanding within this chapter, the applied terminology with its intended meaning is collected in this section.

Geometry Description. A geometry description is a *somehow* formalised description of a geometry. The geometry can be conceptual, digital or existing. A geometry description can be a file, code, text or spoken words.

Geometry Schema. A geometry schema is a syntax that specifies terminology to describe geometries. While a geometry schema defines *how geometry can be described*, it does not determine how a resulting geometry description should be serialised.

Geometry Format. A geometry format is a serialisation of a geometry schema. It is possible for one schema to be serialised in multiple, different formats, as is the case for IFC with its STEP, XML and RDF formats. One geometry format can serialise only one geometry schema.

Geometry Representation Context. Geometry schemas can rely on different methods to describe geometry, so-called geometry representation contexts. Within this chapter, five contexts are considered, whereas arguably more contexts exist. The considered geometry representation contexts are:

- Constructive Solid Geometry (CSG): Geometry of the CSG context is described in a procedural manner, in the sense that geometric primitives, such as cubes and spheres, are defined and their processing is documented step-by-step. Processing can include

Boolean operations, transformations, scaling and similar methods until the desired geometry is created. The resulting geometry is defined to be solid.

- **Boundary Representation (BREP):** BREP geometry is described by defining and connecting the surfaces of the geometry object. The description relies on defining faces, edges and vertices.

- Mathematical Models: Mathematical models can be applied to describe geometry as well. Examples for this context are B-Spline, NURBS or Bézier surfaces.

- Tessellation: Tessellated geometries or meshes are described by tessellating the geometry's surface model. Hence, the geometry description consists of connected triangles or quadrilaterals that, together, form the geometry's surface. During the tessellation process, parameters regarding the number of triangles or quadrilaterals and subsequently resulting geometric detail can be adapted.

- Point Cloud: Point cloud geometry consists of a large number of 3D points that each represents a singular point in the geometry's surface. This context is most common in as-built geometry that is digitised through laser scans or photogrammetry.

Geometry Kernel. Geometry kernels are software libraries that comprise several methods and functions to interpret and interact with geometry descriptions and create digital geometries (visualisations) from them. The kernels are based on algorithms that are typically proprietary, since multiple concurring geometry kernels exist. Commonly, a geometry kernel is designed and optimised for certain use cases and geometry schemas. Examples of either open or proprietary geometry kernels are Open Cascade, ACIS and Parasolid [300].

4.2.2   Importance of geometry and geospatial data to AEC

Geometry and geospatial data are relevant to most use cases and domains in AEC. A brief study on use cases and their requirements on geometry descriptions is given by [404]. This study shows that, while the use cases are diverse, geometry descriptions are required. Yet, the expected details, representation contexts and schemas are dissimilar, as well.

4.2.2.1   Integration in traditional BIM

Traditional BIM models intertwine geometry descriptions closely with non-geometric (semantic) information. For example, the IFC requires both a geometry description and a basic semantic model in order to be imported into common BIM software applications, such as Autodesk Revit. Based on the historic annotated 2D drawings, 3D geometry models are annotated with semantic information, resulting in geometry being a core part of common BIM schemas.

4.2.2.2   Challenges in traditional BIM

Due to the heterogeneous software landscape of AEC, stakeholders commonly apply different software solutions that oftentimes rely on different geometry kernels for interpreting geometry. Since the kernels' algorithms are not uniform, mathematical rounding errors occur, producing inaccuracies when geometries are produced by different geometry kernels based on the same geometry description.

These inaccuracies hinder interoperability, since it complicates round-tripping[1] geometry between different software applications. Moreover, misunderstandings between different domains and stakeholders are increased if, for example, calculated surface areas differ. When parametric geometry descriptions are considered, this situation intensifies, as implicit geometric information serves as a basis for calculating non-geometric properties, e.g. U-values.

To solve issues related to heterogeneous kernel algorithms and subsequent inaccuracies, a uniform geometry kernel would be required. As geometry kernels are mostly proprietary and core of competing software applications, an agreement on a uniform geometry kernel is highly unlikely. Alternatively, a master geometry schema could be defined, e.g. within projects, together with conversion processes to other geometry schemas. Then, software applications can be fed their optimised native geometry schemas and anticipated rounding errors, as the conversion processes are standardised within the considered environment. But since traditional BIM approaches understand geometry as the BIM model's core that is annotated with semantic information, attaching multiple geometry descriptions to the same object produces redundant semantic information and potentially overly complicated conversion processes as semantic information would have to be converted, too.

4.3 INTEGRATING GEOMETRY AND GEOSPATIAL DATA IN A WEB CONTEXT

When integrating geometry and geospatial data in a Web context, one can differentiate between rendering geometry for Web and Cloud applications and publishing data in a semantic web and linked data context. While several solutions exist for rendering geometry in a web browser[2], mostly based on the glTF[3] data schema and format, this chapter focuses on publishing data on the Web.

By integrating geometry and geospatial data in a semantic web or linked data context, it is possible to extend semantically oriented Linked Building Data models, apply extensive spatial querying methods, define (automated) conversions of geometry schemas and provide fully parametric models. However, not all of these functionalities are available to any geometry description[4] and some are still in research at the time of writing.

In general, [404] distinguish four approaches to integrate geometry in a semantic web context, see Figure 4.1. These approaches are briefly discussed in this chapter. For more details and a file size analysis, see [254,404]. Each approach is linking the geometry through semantic web technologies, but the applied technologies for storing the data or formatting the content differ. The approaches have been introduced by [404] and the upcoming sections are based on this work. For more details on the existing geometry schemas and a more thorough discussion and comparison, see the original source.

4.3.1 Approach I: RDF-based geometry descriptions

The first approach relies on semantic web technologies to store and format the geometry description. Thus, this approach comprises ontologies that describe geometry, see schematic of Figure 4.2. Thereby, all geometric properties are available in RDF and can be linked directly. This can be used to reduce redundant information, i.e. when a non-geometric property represents a geometric feature, such as the height of an object. In the case of indirectly redundant information, e.g. volumes, the geometric properties can be referenced with the algorithm that results in the desired value. Hence, the integration

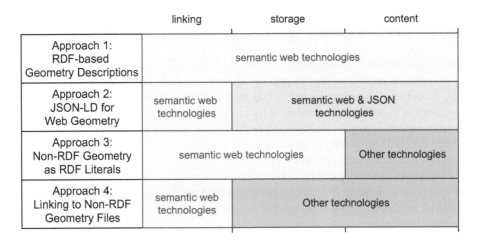

	linking	storage	content
Approach 1: RDF-based Geometry Descriptions	semantic web technologies		
Approach 2: JSON-LD for Web Geometry	semantic web technologies	semantic web & JSON technologies	
Approach 3: Non-RDF Geometry as RDF Literals	semantic web technologies		Other technologies
Approach 4: Linking to Non-RDF Geometry Files	semantic web technologies	Other technologies	

Figure 4.1 Overview of distinguished approaches of [404].

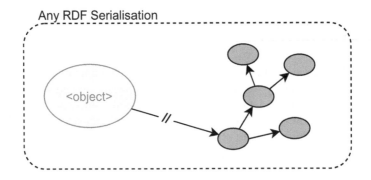

Figure 4.2 Schematic visualisation of Approach I by [404].

of completely RDF-based geometry descriptions into Linked Building Data offers new possibilities for equation-based parametric descriptions, see [403].

Implementations of this approach are not as manifold as common geometry schemas and formats. A selection of current implementations include[5]:

ifcOWL which is not stripped of the IFC geometry part [299]

GEOM ontology which simplifies the IFC geometry part[6] [45,46]

OntoSTEP an RDF translation of any STEP Application Profile (AP) [22]

OntoBREP describing BREP geometries in RDF [304]

OCC/OOP Using the Ontology for Object-oriented Programming[7] with the OCC ontology[8] that maps the OpenCASCADE kernel functions to create an RDF representation of source code to create geometry

Neither ontology is supported by common software applications, and converting geometry from a non-RDF format into these ontologies is only possible in limited combinations, using specified implementations. Moreover, an RDF-based geometry description that follows the same schema as a geometry description that is not serialised in RDF, e.g. OntoSTEP and STEP, will always produce a significantly larger size, as has been shown on the example of (entire) IFC models before [404]. The reason for this circumstance is that geometry descriptions heavily rely on ordered lists to describe coordinates, vectors and matrices, which are complicated to define in RDF.

4.3.1.1 Lists in RDF

While it is possible to add multiple values with the same property in RDF, these lists are not ordered and will be returned in a random order by triple stores. If the order of the list is relevant, which is the case for coordinates in geometry descriptions, dedicated methods have to be applied [296].

4.3.1.1.1 RDF lists The most straightforward option to model ordered lists in RDF is to use the concepts that are part of the RDF specification already. RDF lists introduce a class (`rdf:List`), two properties (`rdf:first` and `rdf:next`) and an individual (`rdf:nil`). With these concepts, it is possible to create a chain of value pairs (`rdf:List`) that will point to the value at the current position in the list (`rdf:first`) and the next value pair in the chain (`rdf:next`) until there are no more entries in the list (`rdf:nil`), see Figure 4.3. As becomes visible in this figure, the application of RDF Lists results in a rather verbose data structure, especially if compared to non-RDF-based data structures, such as an array in JSON (`["first value", "second value", "last value"]`).

4.3.1.1.2 Dedicated properties If the number of entries in the list is known and does not vary, it is also possible to introduce dedicated properties, reducing the amount of needed triples immensely. In the case of coordinates, it is feasible to follow this approach, as normally only two or three dimensions are considered. For example, the GEOM ontology [45,46] introduces dedicated `geom:x`, `geom:y` and `geom:z` properties to go without ordered lists for coordinates. With limited coordinate dimensions, this approach is feasible. But considering polygons with potentially unlimited number of points, queries that extract values from all dedicated properties become overwhelmingly complex.

4.3.1.1.3 Using indices Another commonly applied solution to ordered lists is to apply value pairs in an unordered list, where the pair consists of the value and the value's index, see Figure 4.4. However, this approach is prone to errors, since the index has to be checked and updated every time a value in this list is added, moved or deleted. Nonetheless, this approach reduces the amount of needed triples to describe ordered lists noticeably in comparison to the RDF List approach while maintaining more flexibility than introducing dedicated properties.

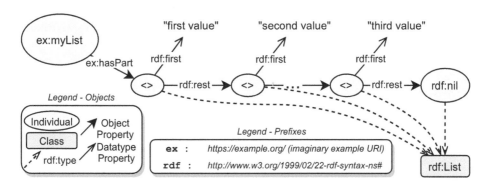

Figure 4.3 Modelling lists in RDF.

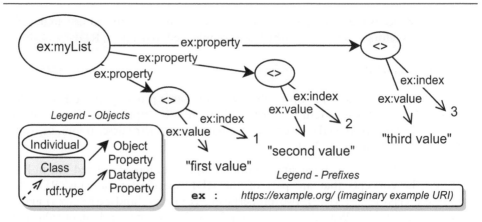

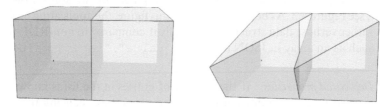

Figure 4.4 Using indices for ordered lists in RDF.

Figure 4.5 GEOM geometry (a) before and (b) after changing order of vertices [404].

4.3.1.1.4 Faulty implementations   Whereas faulty implementations for ordered lists are an issue in disregard of the considered domain, when considering geometry descriptions such faults become more visible. For example, the GEOM ontology does not apply ordered lists as intended in RDF and neglects the issue that property order does not remain stable for RDF data stored in triple stores. The ontology connects several vertices to a BREP geometry using the same property and trusts that the order of the vertices will not be changed. Figure 4.5 shows how the geometry subsequently changes after the order is rearranged for one vertex.

4.3.2   Approach 2: JSON-LD for web geometry

In contrast to Approach 1, the second approach imposes some limitations on users. While all three observed parts of a geometry description (linking, storing and formatting) are still relying on semantic web technologies, the storing and formatting part specifically require JSON-LD [368] as hybrid between Semantic and common Web technologies, see Figure 4.6. For this approach, only one implementation exists to the authors' knowledge, namely **GeoJSON-LD**[9]. Since the release of JSON-LD 1.1, this approach also supports the application of nested lists, which previously was criticised to be lacking [404].

4.3.3   Approach 3: Non-RDF geometry as RDF literals

In the third approach, only linking and storing the geometry descriptions relies on semantic web technologies, whereas the formatting can apply any technology that can be stored or encoded as a string, see Figure 4.7. As a result of allowing different formatting

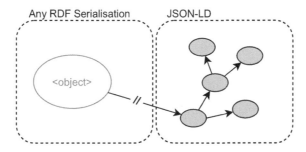

Figure 4.6 Schematic visualisation of Approach 2 by [404].

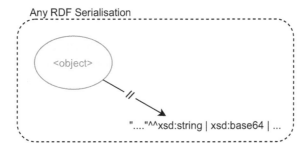

Figure 4.7 Schematic visualisation of Approach 3 by [404].

for geometry descriptions, a much wider range of applicable geometry schemas can be applied within this approach than the first two approaches. All common geometry schemas, including STEP, OBJ, glTF, DWG, STL, IGES, X3D, etc., can be utilised to include so-called snippets in RDF graphs, thereby making them available in a semantic web context. Furthermore, this approach includes the application of the Simple Feature Access (SFA, [174]) with its implementations of Well Known Text (WKT) and Well-Known Binary (WKB), as well as the Geography Markup Language (GML, [285]). For these latter descriptions, more elaborated implementations exist that also allow spatial querying, see Section 4.5.1.

By formatting the geometry description in their native and optimised formats, the size of the descriptions is reduced in comparison to the first approaches. However, by moving the descriptions themselves out of a linked data context, it is not possible to link to geometric properties or objects for parametric dependencies or object-oriented modelling out of the box. If it is required to refer to a part of the geometry description, e.g. the geometry representation of a door which is placed several times within a building model, an additional snippet would have to be created and linked from any related object. Yet, as the snippet is integrated as a literal, it is not possible to have multiple references to it because it is lacking an identifier. Hence, a utility ontology is needed to implement such references, see Section 4.4. Apart from the referencing, the transformation of the geometry object, e.g. when placing it at different positions, has to be considered and modelled, too.

Regarding available software applications that can use implementations of this approach out of the box, more tools are available, especially for web-oriented geometry formats, such as glTF and X3D. Geometry snippets can be easily integrated into web applications. But considering common CAD and BIM geometry formats or formats that

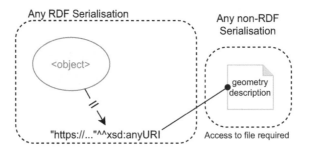

Figure 4.8 Schematic visualisation of Approach 4 by [404].

require encoding, the snippets may need to be pre-processed before they can be loaded into software applications, e.g. because header information is not part of the snippets or RDF literals support UNICODE characters instead of binary content. In the latter case, a binary-to-text encoding is required to store the content of a binary geometry format in an RDF literal.

4.3.4 Approach 4: Linking to Non-RDF geometry files

This last approach is rather similar to the third one, except that the geometry description, formatted in its native schema, is also stored in its native format and only linked through semantic web technologies, see Figure 4.8. This approach can be implemented using any of the data formats of Approach 3. Accordingly, benefits such as reduced file size and more available support by common software applications in comparison to the first two approaches are similar. More so, file size may be lower and support more extensive than in the case of Approach 3, as no pre-processing of geometry description snippets is required and the reduction in size by omitting header information can be easily lost by encoding. On the other hand, referencing to parts of the geometry descriptions is not possible unless the geometry formats offer permanent and unique identifiers to geometry objects or properties.

4.3.5 Multiple geometry descriptions

Summarising, all four approaches have their reason to exist due to benefits and shortcomings. Similar to in non-RDF-based geometry descriptions, there is not one schema that outperforms all others in any use case. Instead, different use cases may be addressed by different approaches. As an overall conclusion, Figure 4.9 gives an insight where each approach has its strength, not considering their individual implementations, which may vary in their suitability. The figure is adapted from [404], only the graph of the second approach is differing in two aspects: Conciseness and Simplicity, as JSON-LD 1.1 now also supports nested lists[10].

In general, semantic expressiveness, extensibility and portability are higher for approaches closer to the semantic web, as ontologies can be utilised to create links between geometric properties and objects and data can be offered distributively as linked data. On the other hand, common geometry formats are better supported, more concise, simpler to formulate and understand, and offer more implementations to choose from.

With the elementary idea in mind that there is not one geometry schema – or approach – that outperforms all others, the vision of multiple geometry descriptions

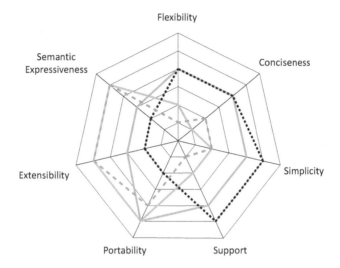

Figure 4.9 Comparison of the approaches, based on [404].

becomes more desirable. And with the linking methods provided by semantic web technologies, dedicated ontologies may support this vision.

4.4 EXISTING IMPLEMENTATIONS FOR INTEGRATION OF GRAPHS

Considering the potential of connecting multiple geometry descriptions of different geometry schemas and even formats, a semantically meaningful linking method has to be applied. Several ontologies define their individual properties on connecting geometry descriptions to non-geometric data. In general, one can differentiate between three types of specialised properties that originate from different ontology types:

1. geometry ontologies, such as GEOM and GeoJSON-LD
2. non-geometry ontologies, such as BOT and ifcOWL[11] [297]
3. annotation and file management ontologies

While **geometry ontologies** provide meaningful and dedicated properties for their individual geometry objects, these properties can usually not be used for other geometry ontologies. More so, these properties are not suitable for non-RDF-based geometry description, as they are used in Approaches 3 and 4.

On the other hand, **non-geometry ontologies** provide properties to integrate geometry descriptions into their specific data schema. These properties may be data properties, allowing to integrate snippets and file references, or object properties to integrate RDF-based geometry descriptions. Yet, not every non-geometry ontology provides both types of properties and since the properties are part of a non-geometry ontology, they commonly reference objects from that ontology. Hence, the application of such properties outside of their origin ontology requires some sort of alignment between the outside ontology and the origin.

Finally, **annotation and file-management ontologies** allow to reference non-RDF resources in a semantically meaningful manner. However, they normally do not consider RDF-based descriptions or snippets that are part of an RDF graph. Hence, properties of these ontologies are not suitable for the Approaches 1–3. However, the idea of managing geometry descriptions instead of just referencing it seems promising. Especially when considering transformation processes, version management and schema conversions, additional metadata for linking is required.

Currently, three ontologies for managing geometry, defining geometry schema and format and describing geometry metadata exist and are primed towards each other. Those ontologies are called *Ontology for Managing Geometry (OMG), File Ontology for Geometry formats (FOG)* and *Geometry Metadata Ontology (GOM)*, and will be discussed in more detail subsequently.

4.4.1 Ontology for managing geometry (OMG)

The Ontology for Managing Geometry (OMG)[12] is highly inspired by the Ontology for Property Management (OPM) [316] and aims to provide generic concepts to manage geometry descriptions in a semantic web context [405]. The ontology does not contain any information about the applied geometry schemas and formats, but focuses on general-purpose functionalities that allow handling of single, multiple and versioned geometry descriptions. By reducing the scope to omit specific definitions, the OMG is supposed to remain in a stable state, where little changes to the schema are required over time. Hence, the ontology is suitable to serve as a reference ontology, whereas concepts that are susceptible to changes are outsourced to dedicated, more flexible ontologies.

With the ontology's scope being the handling of single and multiple geometry descriptions and providing means to version geometry descriptions over time, it serves three different purposes that, potentially, can be combined. However, to reduce schema overhead, if more complex purposes, such as versioning geometry or handling multiple geometry descriptions, are not required, the OMG introduces three levels that cover each purpose. These levels will be summarised here briefly, and they are to inspired and to some extent comparable to the levels discussed in Chapter 2. For a more thorough explanation, see [405] or the online documentation of OMG.

4.4.1.1 Level 1: referencing geometry descriptions in a semantic web context

The first level of OMG introduces two properties: one datatype property for referencing file snippets and external files according to the Approaches 3 and 4; and one object property for referencing RDF-based geometry descriptions as defined in Approaches 1 and 2 (see Figure 4.10). These properties do not contain any information regarding the schema or format of the geometry description, and in the case of embedded or external geometry descriptions, cross-references within the RDF graph are not enabled, as RDF literals for file snippets or locations of external files do not possess identifiers.

4.4.1.2 Level 2: handling multiple geometry descriptions

Since literals cannot be cross-referenced, interlinking of geometry descriptions of any of the four approaches cannot be implemented homogeneously. Thus, the OMG introduces an intermediate omg:Geometry node that can be referenced uniformly

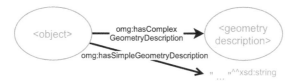

Figure 4.10 Schematic overview of OMG Level 1 [405].

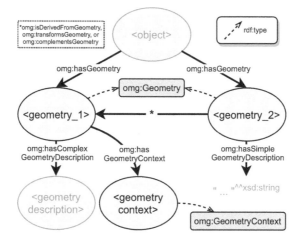

Figure 4.11 Schematic overview of OMG Level 2 [405].

and an `omg:GeometryContext` for grouping geometry descriptions, see Figure 4.11. Apart from the geometry node, relations between different geometry descriptions are part of this level. These relations allow to define derivation, transformation and complementing of geometry descriptions. Again, these relations do not consider any specified information, e.g. transformation matrices or derivation algorithms, but focus on generic links to indicate such relations.

Considering web-based geometry descriptions that rely on multiple input files for rendering, i.e. OBJ geometry that requires a geometry file and a material file that defines colours, textures, etc., the introduced geometry node also allows to attach multiple files through the OMG's datatype property for geometry description at once. If this would be modelled in Level 1 OMG, extracting the correct files from the graph would prove difficult or impossible if multiple OBJ geometry descriptions with material files were attached to the same object.

In the context of compositions, it is also possible to model a geometry description that is part of another (larger) geometry description using the `omg:isPartOfGeometry` between two `omg:Geometry` instances. A prerequisite is that the geometry schema of the larger geometry description supports unique identifiers for the smaller geometries that it includes. The main geometry is modelled using the regular OMG relations, while the geometry description that forms a part of it is described with the `omg:hasReferencedGeometryId` datatype property.

4.4.1.3   Level 3: versioning geometry descriptions

Level 3 of OMG introduces an additional intermediate node, the `omg:GeometryState` that represents a geometry at a certain point in time which is subsequently linked to its

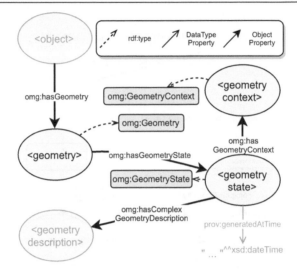

Figure 4.12 Schematic overview of OMG Level 3 [405].

geometry description, see Figure 4.12. In this level, further relations are introduced that rely on OPM concepts to create versions for RDF-based geometry descriptions and infer implicit relations. Moreover, a second derivation property is introduced, which can be used in combination with the derivation property of Level 2 to indicate possible and conducted derivations.

4.4.1.4   *Explicit and implicit dependencies*

Finally, the OMG provides concepts that allow to create links between geometric properties of RDF-based geometry descriptions and their non-geometric duplicates through explicit dependencies, as well as links between geometry descriptions and non-geometric properties that are dependent implicitly, i.e. an object's volume or surface area. These relations are aligned to the concepts of OPM and rely on using `opm:PropertyState` instances.

4.4.1.5   *Summary*

Overall, the functionalities of OMG are summarised in Table 4.1, which also indicates which Level of OMG is required to utilise the considered functionality. For explicit dependencies, any level is suitable, but the geometry description needs to be RDF-based. Implicit dependencies require at least Level 2 OMG, in order to refer to the geometry description in a uniform manner.

4.4.2   File ontology for geometry formats (FOG)

Complementary to the OMG, the File Ontology for Geometry formats (FOG)[13] aims to provide semantically meaningful links to specify the geometry schemas and formats used [43]. FOG is strongly aligned with OMG and extends OMG's generic proper-ties, `omg:hasSimpleGeometryDescription` and `omg:hasComplexGeometry Description` as well as `omg:hasReferencedGeometryId`. The concept of FOG is to serve as some sort of taxonomy of properties for the indication of geometry schemas

Table 4.1 Overview of OMG functionalities

Functionality	Level 1	Level 2	Level 3
Referencing geometry descriptions	✓	✓	✓
Handling multiple geometry descriptions	✗	✓	✓
Part of geometry	✗	✓[a]	✓[a]
Version control	✗	✗	✓
Explicit dependencies	✓[b]	✓[b]	✓[b]
Implicit dependencies	✗	✓	✓

[a] if the larger geometry description provides unique identifiers for smaller geometries within it

[b] if the geometry description provides unique identifiers to its geometric properties (and they are represented by RDF nodes)

and formats, including their release versions, serialisations and specification metadata. For `omg:hasReferencedGeometryId`, the subproperties in FOG specify, besides the geometry schemas and formats, also the type of identifier to avoid confusion by implementations.

In order to take as much advantage from reasoning algorithms as possible, FOG relies heavily on inheritance of properties. For each geometry schema, e.g. OBJ, glTF or GEOM, a subproperty of the suitable OMG property is created, e.g. `fog:asObj`, `fog:asGltf` or `fog:asGeomOntology`. If the geometry schema has several release versions, e.g. glTF 1.0 and glTF 2.0, these properties are extended by further subproperties, e.g. `fog:asGltf_v1.0` and `fog:asGltf_v2.0`. Finally, if the schema can be serialised in multiple formats or complemented by additional files, i.e. for materials, the version property is extended once more, see Figure 4.13.

By structuring its properties strictly hierarchically, FOG allows reasoning engines to generalise highly specific properties. Hence, a geometry description that is related through the dedicated property for DWG geometry descriptions using the 2018 specification can be extracted by querying for DWG descriptions or even generic non-RDF-based descriptions, see Figure 4.14.

In addition, FOG provides refined concepts to group multiple files that can be used to complement a geometry description, i.e. material descriptions. For this purpose, an additional node is introduced, `fog:ReferencedContent`, which is equipped with the referenced part and the original file name, if needed, see Figure 4.15. Concluding, FOG serves as taxonomy-like structure of dedicated properties for specific geometry formats, their versions, serialisations and complementary files. However, since novel geometry formats arise and existing formats are revised and released as newer versions, FOG is not intended to be complete at any point. Instead, it is understood to be a community effort to extend the ontology when new properties are needed, following the proposed architecture and rules[14].

4.4.3 Geometry metadata ontology (GOM)

With OMG and FOG concerning relations between non-geometric objects and their geometry descriptions together with their geometry format, metadata is omitted so far. Therefore, the Geometry Metadata Ontology (GOM)[15] has been introduced [44]. GOM aims to extend functionalities of the OMG/FOG methodology to depict additional

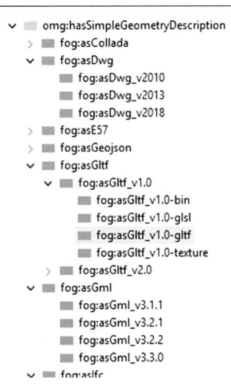

Figure 4.13 Overview of FOG's inheritance structure.

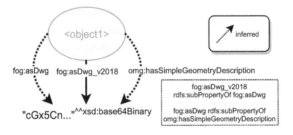

Figure 4.14 Inferring more generic properties via FOG [43].

geometric information regarding coordinate systems and transformation, geometry representation context, represented accuracy and other geometry metadata.

For instance, Figure 4.16 shows how GOM concepts can be used to attach coordinate systems to geometries, including an affine coordinate system transformation that defines the relative position of the two Cartesian coordinate systems to each other. GOM introduces both the concept of coordinate systems, as well as a selection of coordinate system types as subclasses. Due to the verbose description of ordered lists in RDF – and matrices can be represented as a list –, GOM was designed so that regular JSON arrays can be used inside RDF literals connected to a transformation instance. GOM proposes a datatype property (`gom:hasTransformationMatrix`) and two custom datatypes, `gom:rowMajorArray` and `gom:columnMajorArray` for this datatype property.

Another example for GOM concepts is the applied length unit of a custom Cartesian coordinate system, as shown in Figure 4.17. Certain geometry schemas, such as OBJ,

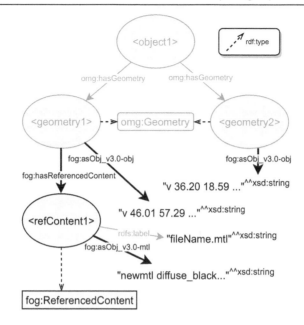

Figure 4.15 Referencing additional content via FOG [43].

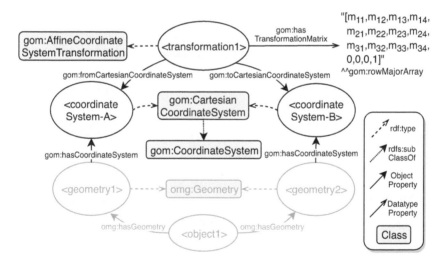

Figure 4.16 Coordinate systems and their transformations in GOM [44].

PLY and PCD, do not have defined units, hindering the analysis of geometry in custom coordinate systems. If no length unit is explicitly modelled, metre is to be assumed. Other metadata covered by GOM is aimed for to enable querying for geometry without having to apply spatial querying, inspired by [359].

Metadata, such as the number of vertices, edges, faces or the geometry's file size (if the geometry is attached according to Approach 4), can give insights on the detail of the respective geometry. Other metadata, e.g. (different types of) surface areas or volumes (geometry representations), can simulate spatial querying, if those properties are kept in sync. Finally, GOM also considers metadata that measures the represented

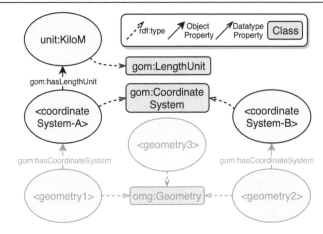

Figure 4.17 Length units in GOM [44].

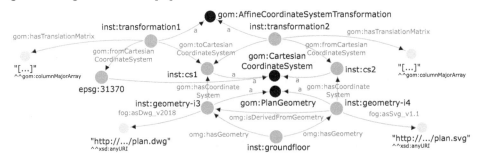

Figure 4.18 Combining OMG, FOG and GOM [44].

accuracy of the geometry at hand, i.e. deviation limits and percentages of surface areas that comply to certain ranges of deviation, which may help searching for high-quality geometry descriptions of surveyed objects.

4.4.4 Summary

The discussed ontologies, OMG, FOG and GOM, can be used together to integrate geometry descriptions into a semantic web context – in disregard of the chosen approach and geometry schema or format. This methodology also allows to manage multiple geometry descriptions of the same object, including dependencies, transformations and version control. Moreover, the used geometry schemas and formats can be distinguished, and meaningful metadata added. An example graph on how those ontologies can work together is given in Figure 4.18, where a non-geometric object (inst:groundfloor) is linked to two geometry descriptions in different schemas and coordinate systems.

4.5 TOOLS FOR INTEGRATING GEOMETRY AND GEOSPATIAL DATA

In addition to the available ontologies to integrate geometry descriptions in a semantic web context, tools exist that allow handling or transforming geometry in said context. This section covers these tools and differentiates between tools for spatial querying that extend the standardised SPARQL query language (Section 4.5.1) and transforming and viewing geometry in web applications or servers (Section 4.5.2).

4.5.1 Spatial querying

Spatial querying combines querying functionalities, e.g. of SPARQL, with geometric processing in order to incorporate spatial analyses into queries. To obtain spatial querying, the query language has to be extended with dedicated terminology, e.g. `squareArea` which is mapped to geometry-processing functionalities, e.g. $width^2$. The extension is best standardised, so any querying engine can implement this extension and spatial queries can be reused across implementations. This standardisation includes not just the terminology but also the geometry-processing functionalities that are mapped to the terminology. Moreover, the analysed geometry schema has to be predefined and standardised, as it needs to be known how geometric properties, e.g. *width*, are defined in the geometry description. Furthermore, the linkage between geometry description and non-geometric objects has to be predefined to ensure that any implementation of the SPARQL extension retrieves the same geometric properties and, thus, returns the same results.

Currently, all standardised implementations for spatial querying relate with geometry description snippets (see Approach 3, Section 4.3.3) that use either WKT or GML as the geometry schema. Also, all implementations are currently relying on their own dedicated linking methods, although the OMG/FOG approach (Section 4.4.1) has been proposed as use case in the latest OGC's Whitepaper [2]. In general, three implementations are widely applied in the domain of Linked Building Data: GeoSPARQL of the Open Geospatial Consortium (OGC) [303], stSPARQL [210] and the domain-specific BimSPARQL [434]. These implementations are briefly discussed in the upcoming sections.

4.5.1.1 GeoSPARQL

The geographic query language for RDF data GeoSPARQL v1.0 [303] is standardised by the OGC and aims to provide RDF structures to describe geospatial data and include their geometry descriptions as RDF literals. GeoSPARQL includes a relatively simple high-level vocabulary for describing geospatial data and an extension to the SPARQL query language for processing that data. The main classes are **SpatialObject** and **Feature**, covering both spatial representations and features respectively (see Figure 4.19). **Geometry** then enables the representation of geometric objects, serialisable as GML and WKT. The query language's documentation includes the expected structure of the considered geometry descriptions. Its main restriction on geometry descriptions is that only the spatial querying of two-dimensional geometry is considered in GeoSPARQL v1.0[16], although there are proposed extensions to GeoSPARQL which will specifically address the need to support querying of 3D geometries [2].

Furthermore, GeoSPARQL provides a terminology to describe topological relations according to *Simple Features*, which is also known as Simple Feature Access (SFA) and is formalised by both the OGC and the International Organization for Standardization (ISO). This is a set of standards that specify a common access and storage model of (mainly) 2D geographic features such as point, line, polygon, multi-point and multi-line. *Region Connection Calculus (RCC8)*, which defines eight basic relations between two regions (in Euclidean, or in topological space) and the *Dimensionally Extended 9-Intersection Model (DE-9IM)*, also known as Egenhofer, which is a

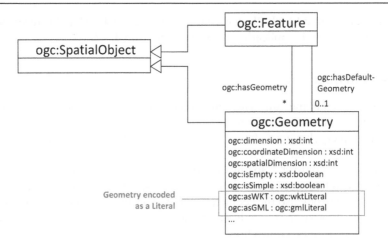

Figure 4.19 Overview of GeoSPARQL. (See also Perry [302].)

standardised topological model also used to describe the relations between two regions (in two dimensions). These include relations for Equality, Disjointness, Intersection, Containment, Overlaps, etc.

For each of these relations, three GeoSPARQL properties are defined to represent the relation itself, a rule describing the relation and query functions to extend the query engine. Additionally, the expected DE-9IM intersection pattern for each relation is specified, ensuring that any implementation of GeoSPARQL has the same understanding of topological relations. Apart from topological relations, further functions are defined in GeoSPARQL to aggregate geometries through intersections, unions or differences, and calculate geometric properties and objects, such as distances or boundaries.

GeoSPARQL is widely implemented in triple stores and query engines; however, most implementations do not consider all aspects. For example, some implementations may cover query functions but not rules, others implement only topological relations. As these implementations are constantly improved, a precise overview of the implemented features is omitted. A selection of currently available implementations include: Apache Jena, Stardog, Virtuoso, Parliament, Strabon and GraphDB.

4.5.1.2   stSPARQL

To query spatial and temporal data in RDF, stSPARQL has been published by Strabon [215]. The query language extension offers mostly the same spatial querying functionality as GeoSPARQL, namely topological relations, aggregations and geometric properties of two-dimensional geometries in GML and SFA (WKT and WKB). However, stSPARQL also considers the temporal context of the spatial data. Therefore, stSPARQL adds a fourth column, extending the RDF triple pattern to a quad pattern, where the last item represents the temporal aspect. Subsequently, it is possible to perform spatio-temporal queries. Yet, in contrast to GeoSPARQL, stSPARQL is not as widely supported; its main application is Strabon.

4.5.1.3   BimSPARQL

Since BIM relies mostly on three-dimensional geometry, the BimSPARQL extension picks up the general idea of GeoSPARQL, namely mapping geometry processing functions for GML and SFA geometry descriptions to a dedicated terminology, and projects it onto three-dimensional geometry. With the third dimension adding a higher complexity, BimSPARQL reduces the scope for geometry descriptions to WKT geometry descriptions only.

Overall, the BimSPARQL vocabulary distinguishes four areas:

- Schema semantics
- Instance semantics
- Product geometry
- Spatial reasoning

Only the last two areas require spatial querying, hence, only those areas are considered in this section. For product geometry, seven general functions are considered to retrieve complete geometries (original, aligned or minimum volume boundary boxes), parts of geometries (surfaces) or properties that are implicitly described by geometries (volume or overall height). Furthermore, specified functions can be defined for individual product types, e.g. space, window or gross wall areas, which rely on the general functions.

In terms of spatial reasoning, *Simple Features* topological relationships are defined in BimSPARQL, i.e. for touching, disjoint, overlapping, bi-directional containment and equality relations. Additionally, three types of distances are specified, three-dimensional, Z-axis only (height) or XY-axes only (plan), and functions that combine spatial and semantic information, e.g. to retrieve whether an object is located within a storey.

While the functionality of BimSPARQL extends the scope of GeoSPARQL and stSPARQL by considering the third dimension, it is combining spatial querying with semantic queries. Yet, BimSPARQL is implemented as a demonstrator only and does not come out of the box with any of the popular triple stores, as is the case for GeoSPARQL.

4.5.1.4 *Geospatial geometric literals*

The geospatial functions supported by standards such as GeoSPARQL and stSPARQL are executed over geometric representations, e.g. the aforementioned literals; Lines, Polygons, etc. Multiple formats for representing geometric literals exist, the most widely used of these are GML and GeoJSON [375]. GML, which is based on the eXtensible Markup Language (XML), is an open source encoding for representing a hierarchy of geometric objects, which include features, collections and geometries, among other structures [285]. A GML schema is used to describe a generic geographic dataset, which contains solids, surfaces, lines, curves and points and can express any type of geometry, and up to three dimensions. More domain-specific representations are also possible; a popular one which is commonly used with Geographical Information Systems (GIS) is CityGML [156] for describing cities (and buildings).

GML literals are represented in XML, with different tags to represent the semantics of a geometry, e.g. a Polygon will consist of multiple tags, with metadata tags providing some string-based "description", an identifier which can be global, or application domain based, to the interior and exterior of the polygon, represented as a "LinearRing" with an x and y position (pos) defined. All points belonging to a geometric object should also share the same coordinate reference system. GML supports the use of any Coordinate Reference System (CRS). While typical CAD tools use a Cartesian coordinate system, in the case of geospatial data, a CRS must take into account the curvature of the Earth. A typical geospatial CRS is the World Geodetic System 1984 (WGS 84 – EPSG:4326), which provides a good approximation at all locations on the Earth, as the most commonly used CRS for spatial data on the Web. In some use cases, more accuracy is required than is available with WGS 84, and so other coordinate systems are preferred. Each country generally has its own CRS, using different projections, e.g. the transverse Mercator map projection.

Given this flexibility, one of the main criticisms of GML is the resulting complexity. For even very simple geometric literals, the number and availability of different tags can make the language overly verbose.

GeoJSON is an encoding based on the JavaScript Object Notation (JSON), developed to support web developers, with the intention of providing a (relatively) simple method to define a geospatial geometry. The basis of an object and the data is given by key-value pairs, and other objects can also form the value for a key, thus enabling nested structures. GeoJSON focuses on a restricted set of features, similar to the OGC simple 2D features, e.g. Point, LineString and Polygon. It also only supports one CRS, WGS 84. GeoJSON does not set out to represent semantics though, and if these are required, it is possible through the use of JSON-LD[17].

Another option for representing geometry is the aforementioned well-known text (WKT) [174], proposed by the OGC. Like GeoJSON, WKT is based on simple features and is thus a less complex solution when compared to GML to encoding geometry. It does though provide more flexibility in that it supports multiple CRS, and so, is suitable where additional accuracy may be required. It is widely used to store geometry data in database systems where large amounts of geometric information must be represented efficiently (using a short representation), but can also be combined in a semantically meaningful manner to support open GIS, e.g. using GeoSPARQL. WKT can represent multiple geometries, as indicated in Figure 4.20, where we see the root geometry class and multiple subclasses, each representing a geometry. WKT can support 0, 1- and 2-dimensional geometric objects that exist in 2-, 3- or 4-dimensional coordinate space, where an additional altitude or elevation may be defined as well as a measurement, e.g. x, y, z and m, where z is elevation and m is a measurement.

4.5.2 Transforming and viewing geometry

Next to spatial querying, there exist implementations to transform geometry for it to be included in a semantic web context or to view geometry in this context. Within this section, a transformation and data service tool is presented, which is part of the ongoing LBDserver project, a demo application to visualise heterogeneous geometries using the OMG/FOG method, and a data service that includes a rendering service for RDF-based geometry descriptions.

4.5.2.1 LBDserver

A transformation tool based on the fourth approach (i.e. linking RDF-based semantics with external geometry files in a non-RDF schema) is implemented as part of the ongoing LBDserver project[18]. The LBDserver provides a modular and extendable infrastructure for handling heterogeneous, federated building data, such as RDF based semantics, different geometric representations and imagery. On top of this service, which can be spread over the Web and reside in diverse databases, client-side applications can be built to to interact with project information in a use-case-oriented manner. At the time of writing, a prototypical Web application for querying and enriching a project

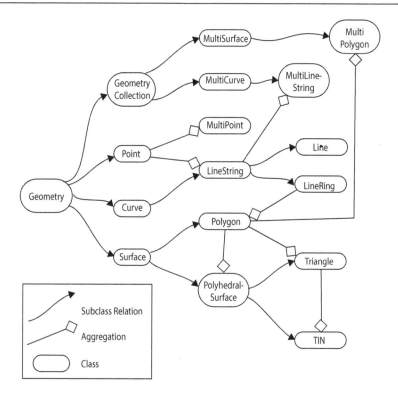

Figure 4.20 Overview of the Geometry class in the WKT schema (see [254]).

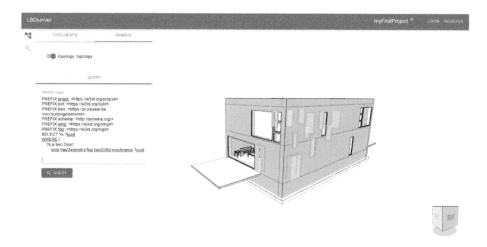

Figure 4.21 Prototype user interface for the LBDserver.

model using glTF-based geometry models and LBD ontologies is available[19], where currently two scenarios are supported (Figure 4.21). This prototype currently supports two 'enrichment' scenarios: the first one handles data which originates from a BIM (IFC) model, the second one considers semantic enrichment of mere geometry, i.e. creating a BIM model 'from scratch' using the geometry as a property to create and access the semantic data.

Using a chain of conversions[20], an IFC file is split into a glTF geometry file and an LBD graph, which are then stored in appropriate databases. Conversions rely upon multiple existing conversion tools such as the IFCtoLBD converter[21], the IfcConvert tool[22] (IfcOpenShell) for the intermediary conversion from IFC to COLLADA, and the COLLADA2GLTF tool[23] (KhronosGroup) for the final conversion from COLLADA to glTF.

Because it is possible to keep track of the original object IDs during the conversion process, an application may perform complex queries on semantic information and use IDs from the results for visualisation purposes by reading sub-documents of the glTF file [243]. Consequentially, the process can be inverted: If an object is selected in a viewer, an ID is retrieved that can be used to query, manipulate and display specific object properties. Because the conversion process does not automatically result in a structure that is compliant with OMG and FOG, an additional mapping needs to take place in order to establish such structure (see Figure 4.22).

A similar scenario is that only CAD geometry is present, which should be enriched with relevant semantic information. This approach aligns with the methodology taken in [410] for enriching existing building geometry with uncertainties, multi-classifications, etc. As there is initially no semantic information at hand, client applications need to construct the OMG- and FOG-based property chain inversely: the CAD object ID needs to be linked to an abstract "object resource", the length of the property chain is depending on the chosen OMG level (as described in Section 4.4.1). After the initial link is created, future geometric representations can also link to this semantic object, which is now instantiated. Similar to the first scenario (Figure 4.22), it is necessary to relate the ID to the complete geometry description file, using the `omg:isPartOfGeometry` relationship.

In other words, both scenarios align with the FOG methodology of referenced geometry IDs within a larger file, although LBDserver implementations are not limited

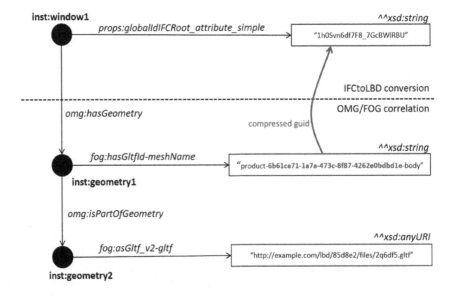

Figure 4.22 Mapping an OMG/FOG structure to IFC-derived triples.

to this methodology: depending on the project and the used (set of) microservices, all four approaches for integrating geometry descriptions in a semantic web context can be implemented.

4.5.2.2 *Visualising heterogeneous geometry descriptions*

As proof of concept for the OMG/FOG method, a web application[24] has been implemented that can read geometries attached to an RDF graph through Approach 3 and 4[25]. Aside the applied integration approach, the geometry descriptions can be available in different schemas and formats [43].

The demo application queries for any geometry descriptions in the connected data sources connected using OMG level 2. If the FOG property allows the application to deduce that the geometry schema at hand is supported, it will load the geometry. To demonstrate to the user that the geometry schemas are different, the geometry colours are distinct to the geometry schema of the visualised object (see Figure 4.23). Furthermore, the geometry descriptions are offered to be downloaded as file, even if they cannot be rendered in the application.

4.5.2.3 *Data service for RDF-based geometry descriptions*

Another tool is the SCOPE data service that includes a Fuseki triple store and interprets RDF-based OCC/OOP geometry descriptions through the OpenCASCADE geometry kernel as a web service [276]. The SCOPE data service combines the BOT, BPO and OCC ontologies through the OMG/FOG method with OMG level 1 and includes a tool that

Figure 4.23 Column Geometries Rendered in the FOG Demo Application [43].

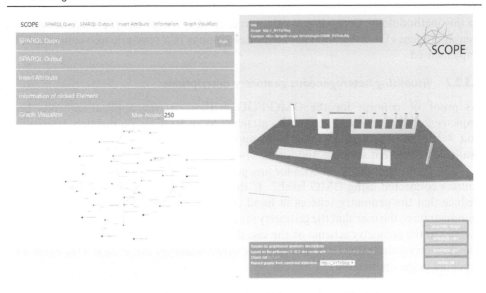

Figure 4.24 Rendered OCC/OOP geometry in the SCOPE data service [276].

allows to export geometric and non-geometric information from Autodesk Revit. This tool is being developed as part of the SCOPE research project[26].

The data stored in the Fuseki database can then be queried and visualised through the service's user interface (see Figure 4.24). The interface allows users to interact with the depicted data bidirectionally. On the one hand, data can be queried through SPARQL and the results are shown as a graph and geometry is highlighted. On the other hand, users can interact with the rendered geometry and clicking on an object will show the graph corresponding to the RDF-based geometry description. Apart from the geometry visualisation, the user interface also provides means to attach non-geometric properties to the selected objects (see also Chapter 2), following the BPO schema.

4.5.2.4   *Integrating geospatial data and building data*

In the Republic of Ireland, the Ordnance Survey Ireland (OSi) maintains GIS data for the country. To improve public access to their spatial data, the GeoHive[27] initiative was created, which allows access to data, but not as linked data. To further facilitate exploration and use of OSi's geospatial datasets, OSi has been publishing datasets [97,98] on data.geohive.ie, as linked data on the Web. This was done first with boundary data, and now more recently over 200 thousand buildings in the county of Galway have been published[28]) to initiate the development of a national geospatial identifier infrastructure based on the OSi building ontology[29] for capturing OSi building data using RDF [250].

The ontology extends both GeoSPARQL to handle geometries and PROV-O to handle provenance. The RDF is generated using an R2RML processor and declarative mappings, which enable the "uplift" of non-semantic tabular data (e.g. relational data) into the RDF. Several examples of interlinking of the authoritative data with other building data standards and datasets using geolocation have been undertaken, and in this section three methods for interlinking the published OSi building data with other datasets through the use of uplift and other data conversion techniques (Java libraries and SPARQL construct) are presented, which are also presented in [250].

4.5.3   Conversion of geospatial data to industry foundation classes

IFC is a standard for data exchange to support interoperability between software applications. A method using R2RML for converting geospatial data into ifcOWL, a serialisation of the IFC schema, is presented. It has been demonstrated that from a set of relatively simply geometric inputs, a skeleton ifcOWL model is generated which includes additional semantics not present in the geospatial data, such as the definition of external walls. To adhere to the IFC schema, the required R2RML mappings are largely due to the many different relations in IFC and also the complex Oracle queries which convert the underlying geospatial data, for example the coordinate system, and to generate geometries. The mapping and a sample data output (both as RDF and STEP) can be found here[30]). The outputs of these mappings have been validated using a combination of constraints in the Shapes Constraint Language (SHACL) and available open source software [371]. The method provides a basis for converting any geospatial dataset into IFC and should be of interest to owners of geospatial datasets as it provides potential new sources of revenue. By taking the linked data approach, this data can then live alongside a growing evolving data space of building data, and through organisations like Ordnance Survey Ireland, authoritative URIs for buildings in Ireland can become available for building data owners to link to and re-use.

4.5.4   Conversion of industry foundation classes to geospatial data

Aligning IFC geometries with existing geospatial geometries remains a challenge, but is essential to support seamless integration of the two domains. In [254,279], the conversion of ifcOWL geometries into GeoSPARQL, as well as other non-geospatial geometries, such as OBJ files, was developed. This approach explored the alignment of an IFC geometry with its geospatial equivalent, enabling the geo-tagging of building elements which have a geometric representation in the IFC model. The challenge remains with IFC that the longitude and latitude are not aligned with the origin of the Cartesian coordinate system used to represent the buildings geometry.

Methods for aligning two WKT geometries have been explored. These make use of a GeoSPARQL function geof:sfOverlaps to match geometries that overlap to support building matching with the combination of additional properties, that is, where two buildings do not align correctly. Using the additional footprint area, a SPARQL filter function returns results based on the filter condition. Two filter conditions specify that the two returned building WKT geometries overlap and that the calculated area of one WKT polygon is within ±25% the area of the overlapping polygon. Both aligned polygons, representing the same physical building, are returned if these two filter conditions are met. The two WKT geometries are taken from an ifcOWL model and another from a Building Topology Ontology (BOT) model. BOT is a developing standard for supporting interlinking of building data [315]. Again, by combining OSi data and IFC, a much richer set of building semantics becomes available for querying, and by aligning existing IFC models' geometric representations, the possibility exists to run geospatial functions over this data.

Ultimately, the goal is to develop a method for querying large datasets of buildings and building-related data, based on the authoritative geometries and URIs which are available through the OSi. Potentially, agreements could be developed between OSi and IFC model owners, whereby queries could return subsets of open data, along with licensing information for data which is considered to have some commercial value.

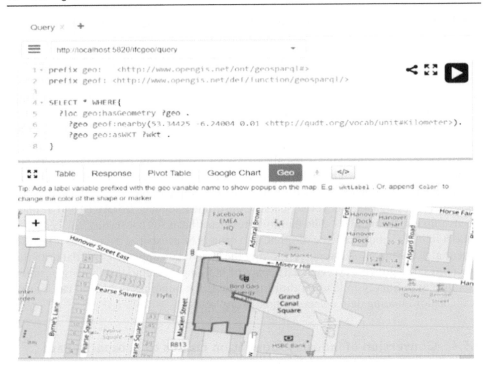

Figure 4.25   GeoSPARQL to locate features near DBpedias "Board Gáis Energy Theater" in Dublin.

4.5.4.1   Interlinking geospatial building data with DBpedia data

[252] demonstrates a SPARQL CONSTRUCT query (4.1) which takes all places from DBpedia in the Republic of Ireland and maps these to GeoSPARQL (adding geolocation). It is then possible to combine DBPedia building data with other building data based purely on geolocation, as GeoSPARQL also provides geospatial functions, such as nearby, which takes a point and returns all points within a certain distance, e.g. 2 km. If one assumes that the location of two buildings will not fall outside a certain distance, data from two sources can be combined based purely on location, i.e. to locate a Feature in the OSi dataset, which is nearby a Place defined in DBpedia. Figure 4.25 shows the result of this query run on YASGUI[31]. This data integration supports the enrichment of OSi building data, e.g. a form and function with other data available on DBpedia, such as the architect of the building, its address, opening date and seating capacity. This simple example, which only looked at one building, shows how data integration using geolocation can support building data enrichment.

Listing 4.1   R2RML Geometry Mapping

```
1  PREFIX geo: <http://www.w3.org/2003/01/geo/wgs84_pos#>
2  PREFIX geos: <http://www.opengis.net/ont/geosparql#>
3  PREFIX dbo: <http://dbpedia.org/ontology/>
4  PREFIX dbr: <http://dbpedia.org/resource/>
5  CONSTRUCT {
6      ?s geos:hasGeometry [ geos:asWKT ?point ] .
7  } WHERE {
```

```
8      ?s a dbo:Place .
9      ?s dbo:locationCountry dbr:Republic_of_Ireland .
10     ?s dbp:latd ?lat .
11     ?s dbp:longd ?long .
12     BIND( STRDT( CONCAT( "POINT(", ?lat , " ", ?long, ")" )
13     , geos:wktLiteral ) AS ?point )
14   }
```

4.5.4.2 Applications to support querying of interlinked geospatial data

As part of Ireland's open data strategy for 2017–2022, the Irish government has launched data.gov.ie which has over 8,800 open data sets published. These can be downloaded in a range of formats, including JSON, CSV and XML. Some of these data sets also have geospatial coordinates provided. In [249], three data.gov.ie open data sets were converted into RDF using R2RML. Next, a WebApp called "GViz" was developed to support the querying of these data sets using configurable GeoSPARQL queries through a WebGL and Google Maps interface (Figure 4.26). This work demonstrated that simple to use interfaces, built upon Web of Data technologies, can provide integrated data on buildings, such as data from data.gov.ie and OSi, which can be queried by non-expert users. Such interfaces can provide a basis for querying and integrating data sets, through processes such as the alignment process described in [251,279].

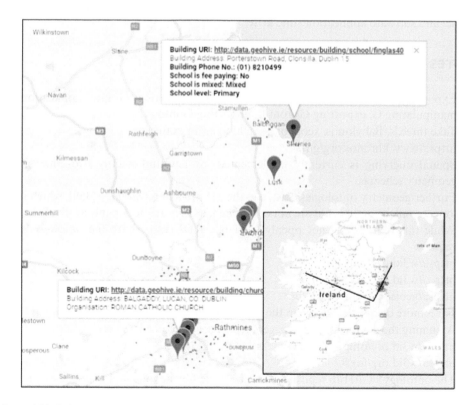

Figure 4.26 Polygon area-based selection using GViz.

In [253], a demo was presented where the OSi authoritative geometry building data set was interlinked with the Irish Central Statistics Office (CSO) data. This work mainly explored mappings to datasets using the RDF Data Cube Vocabulary[32], a W3C standardised vocabulary for the publication of multi-dimensional (statistical) datasets.

4.6   CONCLUSION

Concluding, Web technologies have reached building geometry and open up new possibilities to flexibly attach multiple geometry descriptions to the same object – in disregard of the geometry schema or format at hand. Beyond different overall approaches to integrate geometry descriptions into a semantic web context, uniform linking methods exist that support any of these approaches, OMG and FOG. At the same time, the OMG/FOG method can be used as a basis to attach further geometric metadata, e.g. through GOM, and handle multiple and heterogeneous geometry descriptions, as can be seen in current implementations of various research projects and programmes. More so, query extensions exist and allow for spatial querying within the standardised query language SPARQL. Some of these extensions, i.e. GeoSPARQL, are widely implemented and supported already.

With these approaches, methods and tools, it is possible to describe complete Linked Building Data, including complex geometry descriptions that are chosen individually to suit the use case at hand. By enabling users to attach multiple geometry descriptions of different schemas and formats to the same objects and relate those descriptions to each other, new possibilities to move away from monolithic solutions primed for one geometry kernel and software application suite arise.

NOTES

1 Exporting geometry G from application A, importing G into application B, manipulating G, exporting G from B, importing G into A.
2 Take three.js, Babylon.js, xeogl or Sketchfab as an example.
3 https://www.khronos.org/gltf/.
4 Spatial querying is currently implemented for few and mainly two-dimensional geometry schemas.
5 Further geometry ontologies exist, e.g. the *3D Modeling Ontology* [359], which is no longer available online, but are not considered, as they are not openly available online.
6 While this ontology is not openly available, RDF Ltd. share the ontology upon request.
7 https://w3id.org/oop.
8 https://w3id.org/occ.
9 https://geojson.org/geojson-ld/.
10 For a more detailed analysis on the suitability of the approaches, see [404].
11 Assuming the geometry and non-geometry part are considered separately.
12 https://w3id.org/omg#.
13 https://w3id.org/fog#.
14 The ontology's GitHub repository can be forked by anyone and issues can be raised there: https://github.com/mathib/fog-ontology.
15 https://w3id.org/gom#.

16 A v1.1 backwards compatible revision of the GeoSPARQL standard is currently considered by the OGC and might support the linking and spatial querying of GeoJSON and KML geometry.

17 https://geojson.org/geojson-ld/vocab.html.

18 https://github.com/LBDserver.

19 https://github.com/LBDserver/front-react.

20 https://github.com/LBDserver/converter_backend.

21 https://github.com/jyrkioraskari/IFCtoLBD.

22 http://ifcopenshell.org/ifcconvert.

23 https://github.com/KhronosGroup/COLLADA2GLTF.

24 available at: https://mathib.github.io/fog-demo-app/.

25 Approach 1 is omitted only because the underlying geometry kernel, three.js (WebGL) does not support any of the RDF-based geometry schemas natively.

26 https://www.projekt-scope.de/en/home-en/.

27 http://www.geohive.ie/.

28 http://data.geohive.ie/downloadAndQuery.html.

29 http://ontologies.geohive.ie/osib.

30 https://www.scss.tcd.ie/ mcglink/r2rml/.

31 http://yasgui.org/

32 https://www.w3.org/TR/vocab-data-cube/.

Chapter 5

Open data standards and BIM on the cloud

Pieter Pauwels, Dennis Shelden, Jan Brouwer, Devon Sparks, Saha Nirvik, and Tim Pat McGinley

CONTENTS

DOI: 10.1201/9781003204381-6

This chapter discusses the current and emerging state of open data infrastructure supporting storage, encoding, transmission and exchange of building information, with the goal of supporting cloud-based networks of information systems that are able to communicate with each other. These capabilities point the way to future distributed communication of building geometry, product data, properties, and so forth across the design, building and operate life cycle of a building (see Chapter 6). The chapter describes the existing state of building data standards, focused on the object model, schemas and geometric descriptions currently existing in the Industry Foundation Classes (IFC) open data standard (Section 5.2). Core principles of this existing standard are briefly described, including object and relationship modelling, geometry and metadata class structure, and existing encodings using SPF, XML, JSON and RDF in Section 5.3. We hereby rely on some of the background and detailed methods outlined in Chapters 2–4 and try to point out promising new developments that seek to migrate these existing building information data conventions to the emerging paradigms of data on the cloud. Based on this background, we present three data modelling approaches in Sections 5.4–5.6: backwards-compatible JSON, optimised JSON and JSON-LD. JSON is simply used here as an example serialisation format; choices made in RDF, XML and SPF are very similar. Short samples are given for illustration throughout the chapter, and Section 5.7 finally shows an example project in Rhino Grasshopper[1] that indicates what data service architecture is possible for supporting connected web services using the described representation formats.

5.1   INTRODUCTION

The built environment has a very wide span with several levels of details, application domains, and diverse stakeholders and users acting inside this built environment in all sorts of manners (see also Chapter 1). These levels of details, application domains and stakeholders meet very often, leading to the existence of data about the built environment in all kinds and shapes, being exchanged in a large diversity of ways. Yet, this building data is essential in creating the contextual information required for the planning, design, construction, operation, maintenance and destruction/recycling of buildings. At its most basic form, building data can simply include the address or geolocation of a home required to deliver mail and goods (cfr. Google or Amazon services), or a 2D floor plan of a home used when advertising to buy or sell. When designing a new building,

building data typically includes a full 3D solid model of the building, describing the geometry and semantics of each product (walls, windows, doors and columns), HVAC, the materials used, plumbing and electrical wiring in the building, space boundaries and the relationships between all these. All these data are required to support the wide range of stakeholders involved along the building's life cycle – sometimes in full detail, oftentimes only partial views are needed (see discipline and coordination models in [295]).

5.1.1   Building data interoperability

The state of building information modelling, interoperability, data exchange and data science has evolved a long way in the past decade [3,49,113,115,116,295]. Today we find ourselves in a landscape with a constellation of different and semi-compatible data formatting standards and conventions – both open and proprietary [352]. These standards are diverging in finding specific applicability to quite specific contexts of increasingly sophisticated targeting and granularity. Formats like gbXML[2], Build-ingSync[3], Brick[4] [21] and others target overlapping but distinct applicability in the energy modelling and compliance space, while IFC, BOT, the Digital Twins Definition Language (DTDL)[5] and numerous proprietary but open standards including the Buildings and Habitats object Model[6] (BHoM), Speckle[7], HYPAR[8] and others target specific building systems, professional disciplines or the import, export and functionality demands of specific applications.

As such, the state of building data interoperability has evolved significantly from the early days of IFC, STEP and similar attempts to achieve interoperability [85, 391]. Traditional interoperability initiatives were focused on developing open standards that could more or less completely capture the structure, metadata and geometry representation characteristics of prevalent internal binary representations of specific desktop application software, with the goal of either transmitting or archiving complete, lossless equivalents of these internal binary representations (see also first part of Chapter 9). The more contemporary data encodings presented in this chapter, specifically JSON, JSON-LD and RDF/OWL, either natively offer data translation capabilities, or, as is the case with JSON and to some extent XML, are positioned at the centre of a constellation of toolsets intended to support data querying and manipulation. These foundational capabilities offer significant possibilities for developing flexible interoperability workflows among federated data sets founded on distinct and bespoke data conventions.

5.1.2   Data exchange

The core of the AEC industry relies on data exchange on a daily basis, as information needs to be exchanged continuously between designers, engineers, and contractors, who are often on a tight schedule. Input data from one stakeholder is necessary for others to perform their work, and so it continues in a complex, interconnected information network that constantly undergoes changes. Thus, creating and sharing data in a continuous stream is crucial to deliver the project on time and on budget. This industry, from design initiation, through construction and to the handover of the facility, is highly driven by the need for abundance of data exchanged in a timely manner.

Building Information Modelling (BIM) tools are central in this exchange, leading to plenty of BIM models that need to be combined, merged, exchanged and checked (see also Chapter 1). Most BIM tools have different internal information structures, and even

Standardisation

Flexibility

- agreements among users	- agile data handling
- shared vocabulary	- people responsible for data
- shared language	consumption
- mutual agreement	- less agreements
- scalability	- data transformation
- wider applicability	- tailored and specialised

Figure 5.1 Standardisation versus flexibility in data handling approaches.

the same BIM tool is often used in very diverse ways, with different modelling guidelines, libraries and methods. As a result, it is difficult or near to impossible to define a common and generally applicable data structure in which practitioners define their data about the building.

5.1.3   Standardisation versus flexibility

To be able to deal with the wide diversity in building data structures, two main strategies or approaches can typically be taken: 'standardisation' and 'flexibility' (see Figure 5.1). Standardisation hereby aims at agreements among users and stakeholders in the data structure: as stakeholders use a shared vocabulary, they are better able to communicate with each other and understand each other's content. Such standardisation clearly allows shared and large-scale tools.

Flexibility, on the other hand, assumes that data use and exchange are arranged based on agile data handlers (people) that are able to produce, translate and consume data across projects. This second type of data exchange tends to put less effort on making agreements in data, and puts more effort in using the tools available to transform data into a desired input form. Such systems have much smaller exchanges typically, which have the advantage of being better tailored to one's own project requirements. The disadvantage to this flexible approach is that one has to be agile enough to build, maintain and continuously re-purpose the necessary transformers and convertors.

In this chapter, we aim to give an indication of this diversity of data and try to show how those two approaches (standardisation vs. flexibility) affect choices in the creation of an ifcJSON structure. This chapter is largely based on the work by the ifcJSON group at buildingSMART International[9].

5.2   IFC: THE LEADING STANDARD FOR BIM DATA

A key part of the data exchange strategies in the AEC industry is of course the Industry Foundation Classes (IFC)[10]. As indicated on the BuildingSMART website, IFC is a *standardized, digital description of the built environment, including buildings and civil infrastructure. It is an open, international standard (ISO 16739-1:2018), meant to be vendor-neutral, or agnostic, and usable across a wide range of hardware devices, software platforms, and interfaces for many different use cases. The IFC schema specification is the primary technical deliverable of buildingSMART International to fulfill its goal to promote openBIM*. It can be considered a data model that can be used to represent the built environment. In reality, IFC is mainly used in the AEC industry, in particular buildings; although it can also be used in the other stages in the building life cycle (manufacturing, operational stages, facility management, etc.), and also in practices other than building industry (infrastructure). IFC is closely linked to the vendors of BIM software and the larger design and engineering firms.

5.2.1 IFC data model

Technically, IFC is a data model that can be expressed in a number of schema languages, namely EXPRESS, XSD, OWL and JSON Schema. Although the schema was originally managed and maintained in EXPRESS as the source of truth, buildingSMART is currently shifting to the use of the Unified Modelling Language (UML) as a more technology-agnostic source model[11]. From the UML model (snippet displayed in Figure 5.2, source BuildingSMART[12]) serialised schemas in EXPRESS, XSD, OWL and JSON Schema can be generated.

This IFC data model has always primarily aimed at resolving interoperability between BIM tools (Chapter 1), which is particularly challenging in terms of exchange of detailed 3D geometric data. One of the aims typically associated with IFC is that a BIM tool should be able to export its data, without loss, in a detailed 3D geometric file, so that it can readily be imported into another tool, including its build-up of detailed geometric data. Notwithstanding this great ambition and the value that it would generate, this is of course not possible. As 3D kernels are inherently different between 3D BIM modelling tools, there is always a loss of data, even when limiting to a purely geometric level. This becomes only more apparent when looking into the more semantic data (e.g. product data and properties. as explained in Chapters 2 and 3). Hence, even if standards and agreements are of value (as indicated in previous section), there is always a strong need for flexible data exchange and data handling.

The IFC data model was originally built in EXPRESS [185]. This makes the data model very rich, large, detailed and expressive. In addition to that, it has a clear graph structure, with several `ENTITY` and `TYPE` data types, as well as object properties and data properties, often in forward and inverse directions. Because of this graph-based and non-hierarchical basis, it is perfectly possible to exchange partial data in diverse ways and forms. This can be recognised in viewers (e.g. Solibri Anywhere), in which it is possible to browse an IFC model through the project containment hierarchy, through the element hierarchy or list, through layers, or through levels. In other libraries, for example

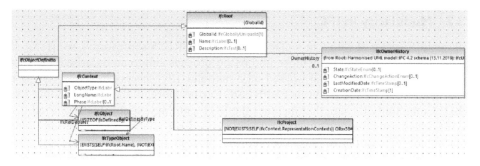

Figure 5.2 UML diagram for a part of IFC.

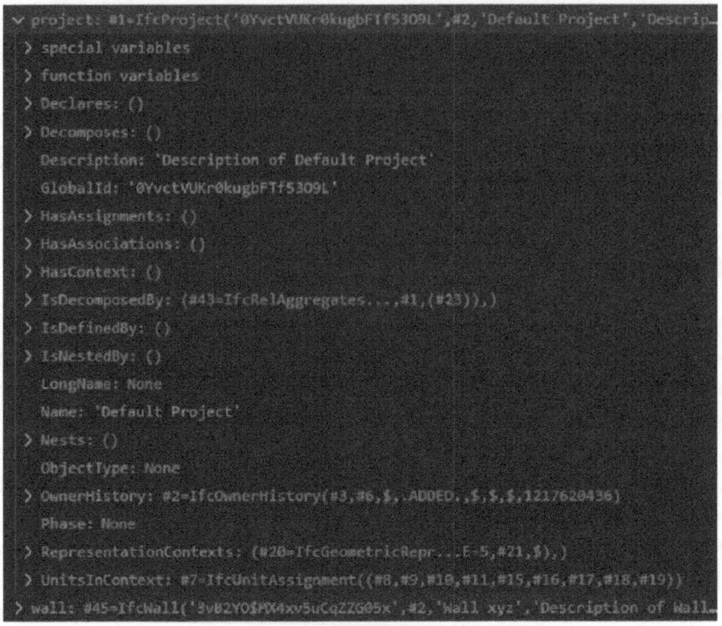

Figure 5.3 A parsed IFC object model including regular and inverse properties in IfcOpenShell.

IfcOpenShell[13], it is perfectly possible to inverse the hierarchies and build upward from a single element to retrieve more generic project data (Figure 5.3).

One certainty in an IFC model is the presence of only one IfcProject entity, as this is one of the requirements included in the EXPRESS schema of IFC (Listing 5.1). As a result, the project decomposition hierarchy (project, site, building, level, space and element) is the most commonly known view for an IFC model.

Listing 5.1 The rule indicating in EXPRESS that there must be no more than 1 IfcProject in every IFC file

```
1  RULE IfcSingleProjectInstance FOR
2     (IfcProject);
3       WHERE
4         WR1 : SIZEOF(IfcProject) <= 1;
5  END_RULE;
```

5.2.2 Modularity in IFC

Although IFC started with a modular structure (EXPRESS short form – see [185]), the data model is not modular any more, and recent projects aim to modularise it again. As an example, Terkaj et al. [383] worked on a modular IFC schema in OWL, relying on the original shortform schema in EXPRESS. Furthermore, the Linked Building Data (LBD) Community Group, that includes the BOT ontology, conceptually started from the IFC schema at large and aimed at making parts of it available as modular ontologies. Conceptually, one can distinguish the following parts in any IFC file:

1. Project structure, which is the building topology in the case of a building, and which could be the alignment part in the case of an infrastructure project.

2. Element classifications and properties, which constitutes the bulk of the semantic data in any IFC file (see also Chapters 2 and 3).

3. 3D geometric data, typically associated with 'spatial elements' and 'built elements' (see Chapter 4).

4. Metadata of all kinds, often aiming to describe the history and authorship of changes as part of the IFC file.

Looking at the ontologies built as part of the LBD community group, (1) the BOT ontology matches with the IFC project structure, (2) the BEO and MEP ontologies match with the element classifications and properties, as well as the BPO ontology, (3) the FOG and OMG ontologies provide an alternative to represent 3D and 2D geometric data and (4) the OPM ontology enables to represent much of the metadata that is also present in IFC.

5.2.3 Partial exchanges in IFC

Although partial exchanges (e.g. elements on a specific floor) are possible in IFC, one typically exchanges a full file in the context of a handover moment. Such handover exchanges thereby fulfil exchange requirements by means of model views (Reference View, Design Transfer View, Coordination View, etc.). It is possible to create Model View Definitions (MVD)[14], yet most software implement only the few major standard MVDs. As such, this MVD approach primarily aims at the standardised approach with little flexibility to, for example, enable to filter and export more custom subsets of an IFC file.

5.3 HOW TO MOVE THE DATA TO THE CLOUD?

Over time, it has been suggested several times to move data exchange and the prevailing industry standard (IFC) in the direction of the cloud. Notwithstanding the very solid data model behind IFC, the information modelling language in which it was built makes it stand apart quite strongly from most data standards used in the web, including XML, JSON and RDF. Hence, several proposals have been made over the last decade to serialise IFC in these alternative languages [31,32,299].

5.3.1 XML

5.3.1.1 XML from IfcDoc

The XML version of IFC has been based on the EXPRESS schema of IFC. This first XML version has been produced as part of the documentation of the IFC standard. This is done using the IfcDoc tool[15], which is originally a library of custom C# code. This IfcDoc tool has many functionalities, one of which is the creation of a large set of C# classes for all object types in IFC. This C# class library would ideally be aligned with a UML Class diagram. In any case, the given C# classes of IFC include bindings with XML, which allows all XML files to be serialised and deserialised directly into these classes (from and to XML). This chapter will not go into detail on how this actually happens, yet the point being made here is that the XML file structure as generated by IfcDoc is a direct derivative from the C# classes.

In addition, an XSD schema is available for the IFC schema, also generated by IfcDoc for the online documentation. Documentation, XML examples and XSD schema can all be found in the online reference documentation[16].

Listing 5.2   Sample of XSD for IfcWall

```
1  <xs:element name="IfcWall" type="ifc:IfcWall" substitutionGroup=
       "ifc:IfcBuildingElement" nillable="true"/>
2  <xs:complexType name="IfcWall">
3    <xs:complexContent>
4      <xs:extension base="ifc:IfcBuildingElement">
5        <xs:attribute name="PredefinedType" type="ifc:
           IfcWallTypeEnum" use="optional"/>
6      </xs:extension>
7    </xs:complexContent>
8  </xs:complexType>
```

Because both XSD and XML are generated using the IfcDoc tool, both are meant to match. Indeed, as can be seen in the XML sample in the below Listing 5.3[17], overall XML structure matches with the schema. As such, a standard language is available that can be relied upon for data exchange.

Listing 5.3   XML sample of IfcWall with opening

```
1  <IfcWall id="i13" GlobalId="3ZYW59sxj8lei475l7EhLU" Name="Wall
       for Test Example" Description="Description of Wall">
2    <OwnerHistory xsi:nil="true" href="i2"/>
3    <IsDefinedBy>
4      ...
5    </IsDefinedBy>
6    <ObjectPlacement xsi:type="IfcLocalPlacement" id="i6">
7      ...
8    </ObjectPlacement>
9    <Representation xsi:type="IfcProductDefinitionShape">
10     ...
11   </Representation>
12   <HasOpenings GlobalId="1nwVYC3VTDeuSc8zbOa89u">
13     ...
14   </HasOpenings>
15 </IfcWall>
```

5.3.1.2 *XML from IFC.JAVA class library*

A similar project is available that achieves the same, but entirely within the Java programming language, instead of C#, and without the complete IfcDoc functionality[18]. Instead, this code repository focuses entirely on providing a stand-alone class library for IFC2x3 and IFC4. This code library similarly has a number of XML annotations, enabled by the Jackson library[19]. This enables serialising and deserialising the ifcXML content directly into Java classes, so one can work with the data in Java-based software. An example output snippet is shown in Listing 5.4. In this case, there is no direct alignment with the XSD schema for IFC. Instead, the XML depends directly on the class structure in the Java class library.

Listing 5.4   Modelling IFC in XML

```
1  <ownerHistory changeAction="ADDED" lastModifiedDate="0"
       creationDate="1217620436">
2    <globalId>d7db3747-44a6-4d92-856e-9979a32bdd69</globalId>
3    <owningUser>
4      <thePerson identification="ID001" familyName="Anon"
           givenName="Anon"/>
5      <theOrganization name="TU/e" description="Eindhoven
           University of Technology"/>
```

```
6     </owningUser>
7     <owningApplication version="version 0.10"
          applicationFullName="My text editor"
          applicationIdentifier="TA 1001">
8       <applicationDeveloper name="TU/e" description="Eindhoven
            University of Technology"/>
9     </owningApplication>
10   </ownerHistory>
```

A number of differences can be found between the XML versions in Listings 5.3 and 5.4, namely:

- Capitalisation is different in the two versions, which can be explained because of differing naming conventions in Java and C#: capitals are common in C#, while Java relies typically more often on camelCase.

- Furthermore, the global identifier is different: the C# version relies on the shortened GUID notation used in IFC, while the Java version relies on the more common and web-friendly UUID notation.

- Example 1 (C#) relies on the id-href mechanism to point to other lines in the XML file. For example, the Wall instance points to an IfcOwnerHistory objects using the notation href="i2". In the IFC.JAVA sample, referencing is done by directly inserting the globalId that was used elsewhere already in the same XML file (not shown in this example), as this is the default serialisation mechanism in the used Jackson library.

5.3.1.3   XML from Autodesk Revit

A third XML version is the XML version of IFC produced by Autodesk Revit. This BIM software enables ifcXML export for any BIM model. Also in this case, the output XML files do not follow the XSD schema supplied by buildingSMART. Below Listing 5.5 shows an example output snippet for a wall.

Listing 5.5   IFC XML output by Autodesk Revit

```
1   <ifc:uos id="uos_1" description="" schema="exp_1" configuration=
        "IFCXML_Official" edo="" xmlns="http://www.iai-tech.org/
        ifcXML/IFC2x2/FINAL" xsi:schemaLocation="http://www.iai-tech
        .org/ifcXML/IFC2x2/FINAL http://www.iai-tech.org/ifcXML/
        IFC2x2/FINAL/ifc2x3g_alpha.xsd">
2   <IfcWallStandardCase id="i1890">
3     <GlobalId>1_6hJIWpzDHurr0YYrwoTw</GlobalId>
4     <OwnerHistory>
5       <IfcOwnerHistory xsi:nil="true" ref="i1678"/>
6     </OwnerHistory>
7     <Name>Basic Wall:Generic - 200mm:293998</Name>
8     <ObjectType>Basic Wall:Generic - 200mm</ObjectType>
9     <ObjectPlacement>
10      <IfcLocalPlacement xsi:nil="true" ref="i1853"/>
11    </ObjectPlacement>
12    <Representation>
13      <IfcProductDefinitionShape id="i1884">
14        <Representations exp:cType="list">
15          <IfcShapeRepresentation exp:pos="0" xsi:nil="true" ref
              ="i1859"/>
16          <IfcShapeRepresentation exp:pos="1" xsi:nil="true" ref
              ="i1881"/>
```

```
17                </Representations>
18              </IfcProductDefinitionShape>
19          </Representation>
20          <Tag>293998</Tag>
21      </IfcWallStandardCase>
22  </ifc:uos>
```

Differences with the sample in Listing 5.3 are clear: other href referencing mechanism (`ref="i4113"`), use of XML Elements instead of XML Attributes, etc. Furthermore, the output XML data tends to be produced in a long list of XML Elements, with a limited amount of hierarchy. This resembles the line-based structure of IFC-SPF, where each line includes one IFC element (Listing 5.6). Similarly, in the produced XML data, every new line includes a new IFC element in a single XML Element. The XML produced by Revit includes exactly as many elements as the IFC-SPF serialisation, making it a direct match with the SPF serialisation primarily. All elements refer to each other using the href-id mechanism, which makes automatic (de)serialisation in custom software difficult. This is very different from the samples that are directly produced by a C# or Java class library (Listings 5.3 and 5.4). XML produced by such class libraries tend to be entirely hierarchically structured, using the XML indenting options available in XML.

Listing 5.6   IFC SPF output by Autodesk Revit

```
1  #42= IFCOWNERHISTORY(#39,#5,$,.NOCHANGE.,$,$,$,1631993784);
2  #217= IFCLOCALPLACEMENT(#161,#216);
3  #223= IFCSHAPEREPRESENTATION(#113,'Axis','Curve2D',(#221));
4  #245= IFCSHAPEREPRESENTATION(#115,'Body','SweptSolid',(#235));
5  #248= IFCPRODUCTDEFINITIONSHAPE($,$,(#223,#245));
6  #254= IFCWALLSTANDARDCASE('3kIOd66qr36u4qFBFSUcYA',#42,'Basic Wall
       :Generic - 200mm:293998',$,'Basic Wall:Generic - 200mm
       ',#217,#248,'293998');
```

The above two examples also include an example of an ordered list, namely in line #248 of Listing 5.6, which lists an ordered list of two shape representations (#223 and #245). This ordered list is encoded using a position identifier (`exp:pos="0"`) in XML (Listing 5.5). As a conclusion, the XML serialisation thus encodes lists using an 'indexed list' pattern to be able to infer order.

5.3.2 JSON

As XML has gradually been superseded by JSON, also the JSON version of IFC has been created later than its XML predecessor. JSON is a widely used data serialisation and interchange format. Its strength comes from its simplicity: it provides a minimal syntax for describing associative arrays (called "objects"), ordered lists, primitive data types (numbers, strings and booleans), and combinations thereof. It is particularly suited for state transfer through web services, as JSON documents are easy to parse, human-readable and support both sequential and random access data structures. In principle, very similar choices exist when serialising an object model to JSON, compared to serialising it to XML. Similar to XML, choices need to be made about:

1. encode information either as objects/elements or as properties/attributes

2. notations, in particular capitalisation

3. use of indentation to clearly indicate full hierarchy, versus a flat list of objects that is cross-referenced as fully as possible

4. how to encode lists and arrays (linked list, indexed list, other), knowing that the majority of IFC definitions assume a list or array or set of objects (cfr. IfcRelationship entities)

The JSON snippets in Listings 5.7 and 5.8 give two example snippets of what this can result in, and similar choices can be seen in both cases when comparing them to the XML examples.

Listing 5.7   Modelling IFC in JSON

```
1   "sweptArea" : {
2        "type" : "IfcIShapeProfileDef",
3        "profileType" : "AREA",
4        "profileName" : "UC305x305x97",
5        "position" : [[ 0.0, 0.0 ], [ 0.0, 1.0 ]],
6        "overallWidth" : 305.3,
7        "overallDepth" : 307.9,
8        "webThickness" : 9.9,
9        "flangeThickness" : 15.4,
10       "filletRadius" : 15.2,
11       "flangeEdgeRadius" : 0.0,
12       "flangeSlope" : 0.0
13   }
```

Listing 5.8   Modelling an IfcIShapeProfileDef in JSON

```
1   "sweptArea" : {
2        "type" : "IfcIShapeProfileDef",
3        "profileType" : {
4            "type" : "IfcProfileTypeEnum",
5            "label" : "AREA",
6            "description" : "The resulting geometric item is of type
                    surface. The resulting geometry after applying a
                    sweeping operation is a swept solid with defined
                    volume."
7        },
8        "profileName" : "UC305x305x97",
9        "position" : {
10           "type" : "IfcAxis2Placement2D",
11           "location" : {
12               "type" : "IfcCartesianPoint",
13               "coordinates" : [ 0.0, 0.0 ]
14           },
15           "refDirection" : {
16               "type" : "IfcDirection",
17               "directionRatios" : [ 0.0, 1.0 ]
18           }
19       },
20       "property" : {
21           "id" : "b7432442-a7be-4474-b52a-b76ad51774dc",
22           "type" : "overallWidth",
23           "value" : 305.3,
24           "unit" : {
25               "id" : "e0687323-b2ef-454c-829f-852ee8a9173f",
26               "type" : "CENUnit" ,
27               "label" : "mm",
28               "symbol" : "mm",
29               "unitType" : "LengthUnit",
```

```
30              "SIUnitType" : "SI derived unit",
31              "fullName" : "millimeter"
32          },
33          "definedBy" : "Mathias Bonduel",
34          "date" : "2020-09-15"
35      }
36  }
```

What is important and interesting to note here, is that JSON is typically used in exchanges with a web server. Most servers and web services nowadays rely on REpresentational State Transfer (REST), often with a backend and frontend in JavaScript (see also Chapter 6). Hence, it is convenient to be able to transfer JavaScript objects from backend to frontend, and back, using an agreed upon JavaScript Object Notation (JSON). This allows communication across the web, from client to server and back, of small or large snippets of data, in which any data consumer is continuously wrapping and unwrapping the data. This web development paradigm (REST) supersedes the SOAP-based web development paradigm, which had a lot more server-side scripting (e.g. ASP, PHP), much less JavaScript and typically an XML-based exchange of data.

Of course, with the sheer number of data exchanges happening every single second across all devices on the planet, this exchange of data across devices using JSON has become critical and a baseline in many if not all web-based applications. In many such cases, the amount of data that is transferred is very small, since further data can be requested at any other point in time. Exchange of data, therefore, is not supposed to be large files with complete models, as the AEC sector has been used to since decades (see MVDs in Section 5.2.3). Instead, the idea is to rely on short data exchanges with small snippets of data, and thus rely on a much larger level of flexibility as opposed to large-scale standardisation.

To return to our distinction between standardisation and flexibility from the introduction of this Chapter (Figure 5.1), this means that, while previous efforts typically focused on standards in the data format, current efforts seem to shift towards standardising a bit less, and enabling more flexible data exchange, in which any data consumer is meant to be able to transform and ingest incoming and requested data.

5.3.3   RDF

A third serialisation language briefly discussed in this chapter is the Resource Description Framework (RDF), of which we will use here the Turtle serialisation format (.ttl). Any IFC model, as well as the schema, is available in this language, which results in an OWL ontology for the IFC schema and a TTL version for any instance IFC file (Figure 5.4).

Also in this case, several choices can be made and also have been made. The object model of IFC has originally fully been translated into OWL and RDF [299], while aiming

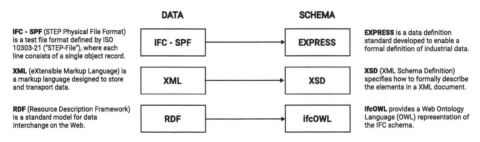

Figure 5.4 Serialisation formats of IFC.

at backwards compatibility into previous versions for IFC. There was also chosen for round-tripping functionality, in which any TTL file for IFC could be translated back into IFC-SPF[20]. In this ifcOWL ontology, almost all content is represented using objects (Elements in XML), and almost none using attributes and datatype properties, thus allowing backwards compatibility. Although this results in a very standard model and representation that can capture anything that is in the IFC model, it is not a very flexible or even short model.

An alternative RDF representation for IFC data can be found in Linked Building Data (LBD) graphs, of which we will only discuss here the simplest form. These LBD graphs rely on a number of OWL ontologies or vocabularies, many of which have been presented already in the previous chapters. The structure is entirely modular, enabling anyone to add their own OWL ontology and use that as well (examples in Schneider et al. [345]). For these LBD graphs, choices other than the ones behind ifcOWL were made, in response to users' requests for:

- simplicity
- modularity
- extensibility

This has by now led to graphs as shown in Listing 5.9. In this example, a number of instances are declared (`inst:site_159` and `inst:building_134`), and for each of them, one can find a reference to one or more relevant classes in the used OWL ontologies (e.g. a `bot:Site`) as well as object properties that relate the object with related objects (e.g. `bot:hasBuilding inst:building_134`), and simple data properties that communicate properties for each building element in the model. As such, results a simple and tightly interconnected graph that in its totality describes a building model.

Listing 5.9   Modelling IFC in RDF (TTL syntax)

```
1   inst:site_159
2       a bot:Site ;
3       rdfs:label "Default"^^xsd:string ;
4       rdfs:comment ""^^xsd:string ;
5       bot:hasGuid "90b511b9-b7c0-465c-8e11-4a4033650f22"^^xsd:
            string ;
6       props:hasCompressedGuid "2GjH6vjy16N8uHIa0pPGyY"^^xsd:string
            ;
7       bot:hasBuilding inst:building_134 ;
8       props:reference "Project Information"^^xsd:string ;
9       props:category "Project Information"^^xsd:string ;
10      props:clientName "Owner"^^xsd:string ;
11      props:projectAddress "Enter address here"^^xsd:string ;
12      props:projectIssueDate "Issue Date"^^xsd:string ;
13      props:projectName "Project Name"^^xsd:string ;
14      props:projectNumber "0001"^^xsd:string ;
15      props:projectStatus "Project Status"^^xsd:string .
16
17  inst:building_134
18      a bot:Building ;
19      rdfs:label ""^^xsd:string ;
20      rdfs:comment ""^^xsd:string ;
21      bot:hasGuid "90b511b9-b7c0-465c-8e11-4a4033650f21"^^xsd:
            string ;
22      props:hasCompressedGuid "2GjH6vjy16N8uHIa0pPGyX"^^xsd:string
            ;
23      bot:hasStorey inst:storey_147 ;
24      bot:hasStorey inst:storey_153 ;
```

```
25      props:reference "Project Information"^^xsd:string ;
26      props:isExternal false ;
27      props:reference "Project Information"^^xsd:string ;
28      props:numberOfStoreys "2"^^xsd:double .
29
30  inst:storey_147
31      a bot:Storey ;
32      bot:hasGuid "90b511b9-b7c0-465c-8e11-4a40cc9af1e7"^^xsd:
            string ;
33      props:hasCompressedGuid "2GjH6vjy16N8uHIa3Ccl7d"^^xsd:string
            ;
34      rdfs:label "8th Floor"^^xsd:string ;
35      rdfs:comment ""^^xsd:string ;
36      bot:hasSpace inst:space_245 , inst:space_404 , inst:
            space_526 , inst:space_649 , inst:space_772 , inst:
            space_892 , inst:space_1023 , inst:space_1144 ;
37      bot:containsElement inst:covering_270426 , inst:
            column_270642 , inst:column_270706 , inst:wall_270788 ,
            inst:wall_271009 , inst:wall_271170 , inst:wall_271267 ,
            inst:wall_271364 , inst:wall_271543 , inst:door_271622
            , inst:door_271998 , inst:wall_272195 , inst:wall_272409
            , inst:wall_272564 , inst:wall_272719 , inst:
            column_20973 , inst:column_21121 , inst:column_21795 ,
            inst:column_21859 , inst:column_21923 , inst:
            column_21987 , inst:column_22051 , inst:column_22115 ,
            inst:curtainWall_28323 , inst:curtainWall_33300 , inst:
            column_38057 , inst:column_38156 , inst:
            curtainWall_38224 , inst:curtainWall_47096 , inst:
            curtainWall_65824 , inst:curtainWall_72641 , inst:
            wall_91464 , inst:wall_91714 , inst:wall_91875 , inst:
            wall_92040 , inst:wall_92205 , inst:wall_92366 , inst:
            wall_92529 , inst:wall_92766 , inst:door_92876 , inst:
            door_93012 , inst:door_93082 , inst:door_93152 , inst:
            door_93222 , inst:door_93292 , inst:wall_93446 , inst:
            wall_93627 , inst:wall_93812 , inst:door_93893 , inst:
            wall_94085 , inst:wall_94279 , inst:wall_94572 , inst:
            wall_94749 , inst:wall_94930 , inst:wall_95099 , inst:
            covering_237115 , inst:covering_237195 , inst:
            covering_237275 , inst:covering_237355 , inst:
            covering_237435 , inst:covering_237515 , inst:
            covering_237595 , inst:covering_237675 , inst:
            covering_237755 , inst:covering_237837 , inst:
            covering_237917 , inst:covering_237997 , inst:
            covering_238077 , inst:covering_238157 , inst:
            covering_238237 , inst:covering_238317 , inst:
            door_245027 ;
38      props:reference "8mm Head"^^xsd:string ;
39      props:aboveGround ".U."^^xsd:double ;
40      props:elevation "0."^^xsd:double ;
41      props:computationHeight "0."^^xsd:double ;
42      props:structural false ;
43      props:buildingStory true ;
44      props:name "8th Floor"^^xsd:string ;
45      props:category "Levels"^^xsd:string ;
46      props:family "Level: 8mm Head"^^xsd:string ;
47      props:familyandType "Level: 8mm Head"^^xsd:string ;
48      props:type "Level: 8mm Head"^^xsd:string ;
49      props:typeId "Level: 8mm Head"^^xsd:string ;
50      props:elevationBase "Project Base Point"^^xsd:string ;
51      props:typeName "8mm Head"^^xsd:string ;
52      props:color "0"^^xsd:double ;
```

```
53    props:linePattern "Centre"^^xsd:string ;
54    props:lineWeight "1"^^xsd:string ;
55    props:symbol "M_Level Head - Circle: M_Level Head - Circle
         "^^xsd:string ;
56    props:symbolatEnd1Default false ;
57    props:symbolatEnd2Default true ;
58    props:category "Levels"^^xsd:string ;
59    props:familyName "Level"^^xsd:string .
```

Also in this case, the original ifcOWL approach is more useful for end users and developers that seek standardisation, while the latter LBD approach is many times more flexible, yet still formally well-structured. This second LBD approach is therefore more of use for developers that are interested in direct, extensible and flexible data use.

One main downside to RDF graphs, in comparison with JSON and XML, is that bindings with object programming languages like C# and Java are much less common. This makes it much more difficult to use this data stored in RDF data stores and load it directly into object-oriented code that can directly consume and use it. Instead, the developer needs to go through either a SPARQL query mechanism, a direct API connection with an RDF-enabled database, or a dedicated software library like RDFLib, Jena, or other. As a result, this language is, for example, less useful for plenty of direct data communications between client and server, in rather large contrast to JSON.

5.4 DATA MODELLING APPROACH I: BACKWARDS COMPATIBLE FILE TRANSFORMATIONS AND DATA EXCHANGES

In all the three listed approaches towards web-based data modelling for IFC (JSON, XML, RDF), there have always prevailed two mainstream stances or approaches for transforming existing IFC data to these formats:

1. Traditional: Focus on backwards compatibility and stability
2. Innovating: Focus on changing the structure in response to newly emerging trends

In this section, we will discuss the first approach (traditional), whereas in the next two sections we will discuss the last approach (innovating), each time focusing on the JSON representation of IFC as a sample case. This distinction can somewhat be compared with the distinction between 'standard' and 'flexible' that is made in the earlier sections of this chapter. In the first approach, a number of criteria are typically listed, namely:

1. Backwards compatibility
2. Data exchange through files
3. Small file size
4. Metadata about files
5. Clear representation of exchange processes

5.4.I Full file serialisations

Traditional and standard approaches often lead to serialisations of data into full files. The reason behind that is unclear. Potentially, file-based exchange has always been the most common method for exchanging data from one person to another, and our tooling and way of working are most accustomed to such an approach. Perhaps, a full file serialisation is easier as it contains all data needed; and that would be easier to request, especially at handover. In any case, a full file serialisation is often the way forward in this setup. A number of file converters have been written, to transfer IFC data from one serialisation into the other (SPF to XML, XML to JSON, JSON to SPF, SPF to JSON and so forth).

5.4.2 File size

Associated with this full file serialisation often comes the focus on file size as a criterion. In that case, the least verbose and the least readable serialisation is often considered 'better' than the more verbose alternative. If file size is indeed a criterion, then it makes sense to make GUIDs as short as possible, and to try to not repeat anything whenever serialising data. In the case of data, this brings about the need for avoiding duplication of data in a single building model serialisation in JSON. This can be achieved using referencing techniques, such as the id-ref, id-href, or any other kind of referencing technique that is seldom standard across JSON serialisations (Figure 5.5).

Two examples of such linking within a larger JSON file are listed in Listings 5.10 and 5.11. Listing 5.10 relies on the `ref` and `type` attributes to point from one entity to the other, which is, in this case, clearly visible for the `ownerHistory` attribute (id: e4980768-d888-49ec-8297-0968dbd994ca). This example lists all entities line per line, while not relying strongly on the hierarchy structure that is available in JSON. Listing 5.11, on the other hand, simply copies the referenced identifier directly as the value of an attribute if possible. For example, in the case of `"ownerHistory"` : `"f974430e-88e1-4772-a724-5ca7d6e0ddb3"`, the referenced ownerHistory object was already declared earlier on in the JSON file. This second approach is made possible automatically by the serialisation and deserialisation options available in the Jackson XML JSON module for Java[21]. In this second approach, file structure does follow a hierarchical structure, starting from the IfcProject object (line 2) and including the entire project tree within (lines 31–44, shortened for illustration purposes).

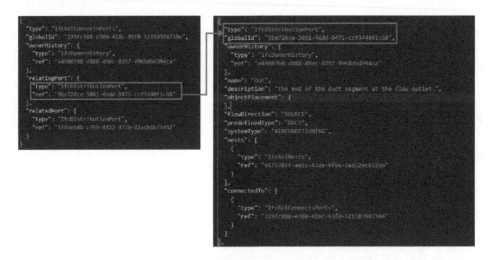

Figure 5.5 Example of referencing from one line number to the other in IFCJSON.

Listing 5.10 Referencing other lines using the href, id-ref, id-href, or similar approach. Example taken from https://github.com/buildingSMART/ifcJSON/blob/master/Samples/IFC_4.0/Building SMARTSpec/air-terminal-element.json.

```
1  {
2    "type": "IfcOwnerHistory",
3    "owningUser": {
4      "type": "IfcPersonAndOrganization",
5      "thePerson": {
6        "type": "IfcPerson",
7        "identification": "Tim"
8      },
9      "theOrganization": {
10       "type": "IfcOrganization",
11       "name": "Tim-PC"
12     }
13   },
14   "state": "READWRITE",
15   "changeAction": "NOTDEFINED",
16   "creationDate": 1321055806,
17   "globalId": "e4980768-d888-49ec-8297-0968dbd994ca"
18 },
19 {
20   "type": "IfcBuilding",
21   "globalId": "6d313a7d-0bf7-4f08-947f-80599436ab30",
22   "ownerHistory": {
23     "type": "IfcOwnerHistory",
24     "ref": "e4980768-d888-49ec-8297-0968dbd994ca"
25   },
26   "compositionType": "ELEMENT",
27   "decomposes": [
28     {
29       "type": "IfcRelAggregates",
30       "ref": "f427ca64-046f-49e9-88d1-b05ac99f5128"
31     }
32   ],
33   "containsElements": [
34     {
35       "type": "IfcRelContainedInSpatialStructure",
36       "ref": "b6d6bd0e-d0c8-44c4-a16b-530f15ecd61f"
37     }
38   ]
39 },
40 {
41   "type": "IfcAirTerminal",
42   "globalId": "703c0f93-955f-4bdc-a8c0-41e62f30f3bb",
43   "ownerHistory": {
44     "type": "IfcOwnerHistory",
45     "ref": "e4980768-d888-49ec-8297-0968dbd994ca"
46   }
47 }
```

Listing 5.11 Referencing other lines using the href, id-ref, id-href, or similar approach. Example taken from https://github.com/pipauwel/Sample_Files/blob/master/IFC%204.0/BuildingSMARTSpec/ air-terminal-element.json.

```
1  {
2    "type" : "IfcProject",
3    "globalId" : "2fb45857-e9ac-4660-8943-604b07b54025",
4    "ownerHistory" : {
5      "type" : "IfcOwnerHistory",
```

```
6       "globalId" : "f974430e-88e1-4772-a724-5ca7d6e0ddb3",
7       "owningUser" : {
8         "type" : "IfcPersonAndOrganization",
9         "globalId" : "5addc637-1158-47df-b45e-4b81ccfc066f",
10        "thePerson" : {
11          "type" : "IfcPerson",
12          "globalId" : "dcda2919-dbcd-41ea-84ef-e67a0499a193",
13          "identification" : {
14            "type" : "IfcIdentifier",
15            "value" : "Tim"
16          }
17        }
18      },
19      "state" : "READWRITE",
20      "changeAction" : "NOTDEFINED",
21      "creationDate" : {
22        "type" : "IfcTimeStamp",
23        "value" : 1321055806
24      }
25    },
26    "name" : {
27      "type" : "IfcLabel",
28      "value" : "Project"
29    },
30    "isDecomposedBy" : {
31      "type" : "IfcRelAggregates",
32      "globalId" : "f427ca64-046f-49e9-88d1-b05ac99f5128",
33      "ownerHistory" : "f974430e-88e1-4772-a724-5ca7d6e0ddb3",
34      "relatedObjects" : [ {
35        "type" : "IfcBuilding",
36        "globalId" : "6d313a7d-0bf7-4f08-947f-80599436ab30",
37        "ownerHistory" : "f974430e-88e1-4772-a724-5ca7d6e0ddb3",
38        "containsElements" : [ {
39          "type" : "IfcRelContainedInSpatialStructure",
40          "globalId" : "b6d6bd0e-d0c8-44c4-a16b-530f15ecd61f",
41          "ownerHistory" : "f974430e-88e1-4772-a724-5ca7d6e0ddb3"
42        } ]
43      } ]
44    }
45 }
```

5.4.3 Round-tripping

A second key topic in these full file serialisations is the notion of 'round-tripping'. When converting data in one format into the other, there almost always exists a request to transfer it back to the original format. This is the case when developers want to test completeness of their conversion algorithms, and often also when external clients require the original format that they are familiar with. Enabling such backwards compatibility or 'round tripping' typically requires the addition of tags or metadata that is needed to return to the original. For example, in the case of IFC, it would be needed at least to include the line number that is used in SPF files (only) to uniquely identify items/lines in the file.

5.4.4 File metadata

When working with files, the metadata is important: who created the file, when, with what tool(s), which version is it, etc. This is available in the IFC-SPF HEADER section. As we are focusing on standard file exchange, it is also necessary to include this in XML and JSON versions of IFC. Hence, the full file in JSON needs to be enriched with a header section that explains what the file is. An example of such header can be found in Listing 5.12.

Listing 5.12   Example JSON snippet with the SPF header metadata included. Example taken from https://github.com/buildingSMART/ifcJSON/blob/master/Samples/IFC_4.0/Building SMARTSpec/air-terminal-element.json.

```
 1  {
 2    "type": "ifcJSON",
 3    "version": "0.0.1",
 4    "schemaIdentifier": "IFC4",
 5    "originatingSystem": "IFC2JSON_python Version 0.0.1",
 6    "preprocessorVersion": "IfcOpenShell 0.6.0b0",
 7    "timeStamp": "2020-10-31T23:51:18",
 8    "data": [
 9      {
10        "type": "IfcProject",
11        "globalId": "2fb45857-e9ac-4660-8943-604b07b54025",
12        "ownerHistory": {
13          "type": "IfcOwnerHistory",
14          "ref": "e4980768-d888-49ec-8297-0968dbd994ca"
15        },
16        "name": "Project"
17      }
18    ]
19  }
```

Valuable about adding this metadata inside the JSON file is that the file becomes comprehensive and can be used as is for hand-over of data. Downsides of taking this approach are that (1) metadata is available in multiple locations, namely inside the file and attached to the file, and (2) any (de)serialiser needs to include also the metadata, or a custom wrapper and unwrapper needs to be coded. Both downsides reduce direct usability of the JSON file content in automated tools.

5.4.5 Inverse relationships

IFC and many other data models are object-oriented models that consist of classes and relationships between those classes. For example, a building is decomposed of many building storeys, which should at least be modelled as two classes (Building, Storey) and one relationship (IsDecomposedOf). In IFC, this is even more complex, as most relationships are ternary or n-ary relationships, often at least with the addition of an OwnerHistory relationship for most classes and relationships. In that case, data modelling consists of four classes (Building, Storey, OwnerHistory and isDecomposedOf) and at least three relationships (relatedObject, relatingObjects and ownerhistoryOf).

An indication of what this leads to in a JSON serialisation is visible in Listings 5.10 and 5.11. Particularly in Listing 5.11, one can find in line 30, 31 and 34 how one `IfcProject` object is linked to one `IfcRelAggregates` object through the `isDecomposedBy` attribute. The same `IfcRelAggregates` object is linked to multiple `IfcBuilding` objects (see array symbols [] in lines 34 and 43) through the relatedObjects attribute. Furthermore, the `IfcRelAggregates` object has a `globalId` literal and `ownerHistory` object linked to it, making it an n-ary relationship object. This structure is also visually displayed in Figure 5.6, also for a second example involving an `IfcWall` and its propertysets and properties.

Complex data models like IFC have plenty of such relationships, often in two directions. In other words, almost every relationship has an inverse relationship, often

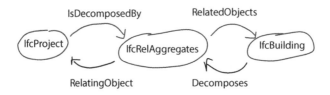

Figure 5.6 Diagram that illustrates the relationship structure in IFC for two example cases: IfcProject - isDecomposedBy - IfcBuilding and IfcWall IsDefinedBy - IfcPropertySet.

including also lists, sets and arrays (not displayed in Figure 5.6). For example, lists of storeys are related with lists of building elements, and vice versa. In Figure 5.6, the RelatedObjects property refers to a set of IfcBuilding objects. The diagram in Figure 5.6 shows the inverse relationships as grey arrows below the objects, from right to left; one grey arrow is available for every forward relationship in the other direction (e.g. IsDecomposedBy has RelatingObject as its inverse).

When serialising such data structure to JSON (or XML), one needs to be careful for cyclic references. It is easily possible to include a relationship from one object to a list of other objects, to then find the inverse of that relation relating further down the JSON tree, but to the original object, and then again the already included relationship, and so forth (see example in Listing 5.13). This obviously needs to be avoided, and that is possible, yet then a choice needs to be made which one of the inverse relationships is not going to be included in the serialised file. For a complex object model like IFC, such choices are often difficult to make and need to be agreed. Listing 5.11, for example, shows that the inverse relationship relatingObject is excluded (missing) for the IfcRelAggregates object.

Listing 5.13 JSON serialisation that does not avoid serialisation of cyclic references (see duplication of IfcProject in lines 2 and 23).

```
1  {
2    "type" : "IfcProject",
3    "globalId" : "2fb45857-e9ac-4660-8943-604b07b54025",
4    "ownerHistory" : {
5      "type" : "IfcOwnerHistory",
6      "globalId" : "f974430e-88e1-4772-a724-5ca7d6e0ddb3"
7    },
8    "isDecomposedBy" : {
9      "type" : "IfcRelAggregates",
10     "globalId" : "f427ca64-046f-49e9-88d1-b05ac99f5128",
11     "ownerHistory" : "f974430e-88e1-4772-a724-5ca7d6e0ddb3",
12     "relatedObjects" : [ {
13       "type" : "IfcBuilding",
14       "globalId" : "6d313a7d-0bf7-4f08-947f-80599436ab30",
15       "ownerHistory" : "f974430e-88e1-4772-a724-5ca7d6e0ddb3",
16       "containsElements" : [ {
17         "type" : "IfcRelContainedInSpatialStructure",
18         "globalId" : "b6d6bd0e-d0c8-44c4-a16b-530f15ecd61f",
```

```
19          "ownerHistory" : "f974430e-88e1-4772-a724-5ca7d6e0ddb3"
20        } ]
21      } ],
22      "relatingObject" : {
23        "type" : "IfcProject",
24        "globalId" : "2fb45857-e9ac-4660-8943-604b07b54025",
25        "ownerHistory" : {
26          "type" : "IfcOwnerHistory",
27          "globalId" : "f974430e-88e1-4772-a724-5ca7d6e0ddb3"
28        },
29        "isDecomposedBy" : {
30          "type" : "IfcRelAggregates",
31          "globalId" : "f427ca64-046f-49e9-88d1-b05ac99f5128",
32          "ownerHistory" : "f974430e-88e1-4772-a724-5ca7d6e0ddb3"
33        }
34      }
35    }
36 }
```

5.4.6 Polymorphism

Many object models, especially also IFC, include inheritance and abstract classes. In such cases, the notion of polymorphism often occurs, which can be defined as the provision of a single interface to entities of different types (e.g. abstract superclass). In that case, an object can 'take multiple forms'. This can easily occur when a relationship from one class points to a parent class, while the data actually includes a child class of that parent class. In such case, the object takes the form of parent class as well as child class, and it is then absolutely necessary and relevant that (de)serialisation mechanisms transform incoming/outgoing data into the correct (child) class.

To be able to support such polymorphism, while still serialising and deserialising data, helper functions or annotations are often necessary. In the case of JAVA and Jackson, for example, the use of appropriate language binding annotations helps to serialise/deserialise into the correct object type. In particular, Listing 5.14 shows how this can be achieved for the IfcBeam class by annotating the class name with annotations that clarify how an object's type information is stored in the JSON or XML serialisation (JsonTypeInfo), and it also indicates which subtypes and supertypes to expect (JsonSubTypes and extends). This leads to the JSON serialisation as shown earlier in Listing 5.11, where the type information is included explicitly in a type attribute.

Listing 5.14 Annotating JAVA class definitions with JSON type information in response to polymorphism requirements.

```
1 @JsonTypeInfo(use = JsonTypeInfo.Id.NAME, include = JsonTypeInfo
     .As.PROPERTY, property = "type")
2 @JsonSubTypes(@JsonSubTypes.Type(value = IfcBeamStandardCase.
     class, name = "IfcBeamStandardCase"))
3 public class IfcBeam extends com.buildingsmart.tech.ifc.
     IfcProductExtension.IfcBuildingElement
4 {
5     ...
6 }
```

5.4.7 Internal and external referencing

An object model often contains references, both within the object model as well as outside the object model. In the first case, objects are related to each other, and that needs to be encoded using a normal JSON referencing technique. In this regard, the id-ref mechanism was already mentioned (see Listings 5.10 and 5.11), yet this is not the most standard or robust method to do this. Good other approaches are unfortunately not available in traditional JSON, and it is advisable to at least use a developer-friendly identification method (uuid, guid and internal numerical integer id).

Referencing to outside object models is even more challenging using JSON, as this requires a file identification mechanism *in addition to* the object identification mechanism. While this is in theory possible using hyperlinks or file paths inside the JSON format, this is seldom done because this is a brittle link to rely on (link breaks as soon as file moves).

5.4.8   Exchange processes

Finally, exchange processes need to be mentioned. These processes are relatively simple considering that this traditional approach assumes files. In this case, exchange processes consist of getting the right file from the right person to the right person. While this can be arranged using traditional methods like email and messaging, recent approaches strongly suggest the use of document-oriented common data environments (CDEs) that are accessible to the correct people for download and use. Such exchange processes need to be formalised in process maps, including information delivery manuals (IDMs) or specifications (ILSs).

5.5   DATA MODELLING APPROACH 2: FORWARD TOWARDS ONLINE DATA LINKING

In the second conversion approach, a more innovating and disruptive stance is taken regarding the original IFC data model. In this case, the following set of criteria is typically followed:

- human readability
- web-oriented data exchange (web services)
- exchange of snippets rather than full files
- simplified file content

These criteria have come about only more recently, and it is also only since more recently possible to address these requirements with appropriate and scalable technologies. In this frame, the main idea is not the direct exchange of files, but rather an 'orchestration of data'. While the problem used to be a matter of sending data from one software to the other, 'data exchange' is shifting into 'data sharing', in which each data element is being consumed, transformed and shared through multiple web-based software applications in the project or data life cycle (see also Chapter 6).

5.5.1   Modular snippets

In this approach, the idea is to avoid full file serialisations, not only because of all the challenges listed in Section 5.4 (e.g. internal and external referencing, inverse relations and cyclic references, file metadata and file size), but also because this approach 2 generally assumes users to request access to precise pieces of data (queries, HTTP requests and filters), instead of complete files and modules. In a way, this requires services to make available 'modular snippets' with data that stands alone and is easy to parse. In such a case, the main idea is to rely on a snippet as listed in Listing 5.15 when a user requests all buildings on a particular site, for example. The `IfcSite` is not included, as it is known by the one that made this request. Also further relations for the `IfcBuilding` (e.g. decomposition) are not included here, yet they can be requested in a next request with different query. Optionally, the `IfcOwnerHistory` could be kept out of this example as well, and as such the complete model (see also in Figure 5.3) can be pruned as preferred.

Listing 5.15   Modular snippets, for example of all buildings on a building site.

```
1  {
2    "type": "IfcBuilding",
3    "globalId": "6d313a7d-0bf7-4f08-947f-80599436ab30",
4    "ownerHistory": {
5      "type": "IfcOwnerHistory",
6      "owningUser": {
7        "type": "IfcPersonAndOrganization",
8        "thePerson": {
9          "type": "IfcPerson",
10         "identification": "Tim"
11       },
12       "theOrganization": {
13         "type": "IfcOrganization",
14         "name": "Tim-PC"
15       }
16     },
17     "state": "READWRITE",
18     "changeAction": "NOTDEFINED",
19     "creationDate": 1321055806,
20     "globalId": "e4980768-d888-49ec-8297-0968dbd994ca"
21   },
22   "compositionType": "ELEMENT"
23 }
```

A good example of how the above-mentioned selective querying of a complete dataset is enabled, is the GraphQL query language[22]. More information is available at Werbrouck et al. [411]. This GraphQL language enables a user to provide a template of what data is needed in a JSON-like syntax, after which the web-based API reconfigures its response to match what is requested. As such, only the requested information is exchanged and a user can almost filter and query on demand (high flexibility).

With this focus on modular snippets and selective querying, it can clearly be noticed that many of the challenges related to the exhaustive listing of all objects and interrelations, as present in Section 5.4, are avoided, simply because the modular snippet remains limited in scope. As a result, cyclic references can almost always be avoided, and also n-ary relations can often be represented in a simplified form, or simply avoided by not requesting the n-th part of a relation. This has motivated many of the simplifications that are suggested as part of the ifcJSON5 work at buildingSMART[23].

The downside of the approach is that many consecutive queries are needed to a server, which leads to overabundant client-server traffic. If the data is stored in a central server, then a lot of Internet traffic emerges, which results in a costly server, too high reliance on a potentially weak network connection, and time delays due to network traffic and congestion. So the above approach is mainly useful if the data does not reside on a far away web server and instead is located close to the point of request, e.g. on the Edge (see also Chapter 10).

5.5.2   Web services and microservices

In this approach, it is important to rely on web services that can respond with JSON snippets for the requests that are made over HTTP. While web services used to be built using the Simple Object Access Protocol (SOAP), which typically responded with XML snippets that include all serialised objects, recent web services rely on a REpresentational State Transfer (REST). While a full explanation of RESTful systems and interfaces[24] is out of scope here (refer to Chapters 7 and 10), it is clear that such interfaces now typically give access to a JavaScript (or Python) backend that stores objects in a key-value formalism more commonly known as JavaScript Object Notation (JSON).

As data is passed on from page to page and from service to service, its structure tends to change often, leading to the request and need to have relatively simple definitions of data. An example worth mentioning here is the IFC.js software or toolkit[25], which parses IFC SPF files into a simplified JSON structure using their browser-compatible library and thus provides the IFC content directly as JavaScript objects to web developers.

5.5.3   What about the 2D and 3D geometry?

This service-based approach of sharing data is considerably more flexible than the file-based approach. In principle, this approach allows a developer to store data on a server in a number of (consistent) formats in the appropriate database(s). This also includes the storage of geometry. For example, if a database stores 2D and 3D geometry for any object (examples in [44] and [410]), then the web service can produce and respond with the representation that is requested by the user. This could be a 2D representation, a faceted Brep representation (3D), a swept solid representation, or even an OBJ representation.

Listing 5.16 shows how each of those could be made available for a wall object in JSON, which can similarly be done in SPF, RDF, and XML. Of course, in reality, not all four of these representations should be returned by a web service, and instead only one of those should be supplied. For use cases that require a simple 3D mesh, the OBJ and faceted Brep representations suffice, while 2D cases need only the 2D representation.

Listing 5.16   Representation of different geometry snippets for a single wall element.

```
1   {
2     "type": "IfcWallStandardCase",
3     "globalId": "f3b7a52f-4eb5-44a8-80e0-87592507aed9",
4     "name": "Wall xyz",
5     "description": "Description of Wall",
6     "representation": {
7       "type": "IfcProductDefinitionShape",
8       "representations": [
9         {
10          "type": "IfcShapeRepresentation",
11          "globalId": "9b76f770-b9ea-4c50-ae00-97b5105644d5",
12          "representationIdentifier": "Body",
13          "representationType": "Brep",
14          "items": [
15            {
16              "type": "IfcFacetedBrep",
17              ...
18            }
19          ]
20        },
21        {
22          "type": "IfcShapeRepresentation",
23          "globalId": "a296a747-eba3-4a96-b20c-ced6dbb0092d",
```

```
24            "representationIdentifier": "Axis",
25            "representationType": "Curve2D",
26            "items": [
27               {
28                  "type": "IfcPolyline",
29                  "points": [ ... ]
30               }
31            ]
32         },
33         {
34            "type": "IfcShapeRepresentation",
35            "globalId": "70a00a01-12ba-44e9-b77e-b990fe550b2f",
36            "representationIdentifier": "Body",
37            "representationType": "SweptSolid",
38            "items": [
39               {
40                  "type": "IfcExtrudedAreaSolid",
41                  "sweptArea": { ... },
42                  "position": { ... },
43                  "extrudedDirection": { ... },
44                  "depth": 2.3
45               }
46            ]
47         },
48         {
49            "type": "IfcShapeRepresentation",
50            "globalId": "70a00a01-12ba-44e9-b77e-b990fe550b2f",
51            "representationIdentifier": "Body",
52            "representationType": "OBJ",
53            "items": [
54               {
55                  "type": "objSTRING",
56                  "value": { "v 0 0 0\nv 0 0 2.3\nv 0 0.3 2.3\nv 0 0.3
                     0\nv 5 0 0\nv 1.25 0 0\nv 1.25 0 2.1\nv 0.5 0
                     2.1\nv 0.5 0 0\nv 0 0 0\nv 0 0 2.3\nv 5 0 2.3\nv
                     0 0.3 0\nv 0.5 0.3 0\nv 0.5 0.3 2.1\nv 1.25 0.3
                     2.1\nv 1.25 0.3 0\nv 5 0.3 0\nv 5 0.3 2.3\nv 0
                     0.3 2.3\nv 0.5 0 0\nv 0 0 0\nv 0.5 0.3 0\nv 0
                     0.3 0\nv 0 0 2.3\nv 0 0.3 2.3\nv 5 0.3 2.3\nv 5
                     0 2.3\nv 5 0 0\nv 1.25 0 0\nv 5 0.3 0\nv 1.25
                     0.3 0\nv 1.25 0.3 0\nv 1.25 0.3 2.1\nv 1.25 0
                     2.1\nv 1.25 0 0\nv 0.5 0 2.1\nv 0.5 0.3 2.1\nv
                     1.25 0.3 2.1\nv 1.25 0 2.1\nv 0.5 0 0\nv 0.5 0
                     2.1\nv 0.5 0.3 2.1\nv 0.5 0.3 0\nv 5 0.3 0\nv 5
                     0.3 2.3\nv 5 0 2.3\nv 5 0 0\nf 4//4 1//1 2//2\nf
                     4//4 2//2 3//3\nf 7//7 5//5 12//12\nf 11//11
                     8//8 7//7\nf 11//11 7//7 12//12\nf 6//6 5//5
                     7//7\nf 10//10 9//9 8//8\nf 10//10 8//8 11//11\
                     nf 16//16 15//15 20//20\nf 13//13 20//20 15//15\
                     nf 14//14 13//13 15//15\nf 19//19 16//16 20//20\
                     nf 18//18 17//17 16//16\nf 18//18 16//16 19//19\
                     nf 24//24 21//21 22//22\nf 24//24 23//23 21//21\
                     nf 28//28 27//27 26//26\nf 25//25 28//28 26//26\
                     nf 32//32 29//29 30//30\nf 32//32 31//31 29//29\
                     nf 34//34 33//33 36//36\nf 35//35 34//34 36//36\
                     nf 38//38 39//39 40//40\nf 38//38 40//40 37//37\
                     nf 42//42 41//41 44//44\nf 43//43 42//42 44//44\
                     nf 48//48 45//45 46//46\nf 48//48 46//46 47//47\
                     n"
57               }
58            ]
```

```
59              }
60          ]
61      }
62  }
```

5.5.4 Exchange processes: open APIs and CDEs

In this approach, data exchange occurs very differently from the file-based approach. It is instead required to have web services available that are available to respond with JSON data that is understandable to their receiver(s). Currently, most representative examples of such software are CDEs and open APIs to data sharing platforms. It is of course important that the output of such services is understandable and predictable, which can be enabled by standardisation on format and notation, or at least modelling agreements and naming conventions.

5.6 DATA MODELLING APPROACH 3: JSON-LD

In addition to the two main modelling approaches listed above, a third option is mentioned here that relies on JSON-LD. JSON-LD is simply another data serialisation format for RDF graphs. As a result, it is readily available for ifcOWL[26]. However, the JSON-LD version of ifcOWL is a fully file-oriented version, in which many of the optimisations mentioned in Section 5.5 are not available. Hence, a new JSON-LD version of IFC is experimented with by the BuildingSMART IFC-JSON Team[27], based on the work in Section 5.5.

This JSON-LD data modelling approach is not entirely aiming to be standard, nor is it fully flexible either, as everything is supposed to be interlinked in JSON-LD. In the below few sections, we will discuss its main features for the case of IFC.

5.6.1 What is JSON-LD?

JSON provides a convenient syntax for data interchange across the web, as indicated at length above, but it requires out-of-band agreements between producers and consumers to ensure that the meaning (i.e., semantics) of transmitted data is preserved. A client receiving a JSON document with a "createdAt" key (Listing 5.17), for example, has no way to know whether the key's value is a timestamp, geofence, or company name, without external information.

Listing 5.17 Simple JSON example.

```
1  {
2    "id":"32uVjsbxhb4",
3    "createdAt":"2012-04-25T01:30:00Z"
4  }
```

This applies strongly to the examples listed in Section 5.5. While the technology (e.g. GraphQL, JSON) enables to change structure and simplify data representation a lot (high level of flexibility), this advantage can also bring uncertainty into an application: the person making a (HTTP) request is not sure in what form and shape one can make a request, and there is also no clarity on how to interpret the (HTTP) response. A total absence of agreements (small-scale standardisation) is detrimental in this case.

/ JSON-LD Input   / Options   ▮ Document URL

{"@context":{"prov":"http://www.w3.org/ns/prov#",
             "createdAt":{"@id":"prov:startedAtTime", "@type":"xsd:dateTime"},
             "@base":"http://server.org/mydata/",
             "id":"@id"
            },
 "id":"32uVjsbxhb4",
 "createdAt":"2012-04-25T01:30:00Z"
}

/ Expanded   / Compacted   ≡ Flattened   ⊞ Framed   •‡ N-Quads   ▦ Normalized   ⊞ Table   ⊹ Visualized   / Signed with RSA   / Signed with Bitcoin

<http://server.org/mydata/32uVjsbxhb4> <http://www.w3.org/ns/prov#startedAtTime> "2012-04-25T01:30:00Z"^^<xsd:dateTime> .

Figure 5.7 More robust data encoding in JSON-LD for the example JSON encoding in Listing 5.17.

JSON-LD[28] addresses JSON's semantic ambiguity by specifying a syntax for annotating existing JSON data with RDF. As such, it brings meaning to the nodes and objects that are being transmitted. Through the use of a single new key, called @context, JSON documents can carry semantic annotations of their contents as RDF (see example in Figure 5.7). Consumers may then use a compliant JSON-LD processor[29] to extract RDF triples from received JSON data. Protocol agreements that would otherwise be out-of-band can now be carried with the data itself. JSON-LD thus makes JSON simultaneously a human- and machine-readable data interchange format, because JSON-LD documents are self-describing. As such, a level of standardisation is brought to the very flexible JSON encodings.

5.6.2   Standardisation inside the JSON specification of data

Many of the features that are present for JSON are maintained in JSON-LD, and it is mainly the addition of contextual data and identifiers (RDF, URIs) that makes a big difference in the interpretation of results and data. Also in the case of JSON-LD, the main purpose is to make data available through databases and web services, and move forward to data sharing using common and accessible data repositories and services, instead of exchanging files in a much less concurrent or live manner.

5.6.3   Unique referencing using URIs

By the addition of contextual data and typically web-based URI identifiers, the JSON-LD documents become part of a global network of data. The web hereby provides a truly global namespace for digital assets. Any asset - physical or virtual, real or imagined - that can be assigned a URI can be described in RDF. If that URI becomes resolvable (i.e., it resolves to a URL), web-based clients can dereference the URI to retrieve machine-readable descriptions of the asset or its representations. Once assets can be identified and retrieved by URI, their physical storage location (e.g. URLs) becomes semantically irrelevant, provided of course that these URIs are dereferenced and do resolve to actual physical locations.

Asset descriptions – like an IfcProject – that would normally depend on identifiers and data local to a file (e.g. GUIDs or even line numbers), can now be distributed over a network arbitrarily. Content negotiation over HTTP replaces what otherwise would require local file reads and writes. JSON-LD supports this model by allowing existing JSON documents to identify themselves, and related documents, using URIs.

5.6.4   Inverse relationships and polymorphism

Through the embedding of the RDF data modelling principles inside the JSON-LD scope, some of the typical RDF features are also available in JSON-LD, often in an easier way compared to XML or JSON. In particular inverse relationships, polymorphism, and classification of data using multiple types can easily be supported, as can be found in the JSON-LD specification[30]. Assigning multiple types is supported using the @type keyword. As shown in Listing 5.18, an element can thus be annotated with a BOT, Brick, and MEP[31] specification simultaneously, for example. The addition of multiple types does *not* inherently require types to be strictly mapped at a high level onto each other (ontology mapping).

Listing 5.18   Annotating a node using three different types in JSON-LD.

```
1  {
2    ...
3    "@id": "http://example.org/VertigoBuilding#sensor1579",
4    "@type": [ "https://w3id.org/bot#Element", "https://pi.pauwel.
        be/voc/distributionelement#Sensor-CO2SENSOR", "https://
        brickschema.org/schema/Brick#Sensor" ],
5    ...
6  }
```

Furthermore, reverse properties can be specified using the @reverse keyword, an example of which is given in Listing 5.19 for one of the inverse relationships in IFC. This example is based on the earlier given example in Listing 5.13. The relevant relationships and their inverses (IsDecomposedBy, Composes, RelatingObject and RelatedObjects) are visually displayed in Figure 5.6. The JSON-LD sample includes the regular property IsDecomposedBy (line 4) that is also used in the example in Listing 5.19. However, instead of declaring the next forward property RelatedObjects (see Figure 5.6), the reverse property is used, namely Decomposes (line 9), which can be more readable in some cases. Certain flexibility is thus available, provided that this also matches with the schema used.

Listing 5.19   Simple JSON example.

```
1  {
2    "@id": "#2fb45857-e9ac-4660-8943-604b07b54025",
3    "@type": "https://standards.buildingsmart.org/IFC/DEV/IFC4/
        ADD2_TC1/OWL#IfcProject",
4    "https://standards.buildingsmart.org/IFC/DEV/IFC4/ADD2_TC1/OWL
        #IsDecomposedBy":
5      {
6        "@id": "#f427ca64-046f-49e9-88d1-b05ac99f5128",
7          "@type": "https://standards.buildingsmart.org/IFC/
              DEV/IFC4/ADD2_TC1/OWL#IfcRelAggregates",
8          "@reverse": {
9            "https://standards.buildingsmart.org/IFC/DEV/IFC4/
                ADD2_TC1/OWL#Decomposes":
10             {
11                 "@id": "#6d313a7d-0bf7-4f08-947f-80599436
                      ab30",
12                   "@type": "https://standards.
                        buildingsmart.org/IFC/DEV/IFC4/
                        ADD2_TC1/OWL#IfcBuilding"
13             }
14        }
15      }
```

```
16      }
17   }
```

5.6.5   Exchange processes and the use of framing

Finally, as was shown before, JSON provides only a syntactic framework for data exchange, including often ad-hoc and flexible or unstable data structure agreements. JSON-LD extends this syntax to support semantic annotations of JSON data. Once all parts of a JSON document have URIs, it becomes possible to restructure a given JSON document based on its content on demand, e.g. to organise elements hierarchically, filter out those containing a specific RDF predicate (similar to what is possible using GraphQL), or flatten all sub-objects. As such, the very diverse choices that were made in the creations and implementations of XML and JSON for IFC (see Section 5.3) can be translated onto each other. This is a very important feature and power of JSON-LD: while JSON flexibility is allowed, there is also room for semantic structure, and even to restructure documents syntactically in support of specific cases.

This particular feature of JSON-LD is called *framing*. JSON-LD framing provides a standardised way by which JSON-LD documents can be restructured on demand without semantic loss. An example of how this framing can be applied to the BOT topology of a building is shown in the below Listings[32].

A simple building can be represented in JSON-LD as shown in Listing 5.20. Then, every JSON-LD Frame is a request, intended for a JSON-LD Frame Processor, that instructs it how we would like the underlying JSON-LD graph presented. Every Frame defines the requested node types (e.g. bot:Building, bot:Storey, bot:Space), along with a description of the "shape" that these nodes should follow (very similar to GraphQL). To restructure the example given in Listing 5.20 into a more tree-like structure, it is possible to define the Frame in Listing 5.21 and arrive at the result in Listing 5.22. As such, flexibility is available, yet maintaining the links to the standard terminology of IFC.

Listing 5.20   JSON-LD example for a simple two storey building.

```
 1  {"@context":[{"bot":"https://w3id.org/bot#",
 2                "hasStorey":{"@type":"@id",  "@id":"bot:hasStorey"}
                  ,
 3                "hasSpace":{"@type":"@id",  "@id":"bot:hasSpace"}},
 4                {"@base":"http://myproject.org#"}],
 5
 6    "@graph":[
 7      {"@id":"mybuilding",  "@type":"bot:Building",
 8         "hasStorey":["level-1",  "level-2"]},
 9
10      {"@id":"level-1",  "@type":"bot:Storey",
11            "hasSpace":["room-1.1",  "room-1.2"]},
12      {"@id":"level-2",  "@type":"bot:Storey",
13            "hasSpace":["room-2.1",  "room-2.2"]},
14
15      {"@id":"room-1.1",  "@type":"bot:Space"},
16      {"@id":"room-1.2",  "@type":"bot:Space"},
17
18      {"@id":"room-2.1",  "@type":"bot:Space"},
19      {"@id":"room-2.2",  "@type":"bot:Space"}
20    ]
21  }
```

Listing 5.21   JSON-LD example for a simple two storey building.

```
1   {
2     "@context": {
3       "@vocab": "https://w3id.org/bot#",
4       "@base": "http://myproject.org#"
5     },
6     "@type": "Building",
7     "hasStorey": {
8       "@type": "Storey",
9       "hasSpace": {
10        "@type": "Space"
11      }
12    }
13  }
```

Listing 5.22   Restructured JSON-LD example for a simple two storey building.

```
1   {
2     "@context": {
3       "@vocab": "https://w3id.org/bot#",
4       "@base": "http://myproject.org#"
5     },
6     "@id": "mybuilding", "@type": "Building",
7     "hasStorey": [
8       {
9         "@id": "level-1","@type": "Storey",
10        "hasSpace": [
11          {"@id": "room-1.1", "@type": "Space"},
12          {"@id": "room-1.2","@type": "Space"}
13        ]
14      },
15      {
16        "@id": "level-2", "@type": "Storey",
17        "hasSpace": [
18          {"@id": "room-2.1", "@type": "Space"},
19          {"@id": "room-2.2","@type": "Space"}
20        ]
21      }
22    ]
23  }
```

5.7   EXAMPLE APPLICATIONS AND CONSUMING WEB SERVICES

The above sections (5.3–5.6) have given a diagonal overview not only of the diverse data modelling opportunities for building data, while focusing on the IFC standard, but also of the different data modelling choices (flexibility vs. standardisation) that are present throughout these different languages and data models. These data modelling choices are shown using JSON and JSON-LD examples, yet they can often be ported to other languages (RDF, XML) as well.

While there could be many examples named and discussed here in this last Section, we limit ourselves to one example in Rhino Grasshopper that shows how one can build a relatively straightforward script to load the JSON data (Section 5.5) inside the Rhino-Grasshopper 3D modelling software. Having direct access to JSON data on the cloud, without having to request, upload, parse and download SPF files, is one of the main added values explored here. This example focuses mainly on the possibility of making this connection and recreating the available geometry and semantics, while further detailed implementations as well as Graphical User Interface (GUI) improvements are out of scope.

5.7.1 Convertors, translators and transmuters

The IFC-JSON development initiative[33] has included a suite of utilities, including translators, schema generators, data transmuters, query processors and example end applications. Among others, this includes a Grasshopper importer/exporter tool[34] that is able to handle the ifcJSON data samples. We will use this tool here to briefly indicate how the data can be used.

5.7.2 Rhino and grasshopper scripting

Rhinoceros 3D is a widely used 3D modelling application, both for authoring of geometry and as a sort of universal converter among 3D geometries and applications, while Grasshopper is an associated low code/visual programming add-on application with native parametric modelling capabilities. As part of the suite of IFC-JSON support tooling, a set of demonstration scripts have been developed for reading, processing and visualising IFC-JSON data from files in Grasshopper.

One example script to load input, link to objects, and display content is given in Figure 5.8. The application allows a user to select an IFCJSON resource, process the content in order to associate geometry with building elements, and filter, colour and annotate by parameter values found in the IFCJSON file.

The example makes use of the widespread availability of JSON processing tooling to develop the functionality, including basic Grasshopper Python scripts with imported JSON processing libraries. As with many of the examples, the Grasshopper application relies on the mesh-based representations of object geometry (in this case OBJ) that are generated as part of the SPF-JSON conversion, as these can be trivially processed into Rhino/Grasshopper-based meshes for visualisation. An example of the OBJ geometry that is packed into the IFCJSON format was earlier shown in Listing 5.16. Figure 5.9 shows how this particular JSON representation for the Duplex IFC house can thus be visualised into Rhino (filtered for all wall elements).

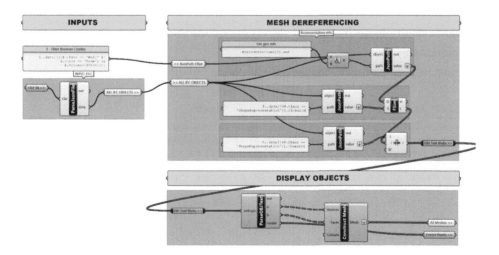

Figure 5.8 Script for handling IFC JSON.

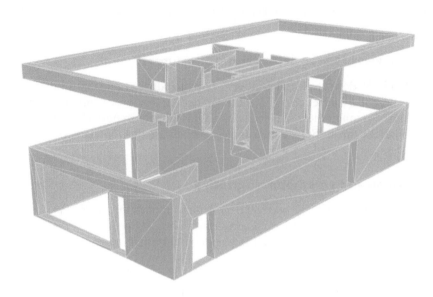

Figure 5.9 Duplex IFC JSON loaded into viewer.

5.7.3 JSONPath-enabled queries

Furthermore, a JSONPath Grasshopper component was developed that can accept a set of IFC-JSON imported building data objects and process these objects against a JSONPath search string. JSONPath[35] is a successor to the widely used XPath libraries, which allows queries formatted as a simple search string to be run against hierarchically structured JSON (or equivalently XML) data. Objects can be retrieved based on the existence of key-value pairs that arise arbitrarily deep in the data as well as conditionals on this data (==, <, >, etc.). The implementation adopts the simplified JSONPath[36] Python-based library developed by Drew Raines and Phil Budne[37].

Listing 5.23 shows a number of JSONPath queries that were used in the creation of this demonstrator. They are used directly in allowing capabilities such as:

- Searching for objects by element class type or parameter value,
- Colouring by class value,
- Annotation of parameter values onto displayed building elements.

Listing 5.23   Example IFCPath search strings on IFC-JSON data

```
1  // find objects of Class type Wall or Beam
2  $..data[?((@.Class == 'Wall' || @.Class == 'Beam')]
3  // find objects less than a given Volume
4  $..data[?(@.Volume<=20000.0)]
5  // get the first Item of all objects of Class type '
       ShapeRepresentation'
6  $..data[?(@.Class == 'ShapeRepresentation')].Items[0]
7  // find all objects whose Class is not ShapeRepresentation and
       build a list of these objects' classes
8  $..data[?(@.Class != 'ShapeRepresentation')].Class
```

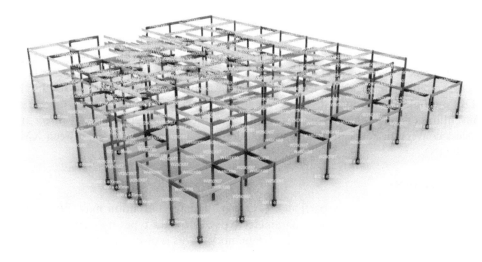

Figure 5.10 Clinic IFC JSON loaded into Grasshopper, including performance of the structural analysis.

While of great utility, these simple path traversal search mechanisms are unable to navigate across relationships by reference – a hallmark of traditional IFC. For the application described, the IFC-JSON converter dereferences some of the relevant relationships, specifically compacting the data tree to attach parameter key–value pairs directly to objects. Similar approaches and solutions were proposed in the past, for example in the creation of L1 properties as explained in Chapter 2 for LBD RDF graphs, but also in JavaScript implementations of 3D IFC viewers such as IFC.js.

A critical remaining task is to associate the OBJ-based ShapeRepresentations to the relevant building elements. This is again approached using JSONPath to find the relevant building element Representations, build a dictionary of ShapeRepresentation objects that contain the geometry meshes, and associate one with the other. Figure 5.10 shows how this is applied to the well-known public IFC Clinic model.

5.8 CONCLUSION: CHALLENGES FOR THE FUTURE

The developments discussed in this chapter offer possibilities for significant advances in building data interoperability, in the processes, supported use cases, and toolset support. The defining trait of flexibility of both data structuring and data use provide some of the greatest avenues for the development of robust data exchange scenarios across multiple applications and domains of applicability. It is clear from the chapter, however, that a good level of standardisation is of big importance as well, in order to be able to understand, interpret and use the data that is either shared or exchanged.

5.8.1 A taxonomy of data representation characteristics

From the review in this chapter, differences among data representations may broadly be characterised using the following characteristics:

5.8.1.1   Encoding

The base-level encoding of the data representations may differ to greater or lesser extent. Binary formats have historically beguiled seamless data manipulation, and many of the languages described have the basic supposition of text-based encodings using UTF character sets. This base level commonality is of course insufficient to support direct interoperability, as the higher level data structural differences of XML versus JSON, RDF, etc. must be addressed. However, at this level, to varying degrees, these data conventions capture common or equivalent data structure concepts whereby direct and lossless conversion between data representation conventions exist (e.g. between XML and JSON).

5.8.1.2   Concepts

The concepts captured in a particular language or data convention may differ. This, broadly, means that entity types, parameter types, and data types may either exist or not exist in a given data convention and data model. This is of course the most difficult aspect of difference to mitigate in a lossless interoperability scenario, since there may be no entity in one frame of data representation to map to a given concept. Hence, this chapter suggests to rely on languages that allow to use multiple types and concepts (e.g. Element (BOT), Sensor (Brick), CO2 Sensor (MEP; IFC)) without those needing to be tightly and restrictively linked to each other on an ontological level.

5.8.1.3   Terminology

The same concepts may exist in parallel standards, but may be represented with different terms. These terminology differences may be minimal – for example in differences of capitalisation, may involve different terms in a common (often English) language, or may be either directly equivalent or relatable concepts with associated differences in text string representation across difference natural languages. The BuildingSmart Data Dictionary (bSDD) is one attempt to develop a universally accessible service offering interoperability of terminology related to building concepts. Yet, also here, the chapter recommends to find a balance between terms that can be more broadly standardised (e.g. bot:Element) and those terms that need more flexible definitions (e.g. CO_2 Sensor or AHU units).

5.8.1.4   Structure

The organisation of data – into conceptual entities or objects, associated parameters, and relationships across data objects – can vary among different data standards. Different encodings (such as JSON) offer widely divergent – but largely equitable – mechanisms encapsulating relationship and organisational structures among data. Some of the base level distinctions among data representations include whether references are made directly between entities through some parent – child relationship in the data directly (a key signature of both XML and JSON), through references among objects (e.g. SPF, RDF and JSON-LD), through primary key identifiers in SQL tabular data, or more or less universal IDs across either a given data set or Internet-wide (URIs in RDF and JSON-LD).

This taxonomy of data representation characteristics is not in any way new, but rather a consistent hallmark of data modelling throughout its relatively short history. Lexical translation and compiler technologies have provided robust mechanisms for bridging these interoperability scenarios for some time.

5.8.2   Flexibility and standardisation

The flexible data approach outlined in this chapter has arisen out of dramatically different use cases than the prior ambitions for full model, bidirectional interoperability between native geometry-based BIM modellers (Section 5.2). These exchange paradigms have emerged specifically from Web applications and microservices, where the underlying data schema and supposed complete and accurate instance information is less the focus than the support of specific, task- or use case-specific interactions among networks of data services. These networks presuppose the availability of required data to service a specific application on demand, but the data may be loosely organised and contain gaps in ways that may only be addressed by merging data from across multiple services.

Underlying conventions in terms of the above (concept, terminology, structure, etc.) characteristics may be existent, but this consistency is not necessarily supposed to be complete across multiple services. Multiple transactions (such as request and response or publish/subscribe) may incorporate additional mechanisms (e.g. JSON-LD framing) to navigate this flexibility, including transmitting or receiving data streams together with validation schemas, dynamic reporting on errors or warnings in human- and machine-readable formats, API signatures that allow customisation of data requests, and – most significantly – on the fly data conversion and filtering utilities.

All of these capabilities have been to some extent available in traditional data exchange technologies (convertors, translators and transmuters), but these capabilities have been handled in code – typically in the importing and exporting software. Furthermore, open data schemes have attempted to universalise consistency across all transactions using a specific data format (e.g. the UML effort for IFC). Yet, most if not all of these efforts need further strengthening and adoption in practice (both flexible data handlers and standard data models), which is also happening.

5.8.3   Towards service-oriented and web-based data handling architectures

In terms of expected implementations, micro-service architectures allow these capabilities to be embedded in the transaction mechanism, allowing more dynamic responses to the specific idiosyncrasies in the pairings of requests and responses. They specifically and often assume and provide mechanisms for handling these idiosyncrasies across larger networked connections among services (e.g. Speckle, IFC.JS). These fault tolerance mechanisms typically come with a premium, both in terms of data storage size as well as processing requirements, but for typical building applications, these costs are manageable and well worth the price in terms of robustness and responsiveness to change among the constituent federated services.

The key characteristics of these increasingly prevalent data structures and perhaps lighter weight standards offer a dramatically revised application architecture – away from the traditional file-based data exchange between desktop applications intended to deliver model development in proprietary applications and formats, and towards an open architecture across light weight applications that process open data natively. This is in fact the larger architecture of the web today, where applications are often light weight client software operating directly in the web browser through languages – predominately built-in JavaScript – and data is passed back and forth between client and web server – and across web services – using the prevalent data encodings of REST APIs and JSON-, XML- and JSON-LD- based encodings. Data transmitted is typically transactional –

subsets of the whole model view of the project and often custom packaged for the given task represented by the application or service.

NOTES

1 https://www.rhino3d.com/6/new/grasshopper/.
2 https://www.gbxml.org/.
3 https://buildingsync.net/.
4 https://brickschema.org/.
5 https://github.com/Azure/opendigitaltwins-dtdl.
6 https://bhom.xyz/.
7 https://speckle.systems/.
8 https://hypar.io/.
9 https://github.com/buildingSMART/ifcJSON.
10 https://technical.buildingsmart.org/standards/ifc/.
11 https://technical.buildingsmart.org/standards/ifc/ifc-tutorials/.
12 https://github.com/buildingSMART/IFC4.3.x-development/.
13 http://ifcopenshell.org/.
14 https://technical.buildingsmart.org/standards/ifc/mvd/.
15 https://github.com/buildingSMART/IfcDoc.
16 https://technical.buildingsmart.org/standards/ifc/ifc-schema-specifications/.
17 from https://standards.buildingsmart.org/IFC/RELEASE/IFC4/ADD2_TC1/HTML/annex/annex-e/wall-with-opening-and-window.ifcxml.
18 https://github.com/pipauwel/IFC.JAVA.
19 https://www.baeldung.com/jackson-xml-serialization-and-deserialization.
20 repository available at https://github.com/BenzclyZhang/IfcSTEP-to-IfcOWL-converters.
21 https://stackify.com/java-xml-jackson/.
22 https://graphql.org/.
23 https://github.com/IFCJSON-Team/IFC.JSON-5Alpha/.
24 https://restfulapi.net/.
25 https://ifcjs.github.io/.
26 https://standards.buildingsmart.org/IFC/DEV/IFC4/ADD2_TC1/OWL/ontology.json and https://standards.buildingsmart.org/IFC/DEV/IFC4/ADD2_TC1/OWL/.
27 https://github.com/IFCJSON-Team.
28 https://json-ld.org/.
29 https://www.w3.org/TR/json-ld11-api/.
30 https://json-ld.org/spec/latest/json-ld/.
31 https://pi.pauwel.be/voc/distributionelement/.
32 based on https://devonsparks.com/2020/10/10/botframes.html.
33 https://github.com/IFCJSON-Team.
34 https://github.com/IFCJSON-Team/IFC.JSON-Grasshopper.
35 https://jsonpath.com/.
36 https://goessner.net/articles/JsonPath/index.html#e2.
37 https://gist.github.com/drewr/783585.

Part II

Algorithms and applications

Chapter 6

Federated data storage for the AEC industry

Jeroen Werbrouck, Madhumitha Senthilvel, and Mads Holten
Rasmussen

CONTENTS

DOI: 10.1201/9781003204381-8

The Architecture, Engineering and Construction (AEC) industry is known for its decentralised organisation; in every stage of a built asset's life cycle, different actors collaborate to design, construct, manage, inhabit or demolish the asset. For every stakeholder, this is just one project in a patchwork with other projects, with other project partners, in another context. With the upcoming of web and cloud technologies, a traditional 'centralised approach' is no longer the only option to manage all building information. Without compromising the 'single source of truth' concept, but reinforcing and validating it indeed, a federated, data-based building project can provide a lot of benefits. A project that is organised as a federated knowledge graph can connect project-specific, contextual, regulatory and product data, but can nevertheless act like a single and coherent data structure if managed properly. In this chapter, basic concepts for the organisation of such federated projects are discussed. The field of federated, data-based construction projects is rather young and aligns with perspectives on the well-known (yet at the same time enigmatic) BIM level 3. However, using international web and AEC standards as a guide, we identify strategies that allow for the setup of such infrastructure.

6.1 INTRODUCTION

Construction projects are inherently decentral: a situation where one player controls the entire process of design over construction and renovation until demolition is extremely rare. Numerous stakeholders participate in a project throughout its life cycle: a complex network of owners, users, contractors and subcontractors, consultants and authorities (see Chapter 1). Each stakeholder is a cog in the overall machine with a predefined set of tasks, typically involved in multiple projects at the same time, with different partners.

For a construction project to be successful, intensive collaboration between all those changing partners is quintessential. The advent of Building Information Modelling (or *Management*) (BIM) [116] brought many benefits regarding information exchange and project management: the object-oriented approach of BIM allowed to attach detailed information to individual building elements, concerning materials, physical properties, construction planning, etc. In this way, a comprehensive and semantically rich building model starts to exist, and information can be shared with whoever needs it for their respective tasks in the project. This holds for the design and construction phases (when BIM is used mostly nowadays), but also during the operational phase, for heritage management [402] or other phases of the Building Life Cycle (BLC). Such continuous usage of the data in the digital building model in an interdisciplinary fashion, throughout the life cycle of the asset, is the core of so-called 'Big BIM'.

A term often associated with BIM (and even more with its recent successor technology, the 'Digital Twin' (Chapter 8)) is its use as a 'single source of truth' (SSOT): if you need a specific property, and your query is precise enough, the answer will be non-ambiguous and trustworthy. In the SSOT-paradigm, data consistency is preserved because it is only stored once; there are no 'duplicates', and each time information is used, it points to exactly this piece of data. As digital twins are used to calculate scenarios for their real-world counterpart, data ambiguity may lead towards mistakes and miscommunications with sometimes costly consequences.

So far the theory: however, the reality is more complex and dishevelled. As mentioned, a construction project is a multi-disciplinary patchwork of (often small- or medium-sized) enterprises (Figure 6.1). In the most general scenario, each one of those enterprises uses different software packages depending on its discipline, resources and

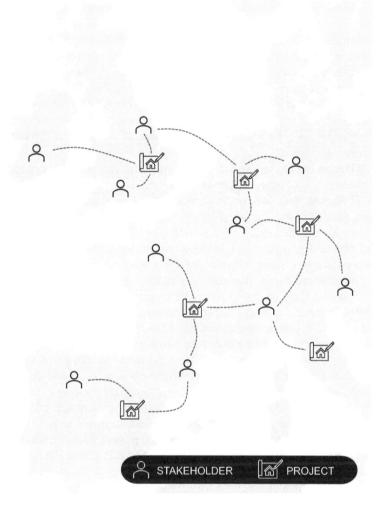

STAKEHOLDER   PROJECT

Figure 6.1 Patchworks in construction consortia.

the specific tasks they have been assigned for a particular project. Since every office in a particular project will be involved in other projects as well, one can easily see that the entire software stack used in a project will probably be an eclectic mix of packages, each one naturally favouring its vendor's proprietary data formats. In such situations, the use of open data formats is essential, a practice often denoted as 'Open BIM'. In projects where all stakeholders use the same ecosystem, a 'Closed BIM' approach may be beneficial as well, as data exchange will most likely be more streamlined. However, at some point in the BLC, it will often become necessary to exchange data with one or more tools outside the 'Closed BIM' ecosystem.

When using a specific tool for a dedicated task, an (encoded) in-memory database is created to allow CRUD (create-read-update-delete) operations on the local (sub-)model and speed up data handling. Saving the result, this data is serialised into a file that can be exchanged with other partners. A new sub-model is now ready to be added to the already present stack of project data. The main problem there, however, is not per se the fact that there is a new file on the pile, but the difficulty to stay synchronised with the other files, or, even more, with the status of the project in reality. Especially during the design stages, where the design changes rapidly, it is often hard to assess what the consequences of a certain change will be. It might seem harmless on first sight to just move a wall, but how does it affect the activities and designs of other project stakeholders? Whenever a change is made in one sub-model, a cascade of actions in the other project files needs to be triggered to ensure data consistency. Automating this trigger is only possible in a highly integrated project environment. For example, in an exchange environment where project documents are exchanged on request and through e-mail, automating such a cascade becomes virtually impossible and the chances of creating ambiguities in the project data become significant.

The advent of cloud platforms changed this in a positive way, and the AEC industry fullheartedly welcomed these as 'Common Data Environments' (CDEs), to improve collaboration, planning, management and automation within construction projects. With the ISO 19650 series (Section 6.5), a 2018 international standard based on the British norms BS 1192:2007 and PAS 1192-2:2013, a clear definition of the technical and practical structure of CDEs became available. In ISO 19650, a CDE is defined as an *agreed source of information for any given project or asset, for collecting, managing and disseminating each information container through a managed process.* Modern CDEs now provide the ecosystems where heterogeneous project information originating from multiple sources can be managed as a more or less coherent shared model. Creating such a model is, of course, most streamlined when a single ecosystem is used, but then the same problem may occur again: software vendors typically (and logically) integrate their products into a single ecosystem to improve the exchange of data between their own solutions, and AEC offices do not have the resources to consider paid subscription to another ecosystem for each and every project they are working on. To maintain the interoperability, mappings between the internal data model of one project file and the data model of the others needs to be present in order to automate this process, even when a single CDE solution is used within a project. Luckily, ongoing initiatives such as the OpenCDE-API[1] (covered by BuildingSMART) gather the large software vendors of this world to align their data models and improve data exchange from one ecosystem to another, so information exchange may also happen when project partners use different CDE ecosystems. Furthermore, bi-lateral agreements between CDE vendors allow direct integration of each other's formats to a certain extent.

The ISO 19650 series focus on CDEs based on the well-known BIM maturity level 2 (PAS 1192), where *file-based* partial models are interlinked into a consistent project model using cloud technologies (Figure 6.2). However, it provides a terminological basis for the envisaged maturity level 3 as well. In BIM maturity level 3, the project becomes entirely data-based: a fully integrated and interoperable web-based management system that allows real-time information management. Ideas of what BIM level 3 are going to look like depend on whom you ask, but a constant is the use of web technologies to facilitate collaboration between stakeholders and integration of external datasets. One of the more popular narratives nowadays is the use of *semantic* web technologies to organise these highly integrated, data-driven environments.

Semantic web technologies or linked data technologies for the construction industry have been extensively covered in the first part of this book. Over the past years, various disciplines concerned with the built environment have been structuring their knowledge in domain ontologies using the W3C (World Wide Web Consortium) standard for linked data; the Resource Description Framework (RDF)[2]. In their use of URIs (Uniform Resource Identifiers)[3] and URLs (Uniform Resource Locators, allowing to retrieve

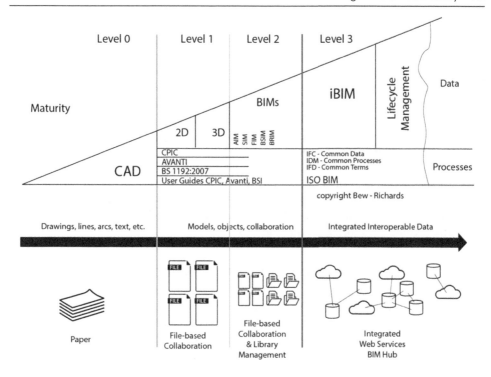

Figure 6.2 The BIM maturity levels as defined by the British Standards Institution (BSI). (PAS 1192-2:2013.)

the identified information)[4], project information can be made accessible through the web and, moreover, need not be centralised on a single server. Corresponding with the situation in reality (Figure 6.1), a stakeholder office can host its own project information on the office server, co-existing with the data it creates for other projects. Using decentral authentication (i.e. identification) and authorisation (i.e. granting specific permissions to identified actors) mechanisms, other collaborators requiring project information can either reference individual data snippets (via their URL) in their own storage server or automatically request specific data from other servers on the web, in order to perform some well-defined actions within a predefined use case. This distributed organisation of project data via stakeholders co-exists with the contextual project information, which is inherently federated, because it is provided by actors who are not actively involved in the project as such. When working with interoperable data models, asset information can be coupled with external data such as open government data [60], geospatial data [251] (see Chapter 4) and product data from manufacturers [406] (see Chapters 1 and 2). Finally, using the infrastructure of the web, not only data itself can be federated but also the tools that do something with this data. The use of microservices based on web APIs (Application Programming Interfaces) is already very common in many disciplines and starts to gain a higher adoption rate in the AEC industry as well. Moreover, when tools and data share common semantics, truly modular and tailor-made project processes become within reach, interoperable throughout the entire Building Life Cycle (BLC).

Bearing this state of the industry and the above-mentioned challenges in mind, Section 6.2 will assess some of the benefits commonly associated with a web-based, interoperable and data-oriented environment in a federated context (i.e. distributed over the web). Section 6.3, in turn, reviews some established concepts for consuming open contextual data as well as interacting with web-based microservices. These technologies provide the foundations of the web, and are thus essential to understand how web-based data exchange takes place. From there on, we move towards project-specific data management. The concept of cloud ecosystems, both general and AEC-specific (the CDEs), will be briefly discussed in Section 6.4. As for the organisation and linking of project resources, an overview of web-based 'information containers' is given in Section 6.5. The storage approach of containers is gaining traction in both industry and academia and allows to organise project data both centrally (Section 6.4) and federated. In the latter case, mechanisms need to be in place to organise access to confidential data in a federated context. Section 6.6 therefore reviews some technologies for federated data storage, authentication and identification. As there are, to the authors' knowledge, no established implementations yet of such framework, we will apply the discussed technologies in Section 6.7 to sketch how a federated, web-based mirror of a project consortium could look like, as a final outlook.

6.2   TOWARDS WEB-BASED CONSTRUCTION PROJECTS

Section 6.1 introduced some of the key challenges of the current AEC practices for the life cycle of a project. This section recalls these challenges while touching upon the potential of (Semantic) web-based technologies to address them. In general, the benefits of a linked data approach in the AEC domain have been pointed out by Pauwels et al. [301], to be summarised as cross-domain collaboration, interoperability and logical inference and proof. Some more specific use cases regarding federation of building data are discussed in the following section. We may summarise them as follows:

1.   Connecting to open datasets for contextual information such as geospatial data and weather data;

2.   Triggering automatic compliance checking based on guidelines provided by the government;

3.   Relating a project's required products with products that are available on the market;

4.   Triggering automatic revision of the federated model and its aspect models whenever a change is made;

5.   Manage real-time data from the building site, including sensor data (IoT), material streams during the construction phase and inhabitant response during the operational phase;

6.2.1   Connecting to open datasets

Apart from project-specific information produced by the project stakeholders, an important role for semantic enrichment of the federated project is reserved for external (open) datasets. These datasets may be provided at web service endpoints managed by governmental agencies, research institutes, universities, etc. They include contextual information such as geospatial datasets, cadaster information or weather. Oftentimes, this information is already exposed through web-APIs, maybe even structured using OWL (Web Ontology Language)[5] ontologies and available in RDF format [60]: such knowledge can be drawn into the project graph whenever necessary. When the data is not available as RDF, technologies to either convert or integrate this data can be used (Section 6.3.1), or links can be established on sub-document level (Section 6.5.2).

6.2.2   Automatic compliance checking

The previous paragraph mentioned open datasets hosted by governmental agencies. A specific sort of governmental data are regulations. At the time of writing, regulations concerning the built environment are typically not expressed in a machine-readable format such as RDF, which would allow to integrate them in automatic process chains. However, with powerful RDF-based technologies for data validation such as SHACL (Shapes Constraint Language)[6] or N3Logic [12], and OWL (Web Ontology Language) ontologies allowing logical inference of implicit information, automatic building compliance checking becomes technically possible. It is, in fact, one of the core incentives for the AEC disciplines to move towards linked data based models [301]. A government, publishing its regulations on the web, e.g. as 'open shapes', becomes an external partner in the project network, similar to the other open datasets that are referenced within this project.

Note that not only governmental compliance checking could play a role: project-specific requirements and ambitions could also be expressed in a machine-readable format and automatically checked on a regular basis. This would ensure that the design and the demands stay aligned over the federated model. Rules could also be used internally at one company to formally express what information should be provided at different project stages. At the component level, this could be expressed as a formalisation of the LOD (Level of Development) specification.

6.2.3   Relating project specifications to products on the market

As the design progresses, the client requirements are translated into placeholder products in the model that will eventually be materialised in real products that can be bought off the shelf. The demand for a certain space with specific physical capabilities originates from a functional demand of the building owner or tenant. For example, 'A two-person office fulfilling indoor climate class B' or 'A laboratory in GMO class 2 with 16 work stations and 8 laminar flow cabinets'. The engineer uses functional requirements of the spaces as boundary conditions for the technical systems serving the spaces and thereby derives the physical requirements for the products that make up these systems. In order to maintain a sustainable competitiveness in the construction industry, the European Union defines legislation for public buildings that prevents the consultants from specifying a specific product for the task. Therefore, the product requirements must also be described in precise, measurable terms (ISO 23386/ISO 23387). If product manufacturers published their product sheets as RDF data (e.g. as RDFa embedded in a HTML page), the requirement model of each product could easily be compared to what is available on the market. This would allow engineers to continuously evaluate the availability of products that meet the requirements, while contractors could later use a similar service to find and buy the products. BIM object databases exist today, and there is a business case in modelling these objects for manufacturers and selling them in a closed platform. In a federated AEC project, the manufacturer would publish the information themselves and host it just as they host their human-readable webpage. Especially for product data that is dependent on the individual manufacturer's current production methods (e.g.

data for Life Cycle Assessment (LCA)), it will significantly enhance the possibilities for performing accurate analysis if there is access to up-to-date data.

6.2.4   Automatic revision of the federated model

The need for reliable consequence assessment of changes made to the model has been partly covered in Section 6.1. Recycling the example from this chapter's introduction: how does moving a wall in one partial model affect other disciplines? Clash detection software allows to detect if the wall now collides with building installations, but other things may be affected as well. Will the HVAC demands of the zones change? Will the structural system be affected? Until a certain threshold is reached, an increased heating demand might be harmless, but when it exceeds the capacity of the designed system, the time and resources needed to remodel the system may increase significantly, especially when the system has already been ordered – or even worse, installed. A granular, data-based CDE could facilitate machine-readable communication between its different nodes, allowing agents to subscribe to changes on specific project information and trigger new calculations (e.g. a thermal simulation) based on the recent adjustments. Minimally, other stakeholders should be informed about the proposed changes (e.g. in a semantic extension of the current BIM Collaboration Format (BCF) specification). New demands could be automatically compared to the capacity of the current design, returning a warning if a certain threshold is exceeded. Many unfortunate changes with comprehensive and expensive consequences could be avoided in such a setup. Of course, since designing a building is an iterative task, it might not be desired to share every single change. Differentiating between work-in-progress, shared and published information (cf. ISO 19650) should therefore be at the core of federated CDE infrastructures.

6.2.5   Managing on-site data streams

The fact that a project features data streams from multiple sources has been mentioned a few times already. Where the above paragraphs focused on information coming from 'core stakeholders', other important information is given by on-site actors and contractors. For example, on-site sensors can monitor the construction process and broadcast to other actors involved (Chapters 3 and 10). In line with the patchwork of stakeholders and projects illustrated in Figure 6.1, projects with an on-site surplus of certain material streams can align with other projects in urgent need of additional resources (e.g. one project that has 2 tonnes of unused sand could redirect to another that needs extra sand because of on-site decisions). Finally, in the operational phase, inhabitants (e.g. of large building blocks) can provide real-time feedback that can be used for fine-tuning installations, damage monitoring or other topics related to managing the facility. Sensor observations can also be used to evaluate the indoor climate, and in combination with a rich semantic model of the technical systems, these observations can be used for automatic fault detection and feedback for tweaking the systems. It might even be possible to feed this information back to the engineers' models for future enhancement.

6.3   INTEGRATING CONTEXTUAL DATA AND MICROSERVICES

Section 6.2 indicated the potential that lies in connecting the project-specific data with external data on the web, provided by actors who are not directly involved in the project. Many third parties (e.g. material libraries, product manufacturersand governmental institutions) can provide useful contextual information for the project, information which thus resides on different web servers. We may see the integration of such data to be a first step towards decentralisation in the AEC industry. This way of sharing data through the HTTP protocol (Hypertext Transfer Protocol) is one of the main building blocks of information exchange through the web – it has been around since the 1990s [130]. Web services using this protocol to communicate, the so-called web APIs (Application Programming Interface), are thus widely used for use cases in virtually all economic sectors. Because they power the World Wide Web, this section starts with an overview of some core concepts (Section 6.3.1). Building upon this concept of web APIs, Sections 6.3.2 and 6.3.3 respectively discuss more data-oriented APIs, which may be used to access open data on the web, and more service-oriented APIs, providing specific functionality to the end user.

6.3.1   Web APIs

Web APIs provide an interface to communicate with a server over HTTP, and trigger certain actions, given the right conditions. This 'machine-based consumption of web content' [400] happens through a structured system of requests (sent by the client to the server and opening the connection) and response (sent by the server to the client). A HTTP request is sent to a specific endpoint provided by the API (in the form of a URL), which tells the server what to do with the information contained in the request. A request can carry very specific information, where the most visible one is the 'method' (Table 6.1). For example, the GET method is used to retrieve data: when you visit a website through your web browser, a GET request is sent to the server, which then returns the actual content of the website to your browser. Other information carried in a HTTP request is, for example, information needed for authenticating the user (using 'Headers'[7]) or additional data necessary to process the request (i.e. the payload), mostly present in the 'body'. While such body can be structured in different formats, one of the most common ways to express structured data is to send it as JSON (JavaScript Object Notation).

6.3.1.1   JSON and JSON-LD

Originating from JavaScript Objects, JSON is now a standardised and language-agnostic specification for binding variable information to specific keys (Listing 6.1). A good API is a well-documented API, describing precisely what information it needs, what happens with that information, and what information is returned. In the case of JSON, this will include specific key names and value types expected by an endpoint in order to function

Table 6.1 Common HTTP request methods and their meaning

HTTP method	Explanation
GET	Requests a representation of the specified resource. Requests using GET should only retrieve data.
POST	Used to submit an entity to the specified resource, often causing a change in state or side effects on the server.
DELETE	Deletes the specified resource
PUT	Replaces all current representations of the target resource with the request payload.
PATCH	Applies partial modifications to a resource

Source: Fielding et al. [130].

properly. JSON is schemaless by default, but can be annotated and validated using the JSON Schema specification[8].

Listing 6.1   JSON example with key-value pairs

```
1  {
2      "firstname": "Jane"
3      "lastname" : "Doe"
4  }
```

When no mere strings are used for JSON keys, but URIs, these keys can refer to concepts defined elsewhere on the web. In this way, these concepts can be reused and dereferenced; compatible with the RDF specifications. Indeed, as RDF is essentially a data model, while JSON is a syntax, RDF data can be serialised in a JSON format, resulting in 'JSON-LD'[9]. Listing 6.2 shows a proper JSON-LD version of the example from Listing 6.1. The '@context' part first defines two namespaces, xsd and schema, that are used throughout the document. The XML Schema Definition (XSD) is used to formally describe that both 'firstName' and 'lastName' have the datatype 'string' given by the special '@type' key. The other special key '@id' is used to map the keys used in this JSON document to widely established concepts from schema.org. 'firstName' and 'lastName' can thus be mapped to https://schema.org/givenName and https://schema.org/familyName, respectively. The resource is further identified with a proper HTTP URI and the schema.org class 'Person' is assigned to it.

Listing 6.2   JSON-LD annotation of the example from Listing 6.1

```
1  {
2      "@context": {
3          "xsd": "http://www.w3.org/2001/XMLSchema#",
4          "schema": "http://schema.org/",
5          "firstname": {
6              "@id": "schema:givenName",
7              "@type": "xsd:string"
8          },
9          "lastname": {
10             "@id": "schema:familyName",
11             "@type": "xsd:string"
12         }
13     },
14     "@id": "http://company-x/employees/Jane-Doe",
15     "@type": "schema:Person",
16     "firstname": "Jane",
17     "lastname": "Doe"
18 }
```

At the time of writing, the use of JSON-LD in web APIs is limited compared to non-semantic web APIs. However, both can co-exist and contribute data to the project: in a BIM project that does not rely on linked data (RDF), the context linked to a JSON-LD response will be of little use, while in a linked data-oriented project, clients may map plain JSON responses to a useful context themselves, using a mapping language like RML[10] [104].

6.3.1.2   API architectures

Most web APIs tend to conform to the constraints of REST architecture[11] (REpresentational State Transfer). An API is considered 'RESTful'[12] when all of these constraints are met. While a detailed discussion on the requirements for a RESTful API is out of scope, one of the most important ones is to maintain a 'stateless' server architecture (i.e. no information on the 'session' is saved, and each request therefore must contain all the necessary information to correctly interpret it).

The RESTful API architecture is worldwide the de facto standard for web API communication. Using the methods defined in Table 6.1, it can be an additional architectural layer on top of the HTTP protocol. However, it also has its limitations, such as the need to send multiple requests to get related resources (which are exposed through different endpoints on the server), and the concept of 'overfetching': a server response may contain more data than is actually needed by the client. Upcoming technologies such as GraphQL[13] offer an alternative, more efficient approach in these cases. GraphQL essentially is a query language, which allows to query a server's resources as an interconnected graph. This means a request for data can be constructed in a very detailed way, only asking for data that is relevant. A GraphQL-compliant API uses a 'schema' to validate incoming queries against predefined 'types', which are used to classify and structure the data internally.

Although GraphQL and RDF are two separate worlds, they share some interesting features, the most obvious one that data is structured in graphs. Because of that, several approaches exist to 'semantically enrich' GraphQL, e.g. in order to be able to use it to query the semantic web in a more user-friendly way than the query standard SPARQL[14] [374]. A detailed discussion of these approaches is out of scope for this Chapter.

6.3.2   Consuming data on the web

In the previous section, it became clear that web APIs provide a generic mechanism to communicate with servers, independent of the programming language that server is written in. Therefore, they can be used for multiple purposes, which can be roughly divided in (1) communicating with a database (i.e., the server acts as *middleware* between client and database) and (2) providing specific functionality to the client as a *microservice*, for example, by performing simulations or calculations. The first one will be discussed in the following section; the second one is the topic of Section 6.3.3.

When data is made available to anyone, it is called 'Open Data'. This means you do not need to authenticate to be able to 'consume' the data. Open data can be expressed in different ways; as a guide for the 'openness' of data, the 5-star schema[15] by Tim Berners-Lee [34] can be used. The first star defines the basic requirement for exposing open data: there are no conditions to the format, only it must be accessible through the web under an open license. The second one requires the data to be structured – in other words: it should be machine-readable. This can be a proprietary format as well, but not a scanned document for instance (that would be an unstructured image). When such machine-readable data is exposed as a non-proprietary open format (e.g., JSON or CSV (Comma Separated Values)), a three-star level openness is reached. From the 4th star onwards, the use of URIs to denote things becomes mandatory. The last phase of open data requires that such URIs are linked to other data, eventually forming a federated graph of open data on the web, complying to the RDF specifications. It is important to note that we can still communicate with such 5-stars-of-open-data web services using the familiar JSON syntax – as we have already seen, one of the many serialisations of RDF is based on a context-extended JSON structure: JSON-LD.

Nowadays, most open data services do not reach the 5th star. However, this is no requirement to find the use of this data within a project, as benefits can already start by harvesting 1-star open data (e.g. images). Therefore, services that provide access to such datasets are a very important aspect of web-based BIM – independent from whether the project uses semantic web technologies. We can conclude that, by integrating external data (be it using RDF or not), projects can be enriched with data that is not necessarily AEC-oriented, but which relates to relevant adjacent domains for specific use cases. Moreover, if this data is not access-restricted, which is often the case with e.g. governmental and geospatial information, it can be relatively easily integrated into digital construction projects.

6.3.3   Microservices

The second way web APIs can be used, is not to expose data, but as microservices which require certain data input to do their job, and return the outcome of this job as the payload of an HTTP response. For some microservices, it is unnecessary to present the end user with a graphical interface: they are called 'headless' APIs. These are services that require no further interaction after an initial request: a service that combines project and contextual data to draw certain conclusions, or an API using the same data to check the model against a set of regulations. Other services, however, require human input as a human-machine interplay to complete complex tasks. For those services, a graphical user interface is necessary. One can imagine that an interface for specifying connections between building elements looks different from an application that is used for indicating damage patterns or classification of heritage objects. Because the principle here is that the user provides the service with the information it needs, data and applications essentially become de-coupled: within the network of microservices that can be used within a construction project, one may just choose the right application for the right purpose.

Initiatives such as the BIM bots from the BIMserver initiative (Section 6.4) or commercial API frameworks from BIM authoring tool vendors clearly show that the 'microservices paradigm' is increasingly gaining traction in the domain of construction informatics – this in contrast with long dominating practices of using a single (BIM or CAD) authoring tool throughout the entire project. This trend towards microservices holds for proprietary, commercial ecosystems and ecosystems using open data models such as IFC and LBD ontologies; as long as they are interoperable with the decided-upon overall data schema of the project. However, if this data model is both open and extendable, as well as dereferenceable on the web (as are the LBD ontologies), the ecosystem will be able to make far greater use of open web data and integrate contextual data and project data on the data level – semantically clear and independent of any other tools that will base upon this data.

Taking this concept one step further, web services that base upon linked data ontologies may link to those ontologies while communicating their data needs to the client, effectively allowing a generic communication mechanism between different services and APIs, often without the need for human intervention. For example, validation technologies, like SHACL, allow an application to make sure the data it receives is structured the right way, while reasoning engines and ontological mappings may be used to enrich datasets to increase chances of fitting the data in such SHACL-based validations. This way, may it be in the far future, semantically clear project plannings may be used to trigger certain microservices; for instance, when certain conditions are met, when specific types of changes are made to the model or at specific points in time. One can then imagine a chaining of microservices Retaking the example of the movement of the wall from Section 6.1, a workflow could be defined as follows: the first service, automatically notified when the change occurs, calculates the wall (and opening) areas for adjacent spaces. A second one takes these as an input to calculate energy losses, a third one jumps in to check if the changed model is still compliant to energy regulations and a fourth one calculates whether a new HVAC design is required. Project-specific 'orchestrator' services, with a close view on the project data, used data models, stakeholders and project planning, may be used for such microservice triggering and model-checking purposes.

However, for such federated chains of services to be actually implemented in construction projects, not only data compatibility is needed but also mechanisms to restrict and grant access to sensitive data: contextual data may be open, project-specific data most likely is not. Several technologies allow to securely store confidential data. The first one, which is the most widely used solution in the current AEC industry, is to store project data at a single server – the centralised solution: the CDE. As user accounts and access permissions for microservices are also managed within this single central server, access control can happen in a rather straightforward way. In Section 6.4, we will briefly touch upon these existing environments. The second approach is to structure the project in a federated manner and use workflows for decentral identity management and authentication. The pros and cons of such project organisation are still under research; to the authors' knowledge, only experimental implementations have been set up at the time of writing. A discussion of some upcoming technologies in this regard are the core topics of Section 6.5 and Section 6.6.

6.4   EXISTING ENVIRONMENTS

Section 6.1 briefly introduced the concept of a Common Data Environment (CDE) and its international standard ISO 19650. A CDE makes project information produced by members of the consortium available to other members in the consortium who may need that information for their own work packages. It is clear that such project-produced information differs a lot from the information provided by open data APIs discussed in Section 6.3.2 – access control needs to be managed and data availability must be 100%. A reliable access to all (published or shared) project-specific data is ensured because the information resides 'in the cloud', on the servers of the CDE vendor. In many cases, CDEs provide additional functionality to allow to communicate about inconsistencies, collisions or general change of plans, and integrate the service within the existing ecosystem of software solutions. Furthermore, the integration with browser-based viewers allows project members without a license for specific BIM authoring tools to view the model, submit issues and take notes.

6.4.1   Non-specialised environments

At the heart of a common environment for project data, however, still lies the infrastructure of storage functionality and access rules: who gets access to what information. With this in mind, general file-sharing solutions could be seen as non-specialised 'CDEs'. When the right connections for authentication, authorisation and data retrieval are established between authoring tools and storage servers, project stakeholders may communicate with one another using these solutions.

6.4.2   AEC-specific environments

Cloud environments that are specifically targeting the AEC industry add some extra layers upon the classic 'Document Management System' (DMS). For example, they enable versioning of project models and metadata linking (Model Management System, MMS), issue communication (Issue Management System, IMS) or, as indicated earlier, integrate specialised geometric model viewers. Proprietary CDEs natively use internal data models for integrating these systems, although open standards exist and are supported to various degrees.

For example, apart from the Industry Foundation Classes (IFC) (the international standard for open BIM), there exists the BIM Collaboration Format (BCF) for issue management, which is expected to play a major role in issue communication between different CDEs used in a project with different stakeholders[16]. Although standardised communication between commercial CDEs is an important step towards industry interoperability, we consider this only one of many requirements necessary for a BIM Level 3 ecosystem: issue management standards, such as BCF, allow to address issues on a 'sub-file' level, but overall, a file-based paradigm is still prevalent, preventing or hindering roll-out of comprehensive data-based infrastructures and duplicating information over different project files.

There are, however, initiatives that have embraced a data-based approach for collaboration in the cloud. One of the most well-known ones is the open source BIMserver project. The BIMserver (formerly IFC Server) [30] is an IFC standard compliant model server which facilitates storage and management of building information modelling projects. It is an open source project based on a model-driven architecture approach, with the IFC data stored as objects rather than as files. The server itself is envisioned as an object-oriented IFC database, which provides functionality for developers to build their own applications on top of it: the so-called 'BIM bots'[17]. Using an open standard such as IFC (or a set of agreed data models such as the ontologies discussed in Part 1 of this book), external web services (Section 6.3.3) can connect to the BIMserver database through service interfaces such as the already discussed JSON API[18] (Section 6.3), 'understanding' each other because they are built upon the same data model. Specialised microservices can then be used for specific use cases (Figure 6.3). For example, the current BIMserver implementation contains features such as model checking, versioning, project structuring, etc. web applications based on the BIM bots paradigm but applied to Linked Building Data form an important aspect of automated federated project management and synchronisation.

As this section focused on existing centralised data platforms for project organisation, the following section will dive deeper into the internal structure of project data, and how project resources can be semantically related to one another, independent from whether the project is organised centrally or in a federated manner.

6.5   CONTAINERISATION OF HETEROGENEOUS DATASETS IN CONSTRUCTION

In a decentralised ecosystem, management of information requires a structuring system that consistently maintains the links between the information. This information can be stored in files or as graphs, spread across data silos, and needs to be linked and maintained in order to function effectively in the linked data world. While interoperability between

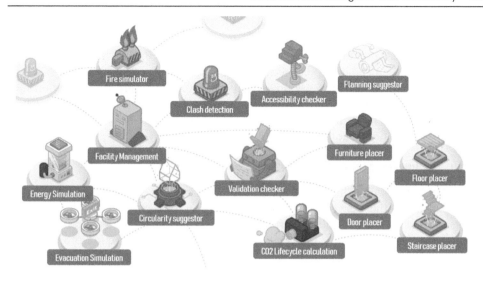

Figure 6.3 Networks of BIM bots on top of the BIMserver. [394]

software platforms and data formats is still a heavily researched area with ongoing efforts, as mentioned in the previous sections, the need remains for an organised structure for collaborating with different stakeholders utilising different data formats. This section discusses the approach offered by 'Containers', which provide a path for collaborative information management by incorporating essential metadata about project information, in a web-compatible way. After an introduction to the concept of containers, some existing specifications that implement variations on the theme are discussed. Building upon the work done in these specifications, the novel ISO standard 'Information Container for linked Document Delivery' (ICDD, ISO 21597) is reviewed as well as its relevance for the concept of CDEs and of its some limitations.

6.5.1   Existing specifications on information containers

Data containers, in their most stripped down form, are enclosures which are used to store information: they are used to partition resources by bundling the relevant resources together, based on the use case for which the container was created. There are no restrictions on what kind of data they can contain. For example, in the context of documenting the damages recorded for a building, the container can consist of image files of the damage, 3D models of the building, PDF reports on the damage, etc. The concept of containers in AEC has been explored previously, notably in two projects. Multi-model Containers (MMC) for the construction industry were originally defined as a part of the Mefisto project[19] [343], to cater to model exchanges wherein varied file formats were used for numerous application domains. This project resulted in the development of a formal definition of a data schema for generic multi-models, which satisfies cross-domain applicability, storing domain models without any change to their original format, and also persistent and with restorable links [342]. Another project defining core concepts for multi-models is the Dutch COINS approach[20] [398].

These initiatives serve as preludes to the 'ICDD' approach for containerisation. It should be noted here that the concepts of containers and CDEs are closely intertwined. Industry-oriented standards such as the PAS 1192, ISO 19650, DIN SPEC 91391-2 and general web recommendations such as the linked data Platform (LDP) all have their own definitions of containers. However, in each of them, varying levels of information such as vocabularies for metadata, types of links between information blobs, etc. are left to the decision of the implementer [348]. This section presents a concise look at how each of the above-mentioned approaches envisions the ideal 'Container'.

6.5.1.1   BS 1192:2007 and PAS 1192-2:2013

BS 1192:2007 was introduced as a set of specifications that supports the information management in the delivery phase of construction projects and the life cycle of assets beyond the design and construction phase. PAS 1192-2:2013 (Specification for information management for the capital/delivery phase of construction projects using building information modelling) expands on the scope of CDE defined in BS 1192:2007 by defining specific status codes and file and layer naming conventions for document management. In the latter, it addresses information on project location, document author, use case, discipline, etc. While it does not explicitly mention containers, it describes four states of data management: 'Shared', 'Work-in-progress', 'Archived' and 'Published'. Files in these states should adhere to the naming conventions in the standard, and additional process specifications in it. This standard has now been withdrawn, replaced by ISO 19650.

6.5.1.2   ISO 19650-1/2:2018

ISO 19650 was already introduced in Sections 6.1 and 6.4 as the main international standard for the definition of a CDE. Both Parts 1 and 2 of this standard talk about CDEs as a technical solution and process workflow, by specifying information container management [187]. While its file naming conventions for container information largely follow the ones defined in PAS 1192:2013, this standard contains additional specifications of metadata assignment for revisions through a 'Revision control system', 'Purpose of Container', 'Issue Date', etc., according to Uniclass 2015 [199].

6.5.1.3   DIN SPEC 91391-2:2019

The German DIN SPEC 91391-2 standard specifies OpenCDE, a list of conceptual requirements for the interface for communication between CDEs, and their interaction with various software applications. In this standard, 'Multi-containers' are specified as well-defined container types. It also contains concepts for building projects (such as the project identifier) and an explicit version handling for documents. A detailed look at these container concepts is provided in Senthilvel et al. [348].

6.5.1.4   Linked Data Platform

Independent from the AEC industry, the W3C specification of the linked data Platform (LDP)[21] defines guidelines for interacting with web resources (both RDF and non-RDF), as to provide an architecture for a read-write Web of Data. In the LDP specification, containers play a very important role. They are considered a very specific web resource, which has the capability to respond to HTTP requests (HyperText Transfer Protocol, see Section 6.3) relating to the creation and modification of resources. Two types of containers are defined, based on how resources are contained, created and deleted: a basic container and a direct/indirect container[22]. In the former, links to a 'document resource' (which can be both RDF and non-RDF, as well as a 'child container') can be defined by using a predefined predicate. In the direct and indirect container, additional relationships such as domain-specific vocabularies can be used for the link relationships, offering more flexibility for the user to define custom resource relationships, beyond LDP definitions.

In brief, in the LDP approach 'anything can be a Container', and these containers can have corresponding RDF links to their resources in three different ways depending on which container is used. However, when this approach is scaled to real building projects, more complex information is usually encoded using specific definitions for links between files/documents, 'sub-document level linkages', etc. As LDP is domain agnostic, it provides the flexibility to define link relationships from any domain-specific vocabulary – the links in a container are left to the discretion of the creator. Hence, a schema for interpretation of the links also has to be supplied by the creator, e.g. in the form of an RDF vocabulary. Currently, numerous implementations of LDP exist[23]. Moreover, LDP in turn forms the basis for other web specifications as well, for example the Solid specification stack discussed in Section 6.6.

6.5.2   ISO 21597: Information Container for linked Document Delivery (ICDD)

The Information Container for linked Document Delivery (ICDD) is a recently published standard (ISO 21597) which focuses on standardising the approach for implementation of multi-model containers for Linked Building Data[24]. As mentioned, the standard builds upon some of the previously described specifications. It consists of two parts, with Part 1 focusing on definition of container structure and link concepts, and Part 2 delving into specific link types.

In the ICDD context, a container is compressed by default as a ZIP archive and has the '.icdd' file extension. As per the standard, it follows a folder structure (Figure 6.4). This .icdd archive has three major components: an Ontology folder, which contains the schema structuring the files in the ICDD ZIP container format, a Linkset folder containing the links, and finally, a Payload folder which contains the files which are being linked together. Each ICDD container also comes with a mandatory meta-file called the 'index'. This file specifies what documents are in the container's payload folder, and how they are linked to each other. In brief, the ICDD ontologies, together with data types and properties, define what kind of meta-information can be used for a file and the links between different files.

6.5.2.1   ICDD in a CDE

As noted previously, the concept of containers and their corresponding links in this standard is highly relevant to CDEs, and to project management on the web in general. Information from stakeholders can be stored as files (i.e. the default CDE approach) or as Linked data graphs (e.g. using a triple store). In both cases, it is necessary to provide metadata – we have shown that a container-based approach can offer benefits there. Hence, most CDE approaches have some level of definition of containers and their associated properties. The linked data Platform (LDP), which was mentioned earlier, is relevant here since it also specifies container definitions, information access protocols,

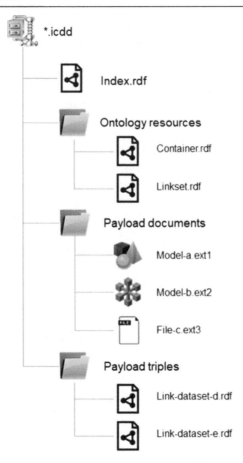

Figure 6.4   Containerisation of ICDD. (ISO 21597-1:2020.)

and a broad definition of links [348]. As noted previously, in LDP, anything can be a container and membership links between resources can be defined ad-hoc by the user. In the case of applying these concepts to (federated or centralised) construction projects, concrete and non-ambiguous definitions for link types are needed, along with an ontology for linking on object/sub-document level (e.g. the ICDD Link Ontology). However, ICDD's primary relevance to CDEs is its definition of the ontologies for organising the container and its contained documents through semantic links, as explained in the previous section. These provide a structured approach for linking heterogeneous building data, which can be used for information exchanges. A mapping between the ICDD and LDP container definitions is shown in Figure 6.5.

While the ICDD standard is intended for specifying information containers for a file-based system, its concepts can be well extended to work on distributed knowledge graphs which are dereferenceable (i.e. information can be found through this URL). These graphs can reside on triple stores or be exposed as machine-readable files on an infrastructure such as LDP. Therefore, the ICDD folder structure can be replicated in federated CDEs as well, by using distributed graphs instead of the payload documents. These graphs can be stored in the containers themselves or be referenced to graphs living outside the container system (e.g. contextual information (Section 6.2)). Thus, a virtual

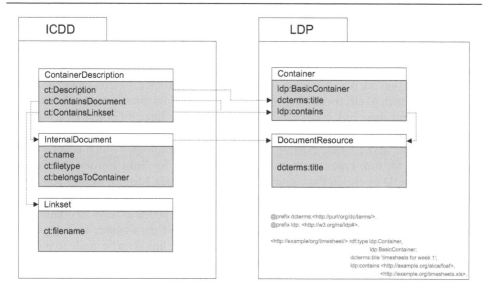

Figure 6.5 Mapping container concepts of ICDD with LDP. (Senthilvel et al. [348].)

project graph can consist of multiple containers, and be linked to different external datasets, thereby creating virtual containers for varied use cases.

An important aspect for implementation of container concepts relies on agreements between CDE vendors on how their systems will exchange data. Such an agreement can be formed based on many overlaid standards and approaches, resulting in a hybrid set of features, implemented and enforced by the CDE vendors. Existing approaches such as the earlier discussed LDP, DIN SPEC 91391, the OpenCDE API, etc. provide a road map for standardising how CDEs should function and the features to be considered when implementing them. The container and link concepts defined in ICDD, along with external vocabularies, can be employed to extend these existing CDE container concepts. Ultimately, containers need to serve the basic functions of querying and data manipulation, while maintaining information uniformity, and integrity.

6.5.2.2 *Limitations and future work*

The current ICDD standard is relatively new at the time of writing this book. Released in early 2020, it constitutes an important step towards establishing standardised implementation approaches for linking building information in a federated environment, although there evidently remain scenarios that are not addressed by the standard. In the current Parts 1 and 2 of this standard, the entities/nodes in a container have meta-information on document history for date of creation and author information. When ICDD is scaled to manage information in larger construction projects, the sheer number of documents, models, etc. would require the development of approaches for automatic creation of links between these resources. In such cases, the need for meta-information on link history also arises. Since the current ICDD schema does not cater to this, an extension to it can potentially aid in this situation.

Furthermore, Part 2 already specifies specialised link-type vocabularies for defining relationships that can be used to define links between conceptually identical documents, such as an 'identical' relationship (two images taken from different angles, which refer to the same wall), conflict indications (different versions of the same document) and resources that are to be considered 'alternatives' (such as an PDF of a plan drawing being an alternative to a snippet of IFC model). However, these links do not cover the entire range of links that are possible in an AEC project (for instance, objects encompassed in other objects, such as electrical wires in walls). An additional question yet to be explored is to what extent these link types can be used for defining links on object level and sub-document level.

In both the above cases, existing or new external ontologies which cover broader terminologies for link types, or meta-data vocabularies for links history, can be used along with ICDD to enrich the container (see also Chapter 1 for an overall context on object types and terminologies).

This section showed that the concept of 'containers' provides ways to organise and interlink project information, both document-based and data-based, in a structured manner. Using the ICDD standard, RDF-based semantic links can be made to information inside the container as well as external information. This property allows to envisage the project as a dynamic subset of the web, rather than living in a central CDE. Section 6.7.1 showed that there are no issues in linking to open data, but indicated some authentication challenges when dealing with confidential data. In the following section, we will discuss some technologies that make such federated organisation of project data possible.

6.6   FEDERATED PROJECT DATA

Nowadays, commercial solutions for sharing confidential data (both AEC-specific and general) collect all project data on a single cloud platform (i.e. a 'centralised' solution). Information (mostly file-based) is uploaded to (or directly created at) the servers of this platform, where it is made available to people with the right credentials. In Sections 6.1 and 6.5, such shared infrastructures were introduced as CDEs. However, one could as well imagine a setup where not only the contextual data is federated, but project-specific data as well, resulting in a 'federated CDE'. After all, as we saw in Section 6.1, the real-world organisation of construction projects is highly decentral: each project is a patchwork of offices and each office participates in multiple projects (see also Chapter 1). In this section, we will discuss some promising technologies for organising digital project data in a federated setup, similar to a real-world stakeholder network.

6.6.1   Decentral identity verification

We have seen in previous sections that web technologies such as HTTP and RDF offer a technological foundation for referencing data on the web. Technologies such as the LDP and Multi-Model Containers (Section 6.5) allow project stakeholders to organise heterogeneous project data in a structured way, maintaining semantic links to external datasets. With this in mind, to put all project-specific data on a single 'central' platform is no longer the only option: stakeholders are not required to upload all their project data to a central project server because RDF allows for cross-server linking of information. However, because this data often is access-restricted, those project servers cannot be organised as if they were 'open' web APIs. Federation of open datasets is so successful because this data is 'read-only': there is no authentication needed to consume the data. However, when considering confidential project-specific data, authentication does play a role indeed. Numerous technologies exist to authenticate clients who request data access – the challenge is to avoid creating accounts for every dataset that one wants to access, or for every single microservice one needs to use throughout the life cycle of a project.

Of course, this is not unique to the AEC industry. Decentral protocols for identity verification and authentication are quite common already in many web APIs with varying scope and functionality. Protocols such as OpenID Connect (OIDC)[25] allow for user authentication and identity checking without the need to create a local account. An external 'identity provider' can confirm the identity of a client to other applications on the web, provided that the client has an account on the servers of the identity provider. This is the background mechanism behind the well-known procedure to 'log in with X', which allows you to identify yourself without creating an account on any website you visit implementing OIDC.

In the case of decentral management of AEC data, an office server that implements OIDC can thus be set up to allow specific external accounts to have access to the data. Most current-day OIDC implementations are provided by large IT companies, but it is, in fact, not necessary to involve those firms in your project. Recent initiatives such as the Solid project[26] [244] eliminate the need to rely on those large enterprises for hosting and sharing data. It is important to note that Solid is not a company, but rather a set of open, discipline-independent web specifications, which, consequentially, may be implemented by anyone. Essentially, Solid just adds an OIDC layer on top of the LDP infrastructure, which encompasses both an identity provider for the owner of the LDP-based data, and a verification mechanism for the identity of visitors requesting access to that data. This infrastructure is generic – it can be used to store and link any dataset from any discipline. To base upon the Solid specifications as the backbone for a federated CDE has been proposed in Werbrouck et al. [409]. The following paragraphs will briefly discuss some of the core Solid specifications, to make the underlying mechanisms more clear.

6.6.2 Federated data storage, authentication and authorisation

As Solid is rooted in the semantic web, it makes extensively use of URIs and URLs to both identify and allocate resources. This way, the primary identification for any actor is not an e-mail address or a username, but a URI, unique by definition. Such URI is called a WebId[27], which has the following structure:

https://bob.example.com/profile/card#me

When this URI is not only used to identify someone, but can also be used to locate actual information about this person on the web (which is mostly the case), it becomes a URL. For example, the above WebId resource is defined in a larger 'card' document where the WebId is defined and linked to additional information about the person. Anyone who looks up this URL in his/her browser will actually fetch the entire card document. If we further decompose the WebId URL, we see that this 'card' resides in an LDP container identified with the string 'profile'. This backwards chain of sub-containers (which is quite short in this example) eventually leads us to the 'root' container. In Solid, this root container is called the 'Pod': a 'Personal Online Data storage' where an individual can manage their data as if it were a regular folder on their computer.

By connecting the WebId to an account on a 'Solid Identity provider', the Solid specifications allow to implement the mechanisms provided by OIDC without the need to rely on large IT enterprises. It thereby combines the benefits of the well-established

OIDC protocol and the decentralised flexibility of the WebId in the WebId-OIDC[28] specification.

With the authentication (and identification) in place, structures can be set up to grant specific access rights to actors (which can be web services as well). In the case of Solid, this is done with the web Access Control (WAC) ontology and specification[29] to relate access rights to certain LDP containers and individual resources. A dedicated ACL (Access Control List) resource points to either an individual resource or an entire LDP container, relating specific actors (via their WebId) to specific access rights, such as reading, updating and deletion of data. An actor may also be authorised to change and grant access rights himself, a privilege mostly only available to the actual owner of the data. Listing 6.3 gives an example of an ACL file granting read access to everyone and full access to the owner, in this case the notorious `https://bob.example.com/profile/card#me`.

Listing 6.3 Example ACL graph (Turtle) granting reading rights to the public and full access to the owner

```
1  # ACL resource for the public folder
2  @prefix acl: <http://www.w3.org/ns/auth/acl#> .
3  @prefix foaf: <http://xmlns.com/foaf/0.1/> .
4  @prefix own: <https://bob.example.com/profile/card#> .
5
6  # The owner has all permissions
7  <#owner>
8      a acl:Authorization ;
9      acl:agent own:me ;
10     acl:mode acl:Read, acl:Write, acl:Control .
11
12 # The public has read permissions
13 <#public>
14     a acl:Authorization ;
15     acl:agentClass foaf:Agent ;
16     acl:mode acl:Read .
```

Based on all the technologies mentioned in this chapter, a workflow can be set up where an AEC office can host a single office server containing both information for internal and external use. Accounts (e.g. WebIds) and passwords for office employees are managed by this office server as well as the data they produce. At the same time, their office accounts can be used to access project information produced by other stakeholders, and in the same way, they may give employees of these other stakeholders access to relevant information. Functionality offered by subscription-based web services can be accessed in the same way. Recommendations such as the ISO 19650 stages of data publication, infrastructures like ICDD Containers (Section 6.5.2) and LDP and decentral authentication mechanisms like (WebId-)OIDC can aid in organising project data in a federated manner.

Envisaging such setup, Section 6.7 will conclude this chapter with some final considerations for the organisation of federated AEC projects.

6.7 COLLABORATION STRUCTURES FOR THE FUTURE

Where the previous sections focused on existing technologies with various scopes, the final section of this chapter will sketch a possible infrastructure for combining them. Because such a framework does not exist yet at the time of writing, the scope of the topics discussed below is largely illustrative.

6.7.1 The stakeholder network

Throughout this chapter, we have illustrated that an AEC project integrates both contextual and project-specific data. In a fully federated project, the collection of services taking care of the former will be quite volatile (because the idea of 'context' is very use-case specific), while the network taking care of the latter will be more stable: it is the network of stakeholders involved in the project. This network only changes occasionally, e.g. when a new phase in the BLC is reached or someone unexpectedly enters or quits the consortium. Figure 6.6 depicts this 'stable' stakeholder network in relation to external contextual data APIs. Data hosted by these APIs can be referred to from within the graphs produced by individual stakeholders, allowing to traverse the graph when additional information is needed. Furthermore, the figure schematically shows how (third-party or project-hosted) microservices, triggered by the responsible stakeholders, may require subsets of the federated project graph in order to function properly.

To allow information exchange and synchronisation between federated datasets in the project, it is necessary to maintain a semantic description of this network. This semantic description serves a twofold purpose: on the one hand, it allows to identify actors in the network, so they can be granted authenticated access to project information on the web. On the other hand, it allows allocating the resources produced by those

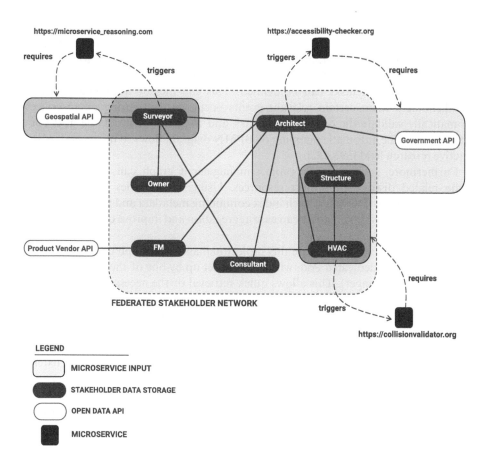

Figure 6.6 Network of project stakeholders, data APIs and microservices.

stakeholders. For the first requirement (identification and data access), a regular e-mail address will in most cases suffice. When the e-mail service is also an OIDC identity provider, authentication of visitors can happen decentrally through this e-mail address, a common procedure on the modern web. The second requirement, however, focuses on allocating information produced by the actor in question. When a regular e-mail address is thus used to conform to the first requirement, it therefore needs to be associated with a data storage endpoint (e.g. provided by a cloud storage provider). However, as we have seen in Section 6.6, both requirements are fulfilled at once by making use of a WebId: because it is a dereferenceable URL, it allows both identification and information retrieval. For illustrative purposes in the remainder of this section, we will therefore assume that all stakeholders have WebId and, consequentially, an associated data store that is compatible with the Solid specs – a Pod.

6.7.2   The project management graph

It is not a strict requirement to represent the stakeholder network using RDF: just a list with WebIds would suffice. However, using a graph-based approach using RDF will yield several benefits, for example, the ability to link roles and tasks to specific stakeholders. Well-known BIM planning processes such as the use of BPMN (Business Process Mapping Notation)[30] diagrams could form the basis for the setup of this machine-readable project planning, combining existing ontological mappings for BPMN [328] with the federated concept of WebIds and the power of automated web services. Such 'project management graph' may then serve additional purposes, for example by providing a semantic project planning document, where dependencies and exchange requirements between tasks can be defined. This can include automatic notification of stakeholders when certain relevant events occur, or configuring microservices to automatically validate the outcome of a task and trigger start-up of the next one (cf. Section 6.3.3). The automatic mapping of BPMN diagram flows to microservices is also an active research field [393,432].

Furthermore, such semantic project management graph can integrate an index of the related distributed project resources. Using vocabularies such as the Data Catalog Vocabulary (DCAT)[31], such index containing metadata and pointers to datasets published by various stakeholders can ease aggregation and improve querying the project for relevant data.

Several options exist to expose this project management graph on the web. First, it may be hosted in a dedicated Pod, which can be set up by one of the main stakeholders (e.g. the process manager). This allows quick retrieval of the necessary data, and offers an unambiguous endpoint to whoever wants to consult it (be it a human actor or a microservice). Another option to organise the stakeholder network is to use distributed, cryptography-based technologies (e.g. blockchain) for maintaining a single source of truth. In this case, everyone has a local copy of this graph which is continuously being synchronised with copies at other stakeholder data stores. Lastly, a 'fully federated' stakeholder graph can be constructed when everyone keeps track of their own sub- and super-contractors, on their own office server, although this scenario introduces additional complexity (e.g. in terms of querying) and will also hinder further enrichment of the stakeholder graph.

Because the stakeholder network graph and its extensions serve as a 'master index' for the federated project, a clear specification for their structure is needed. It is the first step towards the creation of a highly automated construction project, based on the infrastructure of the World Wide Web. As far as we know, a framework that serves such project setup is non-existent at the time of writing, and many hurdles need to be taken before theoretical workflows result in useful implementations for the AEC industry. Taking a step-by-step approach, however, will gradually result in projects that become more and more interoperable, in terms of internal alignment of project resources, but as well regarding external data on the web.

6.8   CONCLUSION

This chapter discussed several technologies that may be used for federated project organisations. As we illustrated throughout this chapter, this is not an all-or-nothing narrative: several actions could be taken to reap the fruits of interoperability in a gradual way. Some of these fruits are quite low-hanging, such as the use of open data formats (Section 6.1) and harvesting contextual data from web APIs (Section 6.3). Linking project information through web-based containers based on industry-oriented standards could be a next action to take (Section 6.5). A final step, to organise project information in a 'federated CDE', where data is not uploaded to a central project server but resides with the stakeholder who produced it, can allow these stakeholders to naturally combine their contributions to multiple projects and draw conclusions for future work (Section 6.6).

The combination of these different steps may lead to new digital workflows that incorporate highly automated information flows (Section 6.7.1), and allow a data model to be used for more than the typical use of BIM models for design and construction. A building model may link to relevant information from various sources and domains: in that case, the entire project is an ever-changing subset of the web – its size depending on the use case. In this very broad interpretation of the 'Big BIM' concept, data can be fetched and reused throughout the entire life cycle of the building: the same model that optimised design decisions and construction site management can be used later on for Facility Management (FM), be an aid in renovation or demolition processes, or, occasionally, be enriched with relevant historical and contextual information. In this way, a multi-disciplinary approach for managing the built environment gradually comes into sight: an automated network of reusable data, a single source of truth embedded in the infrastructure of the World Wide Web.

NOTES

1 https://github.com/buildingSMART/OpenCDE-API.
2 https://www.w3.org/RDF/.
3 https://www.w3.org/wiki/URI.
4 https://www.w3.org/TR/url/.
5 https://www.w3.org/OWL/.
6 https://www.w3.org/TR/shacl/.
7 https://www.w3.org/Protocols/rfc2616/rfc2616-sec14.html.
8 https://json-schema.org.
9 https://json-ld.org.
10 https://rml.io/.
11 https://www.ics.uci.edu/ fielding/pubs/dissertation/rest_arch_style.htm.
12 https://restfulapi.net/.
13 https://spec.graphql.org/.

14 https://www.w3.org/TR/sparql11-query/.
15 https://5stardata.info/en/.
16 https://github.com/buildingSMART/BCF-API.
17 https://www.slideshare.net/berlotti/bim-bots.
18 https://github.com/opensourceBIM/BIMserver/wiki/Service-Interfaces.
19 http://www.mefisto-bau.de/.
20 http://www.coinsweb.nl/wiki2/index.php/Welcome_to_the_COINS_wiki_pages.
21 https://www.w3.org/TR/ldp/.
22 https://www.w3.org/TR/ldp/#ldpc.
23 https://www.w3.org/wiki/LDP_Implementations.
24 https://www.iso.org/standard/74389.html, https://www.iso.org/standard/74390.html.
25 https://openid.net/connect/.
26 https://solidproject.org/.
27 https://www.w3.org/2005/Incubator/webid/spec/identity/.
28 https://github.com/solid/webid-oidc-spec.
29 https://github.com/solid/web-access-control-spec.
30 https://www.bpmn.org/.
31 https://www.w3.org/TR/vocab-dcat-2/.

Chapter 7

Web-based computing for the AEC industry

Overview and applications

*Mohamed Elagiry, Rubèn Alonso, Eva Coscia,
and Diego Reforgiato Recupero*

CONTENTS

The digitalisation process in the highly fragmented architecture, engineering and construction (AEC) industry is essential for improving productivity and profitability. Digitalisation in this industry has led to the adoption of various technologies, which has resulted in interoperability issues among software being used that need to be solved. Semantic web and linked data technologies provide both tools and methodologies that solve such issues. The Cloud has become the new medium for AEC technologies, where all the computing resources are collected and managed automatically [218]. The shift to a cloud-based medium affects information technology (IT) consumption because the provisioning of on-demand services now occurs over the Internet [112]. This chapter introduces cloud computing and web-based applications and highlights their benefits for the AEC industry. Furthermore, it presents two case studies of tools that could support the AEC industry's transition to web-based solutions. Specifically, the EnergyMatching Platform which is a common entry point for stakeholders interested in renewables for energy harvesting to optimise building and district loads, and the Construction Project Quick View platform, which is a mashup solution to integrate different building model formats and information.

DOI: 10.1201/9781003204381-9

7.1   INTRODUCTION

The AEC industry is currently subjected to technology disruption, technology mainstreaming and convergence of digital technologies across the construction life cycle, driven by the rise in mega-trends such as big data, artificial intelligence, Internet of Things and cloud, as well as immersive solutions such as augmented and virtual reality (AR and VR). The main objective of these innovations is to foster seamless synchronisation between assets, models and stakeholders to support decision-making and increase productivity levels.

Cloud computing environments, which describe a variety of computing concepts, provide Internet-based platforms that are used for computer technologies. Cloud computing is increasingly relied on to store both public and personal information and has become the standard medium for relevant hardware, software and service providers according to the needs of users. Cloud computing is an extension of grid computing, distributed computing and parallel computing. Its main purpose is to provide secure, quick, convenient data storage and net computing which is also service-centred.

This book chapter introduces the cloud computing technology definition and requirements, focusing on the AEC industry, through the definition of the five essential characteristics, three service models and four deployment models. The chapter specifically describes the transition to web-based solutions in the AEC industry and the opportunities resulting from that transition such as tools for document management, process tracking or quality control, which are of help to digitise the industry. This chapter also describes the challenges to overcome challenges for the successful implementation of web services in the built environment due to the transition from single-users to multi-tenants, which requires designing user management, administering the authorisation of multiple users to access resources and execute tasks, establish communication mechanisms between the users and ensure the security and privacy requirements.

Finally, this chapter presents two case studies of tools that actors in the AEC industry can use for their transitions to web-based solutions. The first case study is on the EnergyMatching Platform which is a 'web-enabling by wrapping' solution, which resulted from the EnergyMatching H2020 research project. This case study outlines the governance model that supports the adaptive and adaptable envelope solutions for renewable energy harvesting to optimise building and district loads. This platform supports stakeholders through an optimisation tool that suggests optimal configurations of building-integrated photovoltaic (BIPV) systems and provides appropriate examples of active building skin solutions (e.g. BIPV and solar window block). The second case study is on the Construction Project Quick View Platform, which is a mashup solution that integrates different building model formats and information by using functionalities or data from other web services to present those models and information as a new integrated service.

7.2   CLOUD COMPUTING

The National Institute of Standards and Technology (NIST)[1] defines cloud computing as *a model for enabling ubiquitous, convenient, on-demand network access to a shared pool of configurable computing resources (e.g., networks, servers, storage, applications, and services) that can be rapidly provisioned and released with minimal management effort*

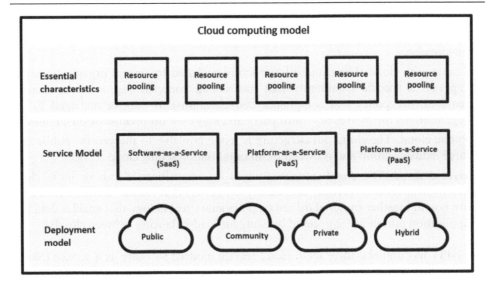

Figure 7.1 Cloud computing.

or service provider interaction. The NIST definition lists five "essential characteristics" of cloud computing: on-demand self-service, broad network access, resource pooling, rapid elasticity or expansion, and measured service (see also Figure 7.1). It also lists three "service models" (software, platform and infrastructure) and four "deployment models" (private, community, public and hybrid) that together categorise ways to deliver cloud services. The definition is intended to serve as a means for broad comparisons of cloud services and deployment strategies, and to provide a baseline for elaborating the could computing definition and its various applications.

A solution must exhibit the following five characteristics to be considered a true cloud solution [330]:

- On-demand self-service: Cloud computing resources can be provisioned without human interaction from the service provider.
- Broad network access: Cloud computing resources are available over the network and can be accessed by diverse customer platforms.
- Resource pooling: Cloud computing resources are designed to support a multi-tenant model. Multi-tenancy allows multiple customers to share the same applications or the same physical infrastructure while retaining privacy and security over their information.
- Rapid elasticity: Cloud computing resources can scale up or down rapidly and, in some cases, automatically, in response to business demands.
- Measured service: Cloud computing resources usage is metered and manufacturing organisations pay accordingly for what they have used.

There are four cloud deployment models: public, private, community and hybrid. Each deployment model is defined according to where the cloud's infrastructure is located.

- Private cloud: The cloud infrastructure is operated solely for an organisation. It may be managed by the organisation itself or a third party and may exist on-premise or off-premise.
- Community cloud: The cloud infrastructure is shared by several organisations and supports a specific community that has shared concerns (e.g. mission, security requirements, policy and compliance considerations). It may be managed by the organisations themselves or a third party and may exist on-premise or off-premise.
- Public cloud: The cloud infrastructure is made available to the general public or a large industry group and is owned by an organisation selling cloud services.
- Hybrid cloud: The cloud infrastructure is a composition of two or more cloud deployment models (private, community or public) that remain unique entities but are bound together by standardised or proprietary technology that enables data and application portability (e.g. cloud bursting for load-balancing between clouds).

NIST has outlined three main cloud service models: software as a service (SaaS), platform as a service (PaaS) and infrastructure as a service (IaaS) [274], but the cloud innovation landscape continues to grow and expand. The SaaS model has been around the longest relative to the other models, dating back to the application service provider (ASP) days. SaaS is a model that enables quick access to cloud-based web applications. The vendor controls the entire computing stack, which can be accessed using a web browser. These applications run on the cloud, and you can use them by a paid licensed subscription or for free with limited access. On the other hand, PaaS is essentially a cloud service model whereby one can develop, test and organise the different applications for a business. Implementing a PaaS simplifies the process of enterprise resource planning software development. The virtual runtime environment provided by PaaS unlocks a favourable space for developing and testing applications. While the IaaS service model is basically a virtual provision of computing resources over the cloud, an IaaS cloud provider offers the entire range of computing infrastructures such as storage, servers and networking hardware alongside maintenance and support.

Each service model has its own benefits and drawbacks and should be chosen based on a comprehensive understanding of the cloud model options, the evaluation of the business requirements and an exploration of how the chosen model can deliver the intended set of workflows and always with careful consideration of the specific use cases and business model. Whether using a cloud provider or not, the maintenance and monitoring of the related systems and applications must always be guaranteed [330].

7.3   WEB-BASED COMPUTING TOOLS IN THE AEC INDUSTRY

The size of the AEC industry is huge. It is estimated to reach US$ 117.59 by 2023. The digitalisation of building infrastructure may save up to 10%–20% of resources involved in the construction workflow, while project time savings could reach 14%. Adopting digital tools increases collaboration, enhances project clarity, reduces the risk for errors and makes construction sites safer[2]. Moreover, the typical work conducted in the AEC industry involves a collaborative approach across different locations such as job sites and offices, and is performed by various stakeholders with disciplines such as owners, architects, contractors and facility managers. On the other hand, user needs evolve rapidly, and the AEC sector is no exception. New demands emerge constantly, such as real-time information sharing, working collaboratively with stakeholders at different phases of a project life cycle, and seamless integration with suppliers or third parties.

That said, cloud adoption is a necessity and not a choice anymore, which has resulted in the evolution from isolated AEC software tools, installed on user devices to web-based solutions where users can work as they would with applications installed locally on a device but incorporating improvements in collaborative working or information availability. Eighty-five percent of construction contractors use or plan to use cloud-based solutions according to a 2017 survey of nearly 1,300 firms performed by Associated General Contractors of America in collaboration with Sage Construction and Real Estate, a leading purveyor of cloud-based solutions. By comparison, a 2012 Sage survey found that only 16% of construction contractors believed that cloud computing was important to their business[3].

The AEC industry must manage increasingly complex data and systems. As traditional data centres are becoming insufficient for the tasks of today's age, more AEC firms have been turning to cloud-based storage solutions. According to a survey by Microdesk conducted with 15,000 AEC companies [261], around two-thirds of AEC firms store data in the cloud. Small and large firms tend to store a high percentage of their data in the cloud, with an average of 51.5% and 55%, respectively. Medium-sized AEC firms are struggling to find cloud solutions that suit their budgetary and technical requirements, storing a much more modest percentage of their data in the cloud, around 17.6% of their data in the cloud on average, which shows that there are significant gaps between how large enterprises, and small- or medium-sized businesses manage digitalisation.

Key AEC vendors (e.g. Autodesk, AVEVA, Bentley Systems, Nemetschek and Trimble) offer different cloud solutions. On the other hand, due to their flexibility and ability to adapt to disruptive technological changes, a large number of small- or medium-sized enterprises (SMEs) are early adopters of innovative technologies such as cloud solutions. SMEs may also fulfil complementary roles for large-sized companies by operating their extensive network systems [271].

As a result, the usage of cloud-based technologies is driven by the need to adopt mobile software on job sites, to perform the following tasks: creating daily field reports (44% of respondents), accessing customer and job information (40%), tracking employee time and approval (40%) and sharing resources such as drawings, photos and documents (38%) [261].

7.3.1 Opportunities

It took a long time for web services to emerge as the new standard for software architectures. Since then, new software applications have been consistently developed, by leveraging data or functionalities of existing applications through the web [373]. In the AEC industry, while it has become increasingly common to see web services, they still rely heavily on totally disconnected desktop-based tools.

While cloud opportunities are certainly limitless, we will concentrate on *four main opportunities* where the transition of the AEC industry to web-based solutions is having a major impact. The first opportunity is the transition to **beyond paper-based solutions**. It may seem that the paper-based solutions phase has long since been abandoned, but the construction sector is still among the least digitized industries [23,225]. Although BIM is now mainstream, the printing, management, usage and storage of paper documents is still a common practice. Web tools for document management, process tracking or quality control are some of the areas where web-based tools are emerging.

New web-based tools also allow **seamless integration between businesses (B2B) or between businesses and consumers (B2C)**. A B2B platform allows the improvement of communication with suppliers and enables the automation of supply provisioning using real-time interaction. On the other hand, having B2C support is beneficial for the image of the institution and offers customers a sense of support and availability. In [171], Dandan He et al. designed an e-commerce platform for industrialised construction procurement supporting both B2B and B2C and considered linked data as an effective means to integrate information.

Speaking of integration, the incorporation of external data or the **use of open data** is facilitating more efficient management and business intelligence, which leads to effective ingesting, storing, organising and maintaining the data created and collected by an organisation. Furthermore, it provides analytical information to help drive operational decision-making and strategic planning by corporate executives, business managers and other end users.

For example, the integration of climate data, for efficient resource management, or event notifications to improve planning and decision-making, are attractive opportunities for web-based applications in the construction sector. The construction data auto-correction system [427], developed by using an open data database made public by the government of Taiwan, illustrates this idea. Other actions like the integration of weather open data and sensor data from construction sites [220] supports better scheduling of workers and resources and helps to improve the safety by reducing worker exposure to extreme weather or hazardous environmental conditions. Another good example is the Augmented-Urban Building Energy Model (BEM) combining open data and simulations to produce synthetic load curves for every building in a city and improving the energy planning of sustainable cities (see also Chapter 3).

Finally, there is an opportunity surrounding the efficient management of knowledge. **Knowledge management** is a topic that has long been among the most important aspects of project and organisational management. In the case of the construction sector, until recently not much progress has been made on this matter, and therefore it is both an opportunity and a requirement. Robinson et al. [325] mention a series of benefits of knowledge management in the construction sector, among them: the retention of the tacit knowledge of key employees, the promotion of continuous improvement, the ability to respond to customers quickly and the need to reduce workload.

7.3.2   Challenges

Of course, these four opportunities in Section 7.3.1 also come with challenges to overcome for accommodating the successful implementation of web services in the AEC industry. As in the opportunities section we will focus on four of the extensive collections of challenges arising from the transition to web technologies for the construction sector within the AEC industry.

The first challenge comes from the **desktop delivery model**, where applications are usually single-user. Therefore, the transformation of applications to becoming web-based requires designing user management, administering the authorisation of users to access resources and execute tasks and establishing communication mechanisms between users (see also Chapter 6). Although security is important in all software applications, the security and privacy requirements in single-user isolated applications and web-based multi-tenant applications differ and have to be taken into consideration. In [25], Basu et al. provide a comprehensive overview of the challenges related to the deploying of solutions in the cloud, considering the different service models (SaaS, PaaS, IaaS) and deployment models (private cloud, public cloud, community cloud and hybrid cloud).

The second challenge is the **re-engineering** of legacy desktop applications and the reuse of services. Beaty et al. [27] mention that moving to the cloud both for server and desktop applications requires good transformation planning to be able to determine resource requirements and the best methods to allocate these resources to the physical hardware which hosts them. Managing computer resources and scaling web applications to meet service level agreements (SLAs) is one of the main challenges when trying to transition to the web, and elasticity is one of the main reasons for attracting web application developers to move their applications to the cloud [312]. Achieving a good balance between cloud infrastructure costs and SLAs, and finding the optimal way to auto-scale the cloud-based web applications is an active research topic.

The already mentioned Qu et al. [312] considers that auto-scaling can be abstracted as a MAPE (monitor-analyse-plan-execute) loop [76] and present a taxonomy focused on aspects like application architecture, adaptivity needs, resource estimation, as well as scaling timing and methods (among others). With a similar objective, M.S. Aslanpour et al. [14] also present a survey and introduce their own auto-scaling method based on the MAPE loop. In this method they target improvements to the last phase of the loop (execute) with a cost-aware approach, obtaining a 7% improvement in infrastructure costs while achieving a 5% reduction of the response times and still reducing the potential for SLA violations.

Speaking of resources, sometimes it is required to transform **simulators or models** to a web-based platform. In some cases, these require high-performance machines or even the use of system software that must be configured and adapted for web use. This can be complex and requires an appropriate design. Christoph Fehling in his book on Cloud Computing Patterns [129] presents several patterns to solve the complex task of creating and migrating to cloud applications. In particular, he considers that certain applications such as those that handle public data or provide information to public users and have been developed based on distributed architectures can be migrated easily, but as a general rule it is necessary to study the requirements in detail and to plan how to integrate and deploy them. Jamshidi et al. [188] also focus on studying the migration of applications to web and cloud environments, and identify the lack of a framework and tools that automatically facilitate cloud migration tasks. Zhao et al. [436] have also investigated this issue in detail, and agree that the ideal strategy depends on the needs of an organisation and the application or legacy system, thus organisations need to choose the most rational strategy before the migration. In general, they believe that cloud computing and service-oriented architecture (SOA) pattern designs are complementary and can be a useful foundation for re-engineering and migrating applications.

Finally, it is important to provide a **data exchange** system, which can be challenging in the AEC industry (see also Chapter 1). That can be done by (1) transferring the files/documents (e.g. information containers in Chapter 6), (2) sharing a storage or data repository (e.g. Pod in Chapter 6) or (3) passing information through messages or application programming interfaces (APIs). Typically these APIs are used to bridge applications and can come in various formats such as web services, remote calls, message passing or even through application-dependent protocols [280].

7.4   USE CASES AND SCENARIOS

In this section, we present two case studies of possibilities for how AEC industry actors can transition to web-based applications, and in turn show how the industry is approaching digitalisation. In some instances, the transformation simply involves adapting the source code and making it usable through a web interface, deployed to a server, but this is not always possible, so it may be necessary to use other alternatives.

For example, projects like CelLEST [372] investigated options such as reverse engineering of a service's executable specification of the legacy application, so it could be accessed through other kinds of interfaces. In certain cases, it is possible to access the AEC legacy application through a command line interface (CLI), interact directly with the business logic, or even include data or functionality from third-party services to generate a new service.

The following two subsections demonstrate two scenarios of transitions being carried out, one through the web-enabling of an application that can receive commands by a CLI and the other through web mashups.

7.4.1   Web-enabling by wrapping

In some circumstances, one can opt for the conversion of a standalone tool to a cloud version. This conversion is not usually trivial but in many cases, it is possible by developing wrappers that expose the functions of the application for use from a web platform, and adapting the backend to perform the operations efficiently.

The EnergyMatching project [247], funded by the European Union under the Horizon 2020 program, is dedicated to the study of "Adaptive and adaptable envelope Renewable Energy solutions for energy harvesting to optimize building and district load" and one of its objectives is the development of the EnergyMatching platform, a public web-based knowledge transfer platform, where architects, engineers and building operators can:

1. find by means of optimisations and simulations the optimal capacity and positions of their photovoltaic system and the optimal storage capacity to maximise the techno-economic performance of the system;
2. share the outcome of the simulations with stakeholders, including their inputs, outputs and even 3D models; and
3. design the solar window block setup by searching and filtering different system configurations obtained from the dynamic simulations to allow the users to customise the system to their needs and understand the configuration performance.

7.4.1.1   *Transforming legacy data and code*

The part that deals with window blocks (part 3) consists of a series of Microsoft Excel files that include the results of dynamic simulations carried out using different system configurations as their source. Making the window block part available through web access was straightforward, requiring only the processing, cleaning, transformation and storage of these data in a cloud database. There are various methods for how to integrate and unify spreadsheets and databases [33,50]. In the case of EnergyMatching being simulation results, linking configuration parameters to a collection of resulting variables, they could be converted to a large CSV file. Using the Python's Openpyxl library[4], EnergyMatching partners iterated using multiple sheets and rows and processed each one of them to be able to insert the information related to the configuration and simulations in the database.

It is worth mentioning that if the spreadsheet can be converted to comma separated values (CSV) file and one wants to keep in the database the same layout as the spreadsheet, the data entry functions of the database can be used. For example, the COPY function of PostgresSQL or the LOAD function of MariaDB allow the loading of CSV files in a simple and immediate way. Once the data is properly stored in the cloud database, a web visualisation and filtering module then allows visitors to choose different building and construction related parameters (e.g. room type, location, orientation) and visualise the results of thermal performance, energy production-consumption matching and daylighting, of three different configurations: BIPV sill, BIPV overhang and BIPV vertical.

On the other hand, the transformation of the part with the simulation of the optimal capacity and position of the photovoltaic system (part 1) was considerably more complex. The optimisation application is desktop software that is composed of several applications resulting from research [231,232]. It mixes code in Python with the use of tools for performing lighting simulation, such as Radiance [408]. In many cases, it is not possible to modify these desktop applications, either because of licensing or intellectual property rights (IPR) issues, affecting the difficulty of modification or compatibility of legacy software, or simply because the software is too complex to be modified. In these cases, a wrapper-based solution may be of interest, as it was in the case of the EnergyMatching Platform.

7.4.1.2   Virtualisation

Hypervisors (or virtual machine monitor, VMM, virtualiser) is a kind of emulator; it is computer software, firmware or hardware that creates and runs virtual machines. Hypervisors and containers are making it easier to migrate legacy applications to the cloud. Hypervisors are ideal when applications require different operating systems (OS), containers are more lightweight and share operating systems and libraries [36], making them interesting when applications share operating system requirements and when a smaller footprint is needed. Migrating legacy applications to a cloud using hypervisor or containers has been an active field of research and engineering in recent years.

In [363], the problem of migrating a legacy web application to cloud is addressed by using a Docker container, by wrapping *existing single-tenant applications and running them as multi-tenant applications*. In [239], Maenhaut et al. also investigate the issues of the multi-tenancy of applications and the migration of legacy software to the cloud, providing their approach in two different migrations of software to their cloud. As a conclusion, they point out that *migrating legacy software to a cloud comes at a cost, and some application components may need to be modified or rewritten* but the use of containers and an efficient migration approach can reduce those costs and the code to be modified significantly. Whether with hypervisors or containers, applications remain isolated hence a light wrapper or API can be developed to interact with the legacy application.

7.4.1.3   Containers and wrapping in the EnergyMatching platform

In the EnergyMatching case, the partners relied on containers and, as mentioned, a wrapper-based solution was chosen to coordinate the execution of the simulations. Broadly speaking we can break down the execution of an application into input, logic and output. If we break the execution of optimisations into three parts: input, simulation and output, we can detail how EnergyMatching addressed each phase.

7.4.1.3.1 Wrapping the input   As far as the *input* is concerned, the original desktop application uses a combination of inputs from the CLI, with files located in specific paths and Excel files with the input data. In the web version, a form is provided where the user can upload the necessary files and indicate the input data, for which examples are given and helps to divide the data into either optional or mandatory. To facilitate some of the input files, some of them can be generated using the data obtained from the form, so they have not requested them from the user, and for other mechanisms are provided to retrieve the data. For example, in the case of the EPW files (EnergyPlus[89] Weather file format) needed for the climate information, the web version offers an embedded map where the user can search or select a location and the backend makes use of third-party services to obtain the most recent EPW file with the closest location.

The objective is twofold, on the one hand, to simplify the use of the tool and on the other hand to reduce the potential problems of data entry by the user. The original application was intended for use in research environments and was complex to parameterise, therefore it was decided to reduce the number of interactions required and to obtain the missing data from other online sources. As a general rule, it is necessary to plan the design of the user interface that the new application will have, since many of the assumptions that are made in desktop applications are not fully valid on web platforms, especially if the option of making the web application accessible from mobile devices is considered.

7.4.1.3.2 Wrapping the simulation   The data entry wrapping part was less difficult than wrapping the simulation. The main problem was that the application was resource-intensive due to the fact that it had to process 3D models and perform different calculations with Radiance and other programmes. Such resource consumption is not very important in single-user desktop systems but can be troublesome when making it available through the web. In addition, a simple simulation does not produce real-time results but requires several minutes of processing. For that reason, an asynchronous task queue system, a mechanism to handle time- and resource-consuming tasks outside the main web application, was used to limit the number of parallel executions and queue them for further processing. In order to keep the user informed, it was decided to send emails when the simulation was ready, including a self-generated report in PDF format.

In general, the use of containers allows for application scaling and the dynamic allocation of resources. Even so, in some cases, as in the EnergyMatching Platform, the time needed to perform a simulation goes beyond the two seconds suggested by Google representatives in the widely used phrase *Two seconds is the threshold for e-commerce website acceptability. At Google, we aim for under a half-second.* In these cases, scaling may not be the only solution and it is necessary to design additional asynchronous mechanisms to: (1) not exhaust computing resources, (2) not keep the user waiting when the response will not be immediate and (3) allow the user to continue performing other tasks on the web site in question.

7.4.1.3.3 Wrapping the output   The data output also needed some adaptation. The simulation software generated CSV files, some graphics in PNG format and several 3D models in OBJ format. This information needed to be displayed through the browser and included in the PDF reports. For that purpose, the data was processed and a template-based system was used to generate the web and PDF reports. In particular the Jinja[5] template engine for Python was used for this purpose, since it is well integrated with the Python Framework we used, enables sandboxed execution, and is easy to use from the user interface (UI) designer perspective. In order to provide more visual information and an ad-hoc developed 3D viewer, Three.js[6] JavaScript library was also included so that users could view the OBJ files without having to leave the platform.

In general, the design of the data output must follow some basic design and compatibility rules. On the one hand, from the point of view of the device from which the content will be accessed, offering responsive and mobile-friendly content if accessed through mobile devices is expected [203]. On the other hand, from the user experience (UX) and human–computer interaction (HCI) perspectives, it is important to optimise how information is displayed while reducing potential issues such as cognitive load [206].

7.4.1.4 Platform architecture

In Figure 7.2, all the components and elements of the EnergyMatching Platform architecture are presented, showing in the red box (BIPV Optimisation Tool) the original desktop application. As described above, EnergyMatching partners containerised the solution, developed a wrapper for the tool and enabled a job scheduler and an asynchronous job queue software, making the platform not only scalable but also avoiding problems related to the resource consumption of the original tool.

For the input, they did some pre-processing, simplification and adaptation of input data, which allows users to make use of external tools to retrieve missing data. In order to simplify the visualisation of the output data, they relied on templates both for web outputs and PDF reports, and provided some useful modules for viewing the 3D models, filtering information and for getting information presented superimposed on top of maps. With all these changes, the desktop software became a web-based solution that allows stakeholders to perform optimisations, share information and visualise data in 3D, all from their web browser on the user's PC or mobile device.

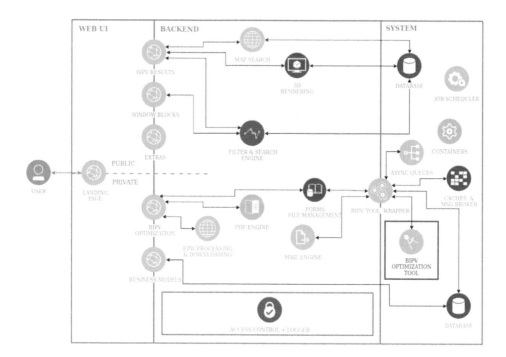

Figure 7.2 EnergyMatching platform architecture and components.

7.4.2   Mashups

A second commonly used solution to transition software to a web-based solution is to rely on mashups. Mashups are essentially web services that use functionality or data from other web services and present those resources as a new integrated service. As Yee [423] mentions, *"how many times have we seen a website and thought, it would be great if we could get this and that in addition to what it already provides."* That is exactly what mashups are all about, complementing different sources of information to generate a solution that meets the user's expectations.

For example, imagine a user searching for buildings currently under construction. This user might be interested in having information about the progress of construction displayed on a map while also having information about nearby transportation and services displayed on the same map. DiFranzo et al. [102] detail a similar application making use of linked open government data and displaying transport or crime information on maps. Or she might be interested even to integrate information on construction, resources and weather in order to plan and schedule the work according to local climatic data [290]. And in addition, she might want to use methods to calculate the acceptable delay duration caused by weather conditions based on weather information available on the web [385]. In such cases of making extensions and additions, a mashup approach is useful.

7.4.2.1   Construction project quick view platform

The following use case can be shown as another example of a web application that uses the mashups approach. This example, unlike the previous case, does not intend to demonstrate the steps, as they depend on the services to be integrated in the mashup, but rather to show the possibilities offered by the integration of services. In this case, an integration was needed of navigable Matterport models, BIM360 documents and maps in a single online platform, which was called the "Construction Project Quick View" platform (see Figure 7.3). The project manager wanted to monitor hundreds of construction projects in different locations throughout Europe and needed a solution that, without going into all the details, would allow him to see all the projects on a single map at a glance, and to obtain preliminary project information through a profile of each project. This profile, in addition to the preliminary information, also had to show the navigable map generated by Matterport while still allowing access to all the project information stored in BIM360.

To meet the requirement and provide information on hundreds of projects, a dashboard was designed in which, in addition to information on the status of the projects, and the possibility of managing the project information, a map showed the position of the projects. The map was based on the OpenLayers API[7], which in turn uses elements from other web-based map services (such as OpenStreetMap or Bing Maps). From each of the markers on the map, it was possible to access the profile of individual projects. The profile supports updating project data and is the one that includes the integration between Matterport and BIM360.

Matterport manufactures hardware and software that allows users to create immersive 3D digital models of spaces. BIM360 is a platform for document management and real-time work, in which teams can collaborate and share data. Both solutions offer APIs that allow the integration of their functionalities in external tools. In the case of Matterport, the interest was that the "Construction Project Quick View" user could have in each profile a 3D/2D viewer of the construction spaces, which was navigable and allowed measurements. Matterport's API and software development kit (SDK) allows

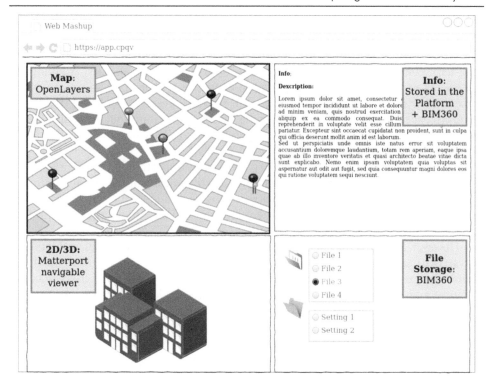

Figure 7.3 Construction project quick view Mashup.

for this integration, so in this way it was possible to obtain more detailed profiles of each project.

In addition, the idea of the "Construction Project Quick View" Mashup was that from this entry point users could access and manage the various displayed construction project documents. These documents were stored in BIM360, which is a cloud tool providing an API to access the plans, models or any other document of the displayed projects. Therefore, through the integration of several APIs and cloud services, a rapid briefing system could be created, providing managers with a quick view of what was going on in the projects, while at the same time allowing them to access information when needed.

7.4.2.2   Extensibility

As an extra advantage, with mashups there is also the possibility to further extend this quick view solution with additional web services. For example, it could be interesting to include an external IFC viewer to display the models, or to use the conversion and visualisation of BIM360 models, to convert and display them. A large number of cloud applications for the construction sector provide APIs that allow the development of new web platforms to make use of them and also integrate with other information such as calendars, weather information, open data or any other relevant information for construction activities. Therefore, designing a mashup merely entails defining what you want to achieve and then researching the information and interfaces available.

7.5 CONCLUSION

The technology offered by cloud solutions is essential to enhancing day-to-day and long-term company processes and protocols for AEC firms of all sizes. Additionally, the pricing of cloud solutions is expected to decline over time [330], so we can expect these solutions to become more easily accessible in the years to come. As the complexity of tasks expected of AEC firms increases as they confront today's challenges of globalisation and urbanisation, cloud solutions are becoming paramount to maximising workflow, costs and sustainability.

The digitalisation of the AEC industry offers many opportunities resulting from that transition such as the adoption of emerging tools for document management, process tracking or quality control, which will help in digitising the industry, enable seamless B2B and B2C integration, and facilitate efficient integration and data management through the use of open data.

However, there are also challenges to overcome for the successful implementation of web services, such as the transition from a single-user to a multi-tenant model, which will require designing user management, administering the authorisation of users to access resources and execute tasks, and even establishing communication mechanisms between those users, while maintaining the required security and privacy requirements. Another challenge is the re-engineering of legacy desktop applications and the reuse of services, which requires good planning to determine resource requirements and the best methods to allocate these to the physical hardware which hosts them.

The presented case studies, the EnergyMatching and Construction Project Quick View Platforms, which are based on web-enabling by wrapping and mashup solutions respectively, highlight the variety of web-based technologies and the resulting benefits of efficient information management, data sharing, accessibility, the support of decision-making processes and the seamless involvement of multiple stakeholders.

NOTES

1 https://www.nist.gov/.
2 https://www.microdesk.com/articles/cloud-usage-in-aec-industry/.
3 https://builtworlds.com/news/cloud-adoptionis-rising-in-the-aec-but-what-are-we-missing-sage-weighs-in/.
4 https://openpyxl.readthedocs.io - A Python library to read/write Excel 2010 xlsx/xlsm files.
5 https://jinja2docs.readthedocs.io/en/stable/
6 https://threejs.org/ - Three.js JavaScript 3D library.
7 https://openlayers.org/

Chapter 8

Digital twins for the built environment

Calin Boje, Sylvain Kubicki, Annie Guerriero, Yacine Rezgui, and Alain Zarli

CONTENTS

The increasing availability of data and new ways to leverage it have pushed engineering domains into investigating a Digital Twin (DT) paradigm, which assumes full integration and cohesion between the physical and the virtual worlds. Although the prospects of digital twinning have been gradually explored within the fields of cyber-physical systems (CPS) and various BIM uses, the next level of integration needs to consider available methods and tools (procedures and technologies) which would give a digital twin more cohesion over the managed information, more adaptability and bring more value from our virtual models. In this chapter, we introduce the recent relevant aspects related to digital twin in research, along with existing initiatives and potential future paths. These are presented through the lenses of various application domains within the built environment by looking at digital twin requirements, technologies, architectures, modularity of services, and the role of semantics in achieving communication between different entities. Although the future of the digital twin paradigm remains uncertain, the posibility to monitor, simulate, optimise and then consequently automate and actuate the real world at various levels remains one of the most valuable prospects for our built environment, thus ensuring lower negative effects on the environment in the long run.

DOI: 10.1201/9781003204381-10

8.1   INTRODUCTION

Since the earliest efforts to design and construct, human imagination and judgement supported by mental models were virtually the only tools to help build artificial habitats and other inventions. In ancient times, this culminated with written manuscripts of the Roman military engineer and architect Marcus Vitruvius Pollio. The treaty titled "De architectura" was considered an abundance of knowledge on architecture and civil engineering works for centuries, describing sophisticated methods for construction on paper. From antiquity to modern day, paper remains an important medium of knowledge storage and exchange, but fails to meet our new requirements for a knowledge-intensive digital era. Our written knowledge and mental models of old have become computer-interpretable and computational, numeric at their core. The term "digital" itself in English refers to representing something using numbers (more specifically digits), related to the modern English word "digit", a number from 0 to 9, rooted in the Latin word "digitus", which our ancient architect Marcus Vitruvius would have understood differently - meaning "finger" or "toe" in Latin, having changed its meaning with time in to "count" or "to number".

Today however, we use computers to count for us at a far greater scale, precision and speed. Digitalisation efforts commenced with the adoption of Computer-Aided Design (CAD) in several engineering domains. It then continued with significant standardisation efforts under the umbrella of Building Information Modelling/Management (BIM), defining common semantics under the Industry Foundation Classes (IFC). This has made modelling, simulating and exchanging information between computer systems much more convenient. Parallel to these efforts, as our built environment grew in size, became more sophisticated and adopted complex manufactured materials, the need for specialised digital tools to design and manage it across the life cycle became evident. Consequently, this has led to an explosion of new models, tools and technologies to help us better design and construct, while understanding the impacts and preserving the natural and built environments.

With an abundance of information on the web and the latest sensing technologies, the next step of digitalisation is now faced with a different challenge - the integration of our virtual space with the physical construction assets on the ground but also with the dynamic social systems that define our communities from building to city levels. Driven by these recent changes, the paradigm of a "digital twin" of constructed assets, and by extension our environment, has become increasingly relevant in academia and several industries which herald it as a way to tackle economic, technological and environmental challenges of our status quo. Within this chapter, we will gradually begin to understand what is the meaning of a Digital Twin (DT), its requirements and expected benefits to our domain, and how semantics plays a key role in its practical application.

8.1.1   The digital twin concept

Although the concept of a "digital twin" may not be something fundamentally novel or revolutionary, the term is actually experiencing a phase of "hype" and resurgence. Recent studies by several authors [140,265,270,381] have already compiled various lists of definitions on what a digital twin is from academia and industry, but we will try to summarise the fundamental concepts here. The mentioned studies have traced the origins of the term to Michael Grieves' 2003 course on Product Lifecycle Management (PLM), which he later published and explained in a white paper [154]. From the academic perspective, the first usage of this term was published in 2011 within the field of aerospace engineering [389], where the authors define it as a "reengineering the structural life prediction process" of an aircraft. The article begins with a thought process of imagining that by 2025, we would be able to have high fidelity finite element models (FEM) and deliver these at the same time with our physical aircraft. The prediction is very close to becoming true, and this has been happening to some degree in many engineering fields, not least with our construction assets where new buildings are being delivered with their BIM models on an almost daily basis. However close the similarities and blurred the boundaries may seem, a BIM and a Digital Twin are different in concept and in function. They are neither the same nor do they exclude each other.

Let's begin by looking at the etymology of the term "digital twin" by examining the meaning of each of the two words. We have previously established that "digital" in our modern context refers to a "numerical representation" of something, which is processed by modern day computers. The second word "twin" comes from Old English "twinn", a cognate to the English word "two", meaning "double" or "two of a kind". Thus, the meaning of the term "digital twin" implies the existence of two entities, one digital and the other physical. This is referred to as the Physical Twin (PT), the thing that exists in reality. The DT by contrast is virtual and can only exist in the numerical representations of our computer models. So far, one would not be wrong in assuming that a BIM is in essence a DT, as it is a numerical representation of a construction asset in reality. However, this would be false as the key difference is the capability to interconnect the twins and keep them in a state of synchronisation whereby if one changes state, so should the other. Within the white paper of Michael Grieves, a first attempt to define a structure for the DT paradigm is made, resulting in three fundamental parts: (1) the physical, (2) the virtual and (3) the data connection [154]. It is important to distinguish between the "digital twin paradigm" – the idea of the process itself, where the three above mentioned components merge, and the "digital twin", which is the mirror model of the "physical twin". In the last one, the "digital twin" consists in essence of the (2) virtual and the (3) data connection. Most often we will refer to a "digital twin" as this representation of its physical counterpart.

Starting from these three fundamental concepts, we can begin to attribute various features, or abilities to the DT paradigm, based on the technologies which make it possible in practice. At the physical level, the ability to sense (observe) is vital; at the virtual level, the ability to represent and interpret the sensed information is key to be able to exploit the data for decision-making and eventual actuation back on the physical. In between these two states, data is collected, transformed and processed in various ways, depending on the application domains. We already have a good idea of the first two components, as we encounter them in everyday life, but it is more difficult to grasp and understand the technicalities of the third. The data connection is envisaged to be a loop (as shown in Figure 8.1), where data is constantly transformed and exchanged between the DT and the PT.

Starting from the fundamental paradigm, Tao et al. [380] proposed an extended five-dimensional model for the DT concept: the physical (1), the virtual - also referred to as models (2), the data (3), the connections (4) and services (5). Models refer to the ability of the DT to represent the physical twin in the virtual world, ranging from properties, behaviours, simulations, etc. Data represents the data gathered from both the physical (real-time capture) and virtual (simulated). Connections refer to the mapping and live links between concepts, which ensure the synchronicity of the entire system. Services are specialised applications where a DT is applied in a very specific context and provides a

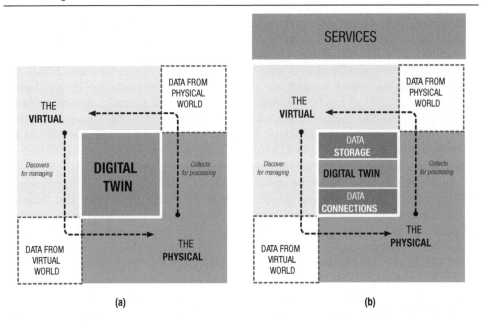

Figure 8.1 The fundamental digital twin paradigm (a) vs the further developed digital twin model (b).

service to its end users. These components will become clearer in the next sections of this chapter, where they are further explained and contextualised.

8.1.2   Related concepts in research

We have previously established that BIM and DT are closely related concepts within the field of construction, but there are several other engineering research domains that are very closely related and are the forerunners of the DT paradigm, most notably: Lifecycle Management (LM), Cyber-Physical Systems (CPS), Industry 4.0 and the Internet of Things (IoT).

The first mentioned, lifecycle management deals with ways to manage an asset from cradle to grave. Within the built environment, the life cycles of our built assets are relatively long compared to manufactured products, ranging from several decades to centuries in the case of heritage buildings. The life cycle of built assets within the construction sector is usually divided into several stages: conceptual design, detailed design, construction, operation, deconstruction and demolition. Within this time, a built asset can be monitored, repaired or renovated using different methods and technologies.

The field of CPS essentially deals with observing the physical world using sensing network infrastructures to send captured data over to the cyber world (virtual world), which is responsible for processing the data, and taking decisions on the next actions of the physical through actuation. This coincides with a DT in almost every way, which can cause confusion. Tao et al. [381] argued that CPS is more conceptual and targeted at researching the possible mechanisms for fusing the virtual and physical worlds, whereas a DT is a more practical application of CPS, directed towards real-world problems. In contrast to this, an editorial from Batty [26] and a response on it from Tomko [386] suggest that the term "digital twin" itself implies an idealisation of perfectly mirroring a built asset, and it may fall short in terms of what the industry and academic communities expect. Nevertheless, a DT is considered a CPS which attempts to represent an entity across its entire life cycle, within the context of increased needs for an interconnected society.

This brings us to the field of IoT, which has experienced an immense success in the last few decades, as it is discussed in Chapter 10 of this book. IoT deals with representing and connecting devices over the Internet and borders the field of CPS where sensors are regarded as "things" which can be connected to and exploited for leveraging big data. The recent success and proliferation of IoT devices has made sensors more affordable and ways to connect and exchange data over the web more convenient, paving the way towards DT implementation [140,236,270].

The paradigm of Industry 4.0 is in a similar place, whereby the development of industry becomes more interconnected through the use of IoT [140], where the digital twin paradigm is seen as an enabler for Industry 4.0 [379]. The specific aspects of Industry 4.0 within the built environment have been recently published in another interesting book [341].

8.1.3 Landscape of nearby engineering domains

There is a series of common uses across various engineering domains that are beginning to explore and adopt a DT paradigm. Since the initial concepts within the aerospace sector for aircraft monitoring [389], this has proliferated to the automotive sector with similar applications, and within the manufacturing sector where products are simulated at very high levels of detail in order for them to be improved in an iterative manner. Other sectors are also exploring the potential of DT, from medicine to agriculture crops monitoring, nuclear power plant structural integrity [246], wind farm monitoring [291], etc.

Within the built environment, we have to consider construction engineering at building and infrastructure levels, where BIM plays a key role but also at a city and districts levels, where BIM is replaced by GIS as a means of information management. As such, the applications of DT within the built environment are quite broad in application and varied in terms of scale. This is in contrast to the aircraft or manufacturing industries, which tend to apply DT to smaller-scale products in relatively controlled environments. Additionally, it is important to distinguish between the motivations which drive DT implementations across the different industries. From their inception, the first DTs were commissioned for economic gains and for iterative improvement of products; this implicitly increased product quality and life expectancy. In contrast to this, within the built environment, the DT implementation is still pushed by economic gains (although branded as savings), but even more so by environmental concerns as the built environment is a main producer of carbon emissions worldwide, a major contributor to land pollution and waste generation. Additionally, our buildings cannot be easily tested and controlled in factory conditions to improve them, but this is instead a longer and more arduous process, whereby buildings can be re-fitted and adapted several times across the life cycle to achieve optimal construction methods, optimal energy use and eventually more conscious recycling and disposal of construction materials.

Several recent studies have already identified the challenges and potential paths to DT implementations within the built environment at building level [68,201], construction site level [41], construction logistics level [153] and city level [135,356]. The motivations behind a DT approach were listed above, with the built environment stakeholders attempting to find common ground on how to proceed in research and practice.

8.2 REQUIREMENTS, TECHNOLOGIES AND ABILITIES

A first step in implementing DT in practice is to identify the requirements from multiple perspectives. The published work of Gharaei et al. [147] attempts to do so from a systems engineering and software development perspective, while articles from Moyne et al. [265] and Shao et al. [350] can offer insights within the manufacturing sector. Although all these are very valid perspectives, we will distinguish the built environment requirements for DT development into two main categories: (1) procedural and (2) technical. The system architecture aspects are discussed in section 8.5 of this chapter in more detail.

8.2.1 Digital twin requirements

Following the recommendations from the literature mentioned above, the requirements for implementing a DT should be rooted in what a DT stands for conceptually – a dynamic environment which synchronises data from real and virtual worlds [350], but it must also consider economic feasibility and the context of each use case [147]. It may not make economic sense to adopt a complete, high fidelity, high-resolution DT, when the nearest acceptable approximation would yield similar outcomes.

Let us first consider the different scales of built environment: we will refer to the building level as the "microscale", the construction site and district levels as "mesoscale", and the city and large infrastructure levels as the "macroscale".

On a procedural level we can begin by asking the question "Why and how should a digital twin be developed for built assets?" in order to begin to understand the process and rationale behind it. For the time being, there is a gap in understanding how DT can be deployed in practice on physical sites at different scales, where its digital components should reside (physical data storage and access) and what precisely are its envisaged uses on and off site. The deployment and management of sensors on the microscale has seen extensive research from smart buildings (see Chapters 3 and 11), and as a result, it has become common practice. Challenges still remain on adapting the sensing to specific use-cases, such as human health monitoring, human behaviour or fire safety, etc.

On the mesoscale, challenges in research and development regarding construction digital twins have been described by Boje et al. [41], also proposing a conceptual framework for its adoption. Methods on instantiating CPS using BIM during construction published by Akanmu et al. [5] is a good starting point which can be easily transposed to the digital twin paradigm. What makes construction sites particular is that the sensing equipment itself is more dynamic than buildings or districts of buildings. The devices used to track the construction site rely heavily on hybrid laser scanning and photogrammetry methods (a fusion of photo/video feeds with point clouds). The challenge lies in deciding on the required sensing on-site, and what is the optimal combination of stationary sensor networks with more dynamic scanning methods (mounted on tripods, drones, etc.). This is expected to vary depending on the construction site (location, size, site availability during the day), the construction schedule and the required level of monitoring quality. A construction site does not live in isolation however, having significant impacts within its proximity in terms of noise, pollution and traffic levels. As such, monitoring health effects of the construction workers and nearby local communities needs to be considered for a more sustainable agenda. In terms of digital twins for districts or groups of buildings, the use case can be narrowed down to the infrastructures which connect them, such as energy [305] and water grids [180], mobility of vehicles and pedestrians and other applications bordering city-scale use cases.

At the macroscale, the prospect of a city or nation-wide digital twin has been discussed for some time, but it may resemble something quite different compared to BIM-based applications. At this scale of magnitude, the level of sensing and its synchronisation pose different technical challenges related to the placement of networks of sensors at greater physical distances, able to capture relatively small parts of information at strategic locations. These need to be in constant communication with sensing of different micro- and mesoscale entities, where a building or site becomes an observer of the events and changes at lower levels.

On a technical level, the question asked is "What and how should existing technologies be used for a functioning digital twin?". The challenges to address here are related to storing and managing the data from sites, its integration with IoT devices, interpretation and subsequent transformations within the virtual-physical loop, which requires a holistic and comprehensive level of cohesion. From this perspective, we can distinguish between different types of needs for our DT, as shown in Table 8.1.

From the built environment there is a bias to associate DT with BIM, and this is perfectly acceptable, as the BIM is the digital representation of a building from design to construction, and in many cases is now being included beyond these traditional stages. However, this poses serious challenges in terms of who deploys the DT in the first place, where is it stored (physically or virtually), who can access it (ownership, security) and in what formats is it kept in across its life cycle (interoperability). When superimposing BIM on the DT paradigm, BIM is part of the models which can define most of our existing built assets, however it is facing challenges in terms of connectivity with the web of things and IoT, and interoperability and extensibility of modelling concepts for different use cases. The same applies on the mesoscale and macroscales with GIS facing similar problems today.

The DT of a built asset will most likely begin with a BIM model. As virtual models start appearing from the designers and constructors, their format and physical hosting

Table 8.1 A list of technical requirements for developing a DT within the built environment

No	Requirement	Description
1	Connectivity	A digital twin needs to be able to connect with physical and virtual resources over the internet.
2	Interoperability	A digital twin needs to speak a universal language which is interoperable with other systems.
3	Extensibility	A digital twin should try to be extensible to new use cases and new additions of data sets.
4	Modularity	A digital twin should be modular from an architecture stand-point.
5	Re-usability (of resources)	Digital twin services and captured data should be easily re-used for new applications or use-cases.
6	Maintainability	A digital twin should be planned for relatively easy maintenance as it develops and changes.
7	Scalability	A digital twin should be scalable to be able to deal with different contexts and resolutions of sensed data.
8	Security	A digital twin must implement the latest proven methods to ensure data privacy and security, and a high level of security at all sub-systems capable of actuation on the physical.
9	Controllability	A digital twin needs to be managed, configured and controlled by its end users.

pass under the responsibility of the facility owners and managers across the life cycle. As a consequence however, the DT is first dependent on BIM-specific data sets (i.e. Industry Foundation Classes). In terms of data semantics and their flexible interchange, a solution at this stage is presented in the form of semantic web and linked data [301], which offers the capability of expressing specific domain knowledge in machine interpretable formats. Formats such as RDF/S (Resource Description Framework/Schema) and OWL (Web Ontology Language) allow more flexibility and extensibility in modelling the semantics. This could allow a more dynamic definition of models at each stage and ways of adapting semantics depending on building type, installed sensors and required DT services at each stage of the life cycle.

As opposed to manufacturing industries, where an asset has a short lifespan and does not change dramatically, the built environment digital twins need to be modular and flexible enough to account for the known data transformation gaps between life cycle stages. Unlike in existing buildings where sensing equipment can be stationary for longer periods, the dynamic nature of construction sites means that sensing devices themselves are in constant replacement and re-calibration – which equates to more time and labour in adapting the site sensing when required. Finally, a maintainability requirement needs to foresee the changing and managing of physical sensing equipment on the ground, the data hosting hardware, as well as the changes of the underlying semantics of the data models.

In principle, a digital twin cannot exist in isolation, as opposed to a BIM model for example. Its live connection to the real world (assets, people) and other virtual entities (internet, systems, other digital twins) needs to consider how the above discussed technical requirements can allow for efficient and effective communication channels. To account for this transition, the white paper from the SPHERE project [229] proposes the concept of a Building Digital Twin Prototype (BDTP) for the construction stage, which is later expected to transition into a Building Digital Twin Instance (BDTI). This is useful when trying to set abstract boundaries for crossing stages, as well as scales. The report goes on to recommend the concept of Building Digital Twin Aggregates (BDTA), which is in effect a grouping of multiple smaller digital twins when attempting to scale from one level to the next. Although these concepts were defined for the building level, they can be generalised over from micro to the meso and macro scales as well. As previously mentioned, the macro scale has to consider an aggregation of multiple BDTIs of built structures as well as more dynamic BDTP for ongoing construction works in the nearby environment.

8.2.2   Digital twin technologies & abilities

After having considered the requirements for developing digital twins at different scales, a more practical approach on adopting different types of technologies is the next logical step. A survey on the existing technologies which have major roles in enabling the development of DTs has been conducted by Fuller et al. [140] from a computer science perspective, but also Qi et al. [309] and Tao et al. [378] from an industrial manufacturing perspective. The key technologies in today's digitalisation process, big data, IoT and AI, are constantly mentioned and discussed as the main enablers of the DT paradigm. The addition of block chain is seen as valuable in the case of supply chain digital twins [24,159].

The white paper by the Industrial Internet Consortium (IIC) [170] provides a list of features that DT should have, such as: modelling (digital representation),

3D representations, visualisation, document management, model synchronisation and connected analytics. Many of these are non-functional system requirements which need to be in place for a DT implementation. Within the built environment, these often fall under the scope of a CDE (see Chapter 6), whereby BIM/GIS data is contextualised with construction processes. Therefore, we can consider these are part of the BIM/GIS integration, which feed into our built environment DTs. A key technology which so far is mostly heralded from the built environment sector is the use of the semantic web ontologies to define domain knowledge and openly share these models as knowledge graphs; combined with a linked data paradigm, multi-disciplinary applications can be semantically aligned and shared over the web. The importance of semantic meta models of a digital twin throughout its life cycle is initially raised by Negri et al. [270], and later investigated at the level of construction sites by Boje et al. [41], which propose a set of so-called digital twin "abilities" which are related to the technologies which should be considered for implementation. These have been slightly adapted and listed in Table 8.2 under the concept of Figure 8.1, as discussed in Section 8.1.

In relation with the more developed view on the DT (Figure 8.1), the abilities facilitate the practical implementation of mentioned technologies, each with a specific role within the system. The abilities that deal with the PT directly have a role in capturing information from the real world and relaying this further down to data and virtual abilities. The actuation plays a distinct key role in that the PT is able to adapt and change state according to the DT reactions (ideally based on simulation, prediction and optimisation). The communication (transfer) of the data itself is achieved through the ability to integrate IoT networks.

Table 8.2 A list of abilities for digital twins, considering the inclusion of available technologies, under the physical-data-virtual paradigm

Part	Ability	Description
Physical	Sense	Observe the physical world in real-time via the use of sensors
Physical	Monitor	Keep track, inform and issue warnings on relevant physical alterations
Physical	Actuate	Control, change, activate or deactivate physical components based on Virtual decisions or stimuli
Data	Integrate	Integrate and consume BIM, GIS and IoT specific data sets in its various formats and standards
Data	Store	Store raw data, processed data and knowledge facts about the system; support rules and reasoning capabilities
Data	Share	Share data, facts and insights in various formats over IoT and semantic web protocols
Data	Link	Map the different resources over the semantic web and verify their integrity regularly
Virtual	Emulate	Represent the physical object along with its properties and behaviour
Virtual	Diagnose	Run scans, detect faults, anomalies and inconsistencies within the system
Virtual	Visualise	Provide data with graphical components for various services
Virtual	Simulate	Apply engineering simulation models from various application domains
Virtual	Predict	Predict the behaviour of the physical based on digital simulations and sensing
Virtual	Optimise	Apply optimisation methods and recommend smart allocation of resources dynamically
Virtual	React	Respond to the physical and digital environments, by delegating AI agents capable of managing (and actuating) the physical based on digital data, following well-defined behaviours, protocols and objectives; this is dependent on the level of autonomy.

Considering the linking ability, linked data plays a key role. This should be based on emulation (representation) of the data, where the semantics of the DT are expressed and leveraged for the use of more abstract concept on the Virtual level. Therefore, the abilities dealing with data need to consider the storage of data coming from sensing (hence the physical), but also from integrated BIM and GIS data sets (considering different scales of application). Additionally, contextual data related to how the DT emulates (represents or replicates) the PT needs to be stored along with any additional virtually simulated data, facts and knowledge about each particular context, which drive the decision support for AI agents and end users. The role of semantics here is a blurred boundary across the three dimensions, as the meaning of the physical, data and virtual abilities need to be expressed (as meta-models), but also the domains which give it specific applications (BIM/GIS, simulation or specialised AI models).

8.2.3   Digital twin levels

The field of BIM introduced the notion of BIM levels [37], a concept which became quickly adopted by industry and academia due to its simplicity when describing the natural evolution of BIM implementation, from paper to web-based BIM. The state of the industry on BIM use however, has been stuck on the BIM level 2 for some time. The status-quo in the industry is in a state of expansion and of BIM use, rather than within a state of BIM evolution. That does not mean however that the industry is not looking at technologies beyond BIM, where the Digital Twin might eventually project BIM (and GIS) to new levels of applications and domains. In a similar fashion, the concept of Digital Twin "levels" has come across as a way to distinguish the level of development and the maturity of the included technologies. Considering the vast number of technologies which are able to fit under a DT paradigm, the combinations of possibilities are very large and diverse. In order to develop a digital twin of a built asset, not all abilities are required, nor should they be adopted in practice straight away. Instead, a DT should be regarded as a progressive evolution [41] from when the asset at any scale is being designed, built, managed and eventually planned for a circular economy.

Madni et al. [236] introduced the idea of Digital Twin levels within the field of aerospace engineering and present four different stages ranging from a pre-DT to an intelligent DT, gaining more abilities and autonomy as it develops and grows. An initial classification of DT levels within the construction industry was proposed by Evans et al. [124] in a white paper, expanding on The Gemini Principles report [42]. Within this white paper, the authors distinguish between 6 different levels of maturity from 0 to 5, with 0 being defined as a way to capture the reality of the physical twin (point clouds, photogrammetry, etc.). Levels 1 to 4 adopt a practical approach to include BIM data (documents, models, etc.), real-time elements using IoT, and a way to integrate and connect them. Level 5 depicts a more complex system with increased autonomy. We can see here very similar understandings of DT levels, from aerospace to construction use. However, the requirements of interoperability and re-usability of the technologies along these levels are not clear at the moment. Based on these views, Boje et al. [41] have proposed three distinct levels, each having to leap over a critical gap (see Figure 8.2).

That said, our current state-of-the-art monitoring platforms (level 1), which use BIM, GIS or a combination of the two for monitoring the built environment, have to cross a semantics gap towards intelligent semantic platforms (level 2). This first step requires that models and knowledge are used and transferred across, ensuring information cohesion. It is at this level that ontologies and semantic web standards such

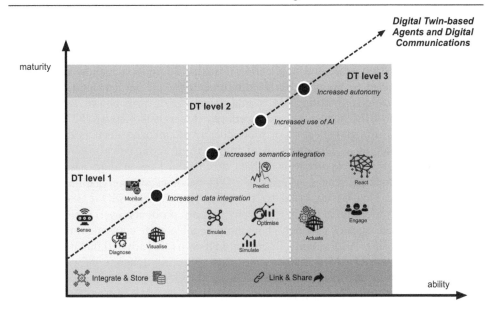

Figure 8.2 Digital twins levels based on perceived abilities. (Designed using resources from Flaticon.com.)

as RDF/RDFS and OWL become crucial in adapting our existing legacy systems to more dynamic web-based ones. The final generation (level 3) should evolve naturally from the previous with the extension to more AI capabilities, but most importantly include the social aspects of the communities interfacing with the DT.

8.2.4   Standards for digital twins

As the subject is still in its infancy, there are no finalised standards specifically designed for digital twin implementation. The newly established Digital Twin Consortium (DTC) in 2020 and the older Industry IoT Consortium (IIC) under the Object Management Group (OMG) are committed to guiding practices on how to deal with the subject of digital twin across several industries. The white paper from IIC [170] already discusses some aspects of existing standards, most notably mentioning the involvement of the Industrial Data committee (ISO/TC 184/SC 4) to develop the ISO 23247 (Digital Twin framework for manufacturing) and the ISO/PRF TR 24464 (Automation systems and integration — Industrial data — Visualisation elements of digital twins) for dealing with the visualisation aspects. The Information technology committee (ISO/IEC JTC 1) has undertaken the task to define the terminology around the subject in ISO/IEC AWI 5618 (Digital twin — Concepts and terminology).

BuildingSmart International has also created a Digital Twin working group in 2020, set on deciding the way forward from BIM and help develop open standards focused around three main areas: data models, data management & integration, data security & privacy. Within their white paper [59], BuildingSmart mention possible links to the ISO 55000 (Asset management — Overview, principles and terminology). On the macro scale, the standards developed by the Sustainable cities and communities committee (ISO TC/268) around the subject of smart cities should also play an important role when adopting a DT strategy.

Considering that a DT is an aggregation of multiple technologies, it is logical to consider existing fundamental standards for implementing these technologies, while combining specific domain standards depending on each context. The context here can be dissected based on scale and application domain. These would allow a high-level description to answer to the procedural requirements for DT implementation. On the micro scale for example, existing BIM standards are the obvious choice, ensuring DT compatibility with BIM data sets and procedures. Rogers and Kirwan [327] proposed related standards from classification of products and dictionaries, such as ISO 12006-2 (Building construction — Organisation of information about construction works — Part 2: Framework for classification) which are a vital step in data harmonisation towards post-occupancy stages, whereby a BDTI becomes an aggregation of several specific products. This is supported by the recently published ISO 23386:2020 and ISO 23387:2020, looking at managing construction products over their life cycle under standardised types or products. This should enable more convenient information gathering and sharing. Additionally, Rogers and Kirwan [327] advised that a DT should enable LEED and BREAM frameworks to start collecting dynamic information and evaluate the performance of the building.

From a technical requirements point of view, the existing standards and guidance on IoT (ISO/IEC 21823-1:2019), data linking, data sharing on the web and APIs should dictate how these are implemented within DT across all industries, thus ensuring cross-domain cohesion. On the European level, there are ongoing efforts by the CEN BIM technical committee (CEN/TC 442) to release prEN 17632 (Semantic Linking and Modelling standard), which will influence semantic web-based BIM and consequently DTs based on BIM. On the macro scale, a good example is provided by Context Information Management working group from ETSI with the already published NGSI-LD API standard, aimed to harmonise the way to interface with not just IoT but also other city databases and common applications for smart cities. The Open Geospatial Consortium (OGC) has several open standards on GIS and related APIs to standardise how data is shared and accessed.

From a system development stand-point, Gharaei et al. [147] considered the ISO 42010 (Systems and software engineering – Architecture description) as a reference standard for defining a DT system in practice, as this document provides detailed meta-models for describing such software systems. Within the specific context of each DT implementation, generic meta-models need to be combined with context domain models. Sabou et al. [333] proposed the Common Information Model schema standard published by the Distributed Management Task Force, which contains various conceptual meta-models for representing complex systems - this was previously explored with smart grids.

As there is no standard to rule them all, Table 8.3 provides an initial starting point of guidance worth considering, some of which have been discussed above. It is common practice that standards are inter-related, and many are often complemented by technical reports on practical issues. The implementation of DT according to existing standards will ensure interoperability with multiple domains, connectivity on and off the web and easier maintenance of the data and models underpinning the DT.

8.3 DOMAINS WITHIN THE BUILT ENVIRONMENT

Now that we have a good idea of what to expect from a digital twin implementation, we will try to compare theory with practice. In this section, we will give some examples

Table 8.3 A non-exhaustive list of relevant standards for DT implementation

Code	Name
ISO 23247	Digital Twin framework for manufacturing
ISO/PRF TR 24464	Automation systems and integration — Industrial data — Visualisation elements of digital twins
ISO/IEC AWI 5618	Digital twin — Concepts and terminology
ISO 55000	Asset management — Overview, principles and terminology
ISO 16739-1:2018	Industry Foundation Classes (IFC) for data sharing in the construction and facility management industries — Part 1: Data schema
ISO 12006-2	Building construction — Organisation of information about construction works — Part 2: Framework for classification
ISO 23386:2020	Building information modelling and other digital processes used in construction — Methodology to describe, author and maintain properties in interconnected data dictionaries
ISO 23387:2020	Building information modelling (BIM) — Data templates for construction objects used in the life cycle of built assets — Concepts and principles
ETSI GR CIM 008	Context Information Management (CIM); NGSI-LD Primer
ETSI GS CIM 009	Context Information Management (CIM); NGSI-LD API
ISO 42010	Systems and software engineering – Architecture description
CIM Schema	Common Information Model (CIM) Schema, version 2.54
ISO/IEC 21823-1:2019	Internet of things (IoT) — Interoperability for IoT systems — Part 1: Framework

of real world case studies and ongoing research, which would qualify as digital twin implementations and initiatives. In the first part, we will attempt to give a (non-exhaustive) list of implemented digital twins across all scales whilst trying to cover the myriad of contexts that the built environment has to offer. From these, as the digital twin concept is taking roots, several organisations and international companies have taken the initiative and begin to give shape to what the digital twin paradigm looks like from a software development perspective.

8.3.1 Digital twin implementation examples

We have already established that there is no need to have a perfect virtual mirror of a built asset as it would not be practical. Instead, the scale, the application domain and the type of sensing equipment will characterise how a digital twin will be deployed and behave in practice. Table 8.4 below lists various implementations across scales (micro, meso, macro) and application domains. The application domain gives the case study the context of the particular problem that it has to solve.

Before taking roots into the fields of BIM and construction, the concept of a DT was quickly embraced by the urban planning community, as a means to achieve a smart city platform. To this end, there are numerous cases of smart city platforms from research and commercial backgrounds. The main challenges and interests on the city scale have usually been the smart use of energy, by balancing the supply and demand, as we can make out from the examples of (1) Cardiff, (2) Helsinki, (3) Rotterdam and (4) Vienna. In these examples, the monitoring of the energy grid and the estimation of the demand lead to an optimisation problem with many dynamic factors depending on the season, population, events, green energy generation potential, etc. The application domains for cities and

Table 8.4 Examples of digital twin implementations across the built environment by scale and domain

No	Case-study	Description	Scale	Domains
1	City of Cardiff[1]	A research-driven platform from several EU and international projects on different use case studies and pilot studies, where the use of semantics was key to achieving interoperability.	Macro	Energy monitoring, optimisation
2	City of Helsinki[2]	3D geometric capture of the city of Helsinki pilot, towards an open data platform under CityGML.	Macro	Open data, end-user involvement, wind and solar simulations, visualisation
3	City of Newcastle (Flood-PREPARED project)[3]	A real-time sensing of the entire city landscape on precipitation in order to simulate flood hazard models with statistical analytics and big data.	Macro	Water flood risk, monitoring, simulation
4	City of Rotterdam,[4] (Ruggedised project[5])	As part of making the city of Rotterdam smarter on many fronts, a digital twin paradigm is considered when integrating data from sensors.	Macro	Efficient street lighting, waste management, transport, energy, etc.
5	City of Vienna[6]	Real-time urban data for smart energy usage.	Macro	Energy
6	City of Zurich [346]	Exploring more detailed modelling of the city built environment and making them open and available for urban planning and citizen engagement	Macro	End-user engagement, simulation, urban planning
7	Virtual Singapore[7]	A national-backed research initiative to create a 3D replica of the city, empowered by IoT, Big Data, but most importantly simulations of various kinds (solar, pedestrian traffic, etc.)	Macro	Solar energy potential, accessibility, urban planning, research, communication
8	Valencia Water Network [77]	A case study on a system monitoring the water usage on a large scale city network using GIS and SCADA	Macro	Water usage monitoring
9	Antwerp Port[8]	Innovation plans for the Port of Antwerp to integrate AI capabilities, augmented reality and blockchain technologies with smart cameras and a 3D map of the port in order to improve operations. A "digital brain" of the port is described.[9]	Meso	Port management, logistics, health and safety, crime safety
10	Queen Elizabeth Olympic Park, UK [96]	Describes the process of implementing a DT on multiple levels, from district down to a building, whilst considering social implications.	Meso	Air quality monitoring, end-user engagement

(Continued)

Table 8.4   (Continued) Examples of digital twin implementations across the built environment by scale and domain

No	Case-study	Description	Scale	Domains
11	Trent Basin, UK (SCENe project)[10]	A real-time monitoring of the energy consumption is made, and it is provided to end-users via an interface (3D model); users are able to learn from their patterns and help optimise their energy usage, as well as learn about energy use in general. Smart homes, IoT. The output is the Community Information Model as an online interactive platform.	Meso	Energy consumption monitoring, end-user engagement
12	Cambridge Campus, UK [311]	A case study intent on deploying a building DT based on a proposed architecture.	Micro	Air-quality, light, motion, vibrations, status of doors
13	Milan Cathedral [8]	A digital twin of a large, complex, cultural heritage building for monitoring structural integrity using finite element simulations.	Micro	Cultural heritage, simulation, monitoring
14	Railway bridge, Stafford, UK [95]	A dynamic BIM-based environment which streams sensor data attached to structural beams, in order to monitor, simulate and help take decisions regarding structural integrity	Micro	Structural integrity, monitoring, simulation
15	Roof structure thermal simulation [235]	Custom built roof of a unconventional shape which needs to be tested and monitored to assess its performance	Micro	Thermal simulation, live labs

[1]https://www.cuspplatform.com/.
[2]https://www.hel.fi/static/liitteet-2019/Kaupunginkanslia/Helsinki3D_Kalasatama_Digital_Twins.pdf.
[3]https://projects.urbanobservatory.ac.uk/projects/flood-prepared.
[4]https://eu-smartcities.eu/news/rotterdams-digital-twin-redefines-our-physical-digital-social-worlds.
[5]https://ruggedised.eu/home/.
[6]https://www.ascr.at/en/.
[7]https://www.nrf.gov.sg/programmes/virtual-singapore.
[8]https://www.portofantwerp.com/en/smart-port.
[9]https://www.portofantwerp.com/en/news/meet-apica-our-digital-brain.
[10]https://www.projectscene.uk/trentbasin/.

districts (see also Chapter 11) range from energy and water usage to transportation networks and disaster and flood management (3 in Table 8.4).

What we can observe from the recent case studies on city and building scale is also the modelling of the end-user engagement and social aspects, which play an important role in gathering data as well as predicting user-tailored solutions. We can see this having an important role to play in defining DT systems in close connection to the PT inhabitants. End-user engagement using advanced visualisation techniques (virtual and augmented realities) is its own established field of research across all scales, many of which are being re-defined as Digital Twins as they attribute the real-time factor along with AI capabilities.

On the meso scale, there are several examples of digital twins, as these usually involve construction sites, specialised districts of buildings (10 and 11 in Table 8.4) or in most cases smart ports (Antwerp and Rotterdam to name a few). These are highly dynamic environments with specialised equipment (vehicles, cranes) and in many cases specialised needs for sensing. These can include cameras and computer vision AI to detect health and safety hazards or in the case of the port of Antwerp (9) - crime detection. The main drivers behind such implementations however are the optimisation of logistics for transportation and construction, which can reduce costs and implicitly have less of an impact on the environment. These are in close connection to supply chain digital twins [24,369], a concept studied under industrial manufacturing.

The micro-scale case studies show us much more detailed technical aspects of DTs, as the scale is more manageable and sensor data is of much higher resolution relative to this scale. Thus, the level of detail and fidelity to the PT is much higher. Energy monitoring and performance assessment of buildings, such as smart homes, is a typical example of DT application. This is a field that has recently converged with BIM on structuring and managing information and sensor data. In all the micro-level case studies in Table 8.4, BIM plays a key role in providing or managing information about the built asset, which is seen as a provider of not just 3D representations but also wider information semantics. The addition of sensing equipment and various IoT devices is coupled here with scan-to-BIM techniques in the cases where a BIM model does not exist such as the Milan Cathedral (13), in order to get the latest virtual representation of the existing assets. Conventional sensing equipment is often deployed to monitor air quality, as it is similarly done in the case study of the Cambridge Campus (12). These are often coupled with end-user engagement services, exposing the DT through web portals or other visualisation methods.

Similar to some cases on the meso scale, there are certain types of assets which require constant monitoring. The case study on the Milan Cathedral (13) is a perfect example on the logistics challenges when dealing with cultural preservation buildings, where a BIM model is usually not available (or hard to model precisely due to the lack of heritage BIM objects) or sensing and simulating experiences various limitations due to the nature of the building itself. Conversely, the application of DT on bridge structural integrity monitoring (14 in Table 8.4) for new or old bridges is a more practical one that can predict the life span of infrastructure assets and issue warnings in real-time in critical conditions. This can help reduce accidents and significantly reduce maintenance costs.

The case studies above gave us an idea of what can be considered for digital twinning the built environment. Although the context is highly specialised and can differ substantially, the fundamental DT paradigm is followed: a physical asset which is equipped with sensing, in connection with its virtual representation on a platform of a software tool which in turn provides a domain-specific service. In many ways, we have been developing Digital Twins for many years. However, as discussed in Section 8.2, there is no standard for DT implementation to date, nor specific tools or platforms which would impose this. Nevertheless, many of their underlying formats and architectures do follow standards and best practices. Additionally, the majority of the case studies above have been conducted in collaboration with industry partners, which are looking to define and streamline digital twinning methods.

8.3.2 Digital twin development initiatives

The technologies used in some of the above projects are various and diverse. Some high-profile companies have taken the lead; however, when it comes to DT solutions - most notably the Microsoft Azure DT and IBM DT, specialised in IoT in a generic manner. Bentley Systems has teamed up with MS Azure DT services to contextualise a built asset

DT, and have launched the open source iTwin[1] platform which was designed to bring cohesion between different types of hybrid data from the built environment models, along with IoT devices.

In a similar way, IBM has partnered with the EU's biggest commercial port of Rotterdam to develop a digital twin solution.[2] For the case of city scales, research platform such as CUSP (City of Cardiff in Table 8.4), Dassault (Virtual Singapore in Table 8.4) and CityZenith[3] have adapted their smart city platforms to follow a DT paradigm. Siemens is another international company looking to implement DT having started from manufacturing to various projects on power grids and water management. Similarly, British Petroleum has taken an interest in deploying and testing DTs for reducing the upkeep costs of oil and gas rigs which have a huge upkeep and renovation cost, as it would completely halt production. The case study on Trent Basin (3 in Table 8.4) was a pilot for IES to expand their energy simulation services into a DT paradigm. Similar solutions will gradually show up as the concept of DT matures in other industries.

Given the examples above, the direction these industries and companies are taking shows that the DT paradigm is considered very seriously. The DT domain is predicted to grow in investment within the next few decades by up to 38%,[4] with the expectations to increase the productivity of companies. This would go hand in hand with recent rises in investment on IoT and Industry 4.0. More important however would be to establish the return on investment and a measure for DT implementation success, such as we have seen with BIM over time.

Apart from international companies which have taken the initiative to adapt to new trends, several nation-wide digital twins have been proposed by the UK [42], Estonia[5] and Luxembourg[6] as ongoing innovation initiatives. Given the scale and the ambition of such developments, a robust IT infrastructure is required to achieve this. The GAIA-X[7] project involving France and Germany is expected to deliver such an infrastructure for handling data at EU level, targeted at multiple industry and environmental domains. For the smart cities domains, the SynchroniCity[8] project aims to find the common ground for requirements and technologies for IoT-enabled smart cities across the EU, and has several pilot sites such as Antwerp, Bordeaux, Helsinki, Porto, etc. More specifically for the construction sector, the DigiPLACE[9] project aims to formulate a common framework and architecture for the digitalisation process of the industry. Similar efforts were conducted by the Interlink[10] project for motorways, which was also a pilot in adopting semantic web technologies for managing large infrastructure assets.

8.4 A REFERENCE FRAMEWORK

Within this section, we will try to give an overview of what a DT architecture might look like within the context of the (semantic) web. After all, the DT in practice is set to be an advanced software system of systems; this means that it needs to include various types of sensors, data and models in order to provide services to its end users.

8.4.1 Conceptual system architecture

Although no common reference architecture for a DT has been defined, some conceptual examples have already been proposed at a generic level [170], at the building level [68,311], at a supply chain level [24] and a construction site level [41]. The common features we can

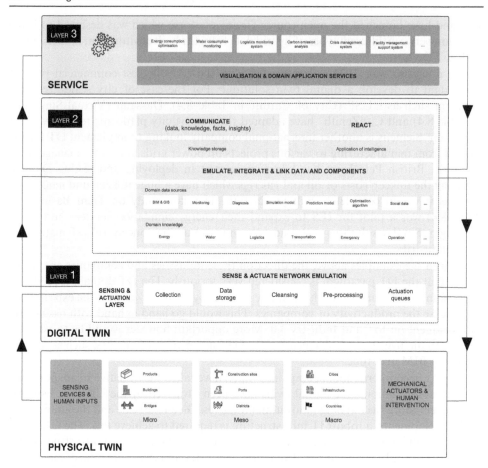

Figure 8.3 A conceptual system architecture framework for digital twins. (Designed using resources from Flaticon.com.)

see from the literature point towards a typical web-based software architecture formed from several layers, as shown in Figure 8.3.

The first layer from the bottom we will refer to as the sensing and actuation layer (1) which deals directly with raw data from the PT, preparing it for the upper layers and dealing with immediate actuation actions back to the PT. Sensors of various types (measurements, 3D scanners, RFIDs, video feeds, human input, etc.) collect the latest information in various formats and store it remotely – usually on cloud databases (see also Chapter 3). The data collected from these sensing networks is remotely stored, cleaned (pre-processed) and later used for processing. A meta-data model needs to be able to emulate the sensing network in place, mapping and linking it to the next layer above.

The second conceptual we will refer to as the computational layer (2) which applies higher level virtual abilities, especially emulation, integration and various forms of AI. This is considered the central processing "brain" of the DT, where the models are located. It is at this level that knowledge modelling, the linking of different virtual components and the application of intelligence are able to facilitate and empower various services. As mentioned in Section 8.2 of this chapter, the models need to be able to represent the physical world that is being sensed. We have termed this "emulation". This means that the modelling process should consider classes and instances of "things" which would describe their identity and location on the web (or the network), their properties and behaviour. The representation of "things" at this level needs to consider not only physical sensors, and virtual objects (such as BIM objects), but also a meta-model describing the interactions between different data models, engineering models and AI. If this layer is to be considered an integrator of information, it needs to know the involved entities, their identities and locations on the web or within a distributed system. It therefore needs to consider the inclusion of the stored data from layer (1), the integration of BIM and GIS databases, the transfer and transformation of information from one model to the next. The use of AI optimisations, machine learning and deep learning at this stage should be triggered in order to predict, react and respond to real world problems. This response can be either a shared recommendation to another system or end users, or it can cause immediate actuation on the physical, depending on the level of autonomy.

The third layer (3) corresponds to the services part of a DT, which is usually described as a web portal, a dashboard, a user interface or an API for other tools to interface with. This layer depends on the use case, but generic functional and non-functional software requirements come into play. We will assume that the DT is exposed to the web at this layer and must follow existing web communication and security protocols. It is at this layer that end users may interact with the DT, where the data and predictions of the DT are visualised (3D, graphs, VR and AR, etc.). Similar to CDEs (see Chapter 6), the access to the DT needs to be defined and configured. Apart from the typical end users, private maintenance access points to experts and auditors need to be considered. Data integrity and checking is a must, which can be achieved with the inclusion of logic rules.

The simplified view of these layers needs to be empowered by semantics at every stage, allowing the interpretation from one layer to the next. The communication or transfer of data between the layers themselves will have to rely on existing protocols. Layer (1) and (2) would have to make use of IoT standards, such as ISO/IEC 21823-1:2019 (covered in Chapter 10), which should cover data collection and possible actuation. The communication between layer (2) components with the third layer (3) is envisaged to be over web protocols, using REST APIs and CRUD operations [170]. When considering the interaction between these components, the already mentioned technical requirements from Table 8.2 need to be considered.

8.4.2 The role of semantics

The challenge in developing DT lies in managing and connecting the dots of all these components, across scales and domains, as we have seen in Section 8.3 of this chapter. Above all, we have mentioned the vital role of ontologies in representing models. Ontologies are best suited to represent complex socio-technical systems, and can represent virtually any "thing", from abstract classes to real-world individuals with the use of the Tbox and Abox statements. The description logics at their core also allow us to formalise rules and validate the integrity of our models and data. The recent releases of JSON-LD (see Chapter 5 and 6) have also made semantic web technologies more compatible with newer web platforms and libraries. Although not all fields have considered semantic web ontologies as formal backend models for their systems, the study from Sabou et al. [333] can provide us with a cross-domain analysis of existing schema models looking to develop DTs.

First, following the described conceptual architecture in Figure 8.3, let us consider the computational layer (2), where the emulation and modelling of the PT take places. On the micro scale, we see that BIM dominates the building representation. The central ontology at this scale is the IFC schema, used in its OWL format (ifcOWL). Several more lightweight and specialised ontologies which are targeted at representing BIM assets at the micro scale have been presented in detail in the previous chapters (Chapters 1–4), covering definitions for geometries, metadata and how they can be linked (BOT, FOG, OMG, etc.). On the meso-macro scales, with the use of GIS, CityGML is a schema which can offer a higher level view of the built environment. For smart grids, Sabou et al. [333] recommend the use of the Common Information Model schema and the IEC 61850 series of standards, while the work of Howell et al. [181] has given us insights into water network ontologies. In addition to emulating sensing, built assets and domain applications, the DT remains in essence a software system. The use of meta-data ontologies to describe the system itself, its components, interactions and events need to be emulated as well. The simulation ability of the DT can be filled by using a set of ontologies from the Aachen University, OntoCape[11] – designed to represent engineering processes. Through linked data and ontologies, "connectivity" and "interoperability" can be achieved at the level of representing (emulating) the physical world and virtual models.

Second, looking at the sensing and actuation layer (1) in Figure 8.3, we can make use of existing ontologies such as the SSN,[12] which is used to describe observations through sensing entities. The use of time series databases to host sensor readings is common practice (Chapter 3), with readings which need to be fetched and contextualised for the layers above. However, sensor observations are not limited to only IoT, but can also envisage software and human agents which can record various types of changes of the physical world state. The Digital Twins Definition Language (DTDL[13]) currently under development could also be used to integrate with IoT devices at a higher level, following W3C standards. As an alternative form of sensing, laser scanning and photogrammetry has become a widely used technology in capturing real world data, which is characterised by a lack of high-level semantics. However, methods such as scan-to-BIM, some using advanced machine and deep learning can help attribute more meaning to these captures automatically. The challenge lies in formalising them in a practical way. The recently published method for Scan-to-Graph by Werbrouck et al. [410] does exactly that by linking modular ontologies such as BOT and FOG. Using these tools we can achieve "connectivity" and "controllability" to the real world objects, retrieving their data and performing actuation when desired. "Interoperability" is achieved implicitly by using the OWL and RDF/S schemas. This ensures alignment of the above layers.

Third, the services layer (3) in Figure 8.3 is characterised by specific applications and interactions with end users, other systems or indeed other DTs. The use of highly specialised ontology models is key here, ranging from costing estimation of construction works, collaboration [40], document management, fire safety simulations [39], waste management to urban design. The introduction of rules such as SWRL, or querying with SPARQL on specific questions can provide means of reasoning and exploring the application context. The use of SHACL is also recommended to validate the underlying models and data [367].

Not all nearby engineering fields have considered the use of ontologies as a prime enabler for DT implementation, but we believe that it is the technology best suited to tackle the "connectivity", "re-usability", "extensibility", "interoperability" and "modularity" mentioned in Table 8.1. Ontologies can easily be extended. The linking of meta-models and domain ontologies ensures some degree of modularity, but care should be taken on which ontologies are best suited for each application. It is often preferable to sacrifice higher level semantics for higher processing speeds. Additionally, maintainability of each ontology should also be put in the balance.

8.5 THE FUTURE OF DIGITAL TWINS

Throughout this book, we have been discussing, among others, the semantic web and ontologies, which at their core were developed within the computer science field to connect the information on the web with a higher degree of meaning. Semantics stands for the "meaning" of the data, its deliberate structure which resembles human knowledge in a formalised way. Before the development and proliferation of ontologies across several domains, semantics were isolated in silos within the construction industry, embedded into closed proprietary software, which prohibited sharing and linking data on the web, and prohibited meaningful connections between knowledge domains.

The proliferation of the semantic web and IoT paradigms has paved the way to a DT implementation, as a practical application of the available tools and technologies developed so far, which have been presented across several chapters of this book. By understanding what a "Digital Twin" paradigm stands for and how it is perceived in research and industry, we can begin to see that in many ways, we have been developing digital twins for a long time. However, the rebranding of our technologies, a more comprehensive and holistic linking of data over the semantic web brings these things closer together, making them machine interpretable and more automatic. The field of DT is going through a phase of "hype" and exploration in practice, but the motivation to better manage our built environment remains a long-term objective, which will contribute to the betterment of our society by reducing our environmental footprint.

NOTES

1 https://www.bentley.com/en/products/product-line/digital-twins/itwin.
2 https://www.ibm.com/blogs/internet-of-things/iot-digital-twin-rotterdam/.
3 https://cityzenith.com/the-world-of-digital-twins.
4 https://www.grandviewresearch.com/industry-analysis/digital-twin-market.
5 https://investinestonia.com/.
6 https://www.list.lu/en/cooperation/innovation-programmes/digital-twin/.
7 https://www.data-infrastructure.eu/.
8 https://synchronicity-iot.eu/.
9 https://www.digiplaceproject.eu/.
10 https://www.roadotl.eu/.
11 https://www.avt.rwth-aachen.de/cms/AVT/Forschung/Software/ ipts/OntoCape/.
12 https://www.w3.org/TR/vocab-ssn/.
13 https://github.com/Azure/opendigitaltwins-dtdl.

Chapter 9

The building as a platform

Predictive digital twinning

Tamer El-Diraby and Soroush Sobhkhiz

CONTENTS

This chapter discusses the need and merit for a possible alternative view on digital twinning. First, the potential that the digital twin will not be IFC-based. The possibility that the digital twin will not be based on any common model of data. Second, and consequently, the fact that the process of digital twinning becomes more important than the digital twin itself. The arguments for these are based on the very role and definition of digital twins. In this chapter, they are not just a repository of data. In addition to data, a digital twin includes two major elements: representation of workflows, which are needed for supporting an automated/algorithmic operation; and simulation models of, for example, energy management and/or user comfort scenarios. Such data will span structured and unstructured data; building, operator and user-generated data; historical, real-time and simulated-futures data. Consequently, the hard-to-achieve interoperability within the traditional structured BIM data will be impossible. Of course, BIM will be a major component in any digital twin, but not the core. The role of a digital twin, in this context, is to discover knowledge: learning what the data is telling us; supporting predictive analysis. This stands in sharp contrast to IFC mentality: compliance to a pre-defined model of knowledge; and achieving interoperability. To this end, the process of composing, analysing and learning from digital twin data becomes the key contribution.

We first present our proposition for predictive digital "twinning", the business case for their existence and value. We situate that against recent advances in IFC-based

DOI: 10.1201/9781003204381-11

twinning and, equally important, the criticism for IFC (model-driven) mentality. To illustrate the arguments made, we showcase an ongoing digital twinning project at the University of Toronto. It is built on top of a non-IFC legacy system for building automation. A no-model architecture is proposed to achieve the following: adding BIM to the digital twin as a component, not as the core element; using machine learning to discover patterns and support predictive analysis and business intelligence applications; and engaging all stakeholders in the learning process.

9.1   INTRODUCTION

Digital Twins (DTs) first became a subject of interest in the architecture, engineering and construction (AEC) industry when efforts shifted to developing live data hosts that can dynamically obtain the building data and manage it throughout its life cycle [234]. The term "live" represents a real-time capture of data but is not only limited to capturing or storing but also includes analysis (sensing) and response. Studies started to integrate new technologies such as IoT and semantic web with Building Information Modelling (BIM). Seamless and practical integration of these technologies seemed to present a new set of challenges, particularly in areas such as data management, interoperability and governance [377]. Digital Twin is the latest resort of the AEC industry to deal with such challenges. However, being a relatively new topic of interest, DT is not perceived the same way among researchers and practitioners.

To this date, numerous definitions exist for Digital Twins such as *A digital replica of a living or non-living physical entity* [117]; or a more detailed one "A digital replica of potential and actual physical assets (physical twin), processes, people, places, systems and devices that can be used for various purposes."[1]. There are many definitions, sometimes contradicting, which can be attributed to the fact that the industry is still shaping the domain [386]. Arguably, these differences originate from how one answers the following question: Is a digital twin an interoperable BIM-based data repository? This is certainly possible as shown in Chapter 8. In this chapter, we present an alternative scenario: a no-IFC (indeed, a no-model) view of digital twins. The need for such varying views is discussed along three main dimensions: technical, business and definitional. In discussing these three dimensions, we hope to lay the foundations for our definition of digital twins.

1. Technical dimension—the fallacy of interoperability
    Achieving full interoperability is a mission impossible [391]. There is a limit to how much IFC can accommodate. The use of ifcOWL and linked data systems can help achieve partial or specific system-to-system agreement. However, given the complexity of intelligent building data, it may not be feasible to develop all the required linked data systems; and it will certainly be very time-consuming to sustain such linkages as the original data models evolve.

2. Business dimension—the objective is analytics
    IFC was developed mainly with engineers (particularly designers) in mind. It was designed to achieve interoperability more than to create an ontological representation of the concepts in AEC—at least beyond product data. Digital twins are different in two main aspects. First, they are directed for use by a broader spectrum of users. Facility operators and asset managers, as well as construction

professionals, are the prime users of digital twins. These stakeholders are business-oriented as much as engineering-oriented. They need to understand what is going on in a site or a building; and how to virtualise future scenarios. They need to harness data and use it to support operational efficiency and asset decision-making. For that, and second, the key interest for these stakeholders is in collecting, analysing, and making sense of unstructured data—mainly text and images. Using appropriate tools (such as machine learning) they can generate business intelligence applications that help them study several scenarios (for design, construction, and operations), detect patterns in facility usage (e.g. occupancy, utilisation and efficiency) and predict future events (e.g. equipment breakdown, occupancy level and energy consumption). The needed knowledge for such analytics is contained in unstructured data. Obviously, a standardised data model is not feasible in this situation. IFC is of less value in this scope—especially that over 90% of enterprise and facility data (what is called dark data) are out of scope for IFC.

3. Definitional dimension—a digital twin is data and models
    A digital twin is not a data repository (Chapter 8). It is meant to be a twin of the building in the virtual world. This must encompass all facility data as well as two types of models. The first is what can be called descriptive models. These are models that attempt to use facility data to understand/capture the current/ongoing conditions of a facility. These models are developed based on an established or pre-defined theory/approach. In other words, we know how the interplay between data sets can be combined to deduct a holistic picture of what is going on in a facility. For example, collating BIM and sensor data to generate an energy model of the facility; using these with occupancy data to develop a model for building utilisation, or air quality and/or circulation in a facility; using these to generate a carbon footprint of a facility; and using all the previous models to generate a sustainability index for a building. The second type of models is futuristic or prescriptive. These are models that aim to experiment with potential scenarios in the virtual world before implementing them in the real world. For example, a proposed operational scheme that can help enhance energy usage; simulating a scheme for data access or use within a facility; replicating maintenance policies under new assumptions or revised technical architecture, or the addition of new equipment.

9.1.1 Digital twins and intelligent buildings

With the above discussion, two key definitions can be provided. First, a digital twin can be defined as a collection of facility data (both structured and unstructured) that enables the development of descriptive models and testing prescriptive models. By developing descriptive operational and engineering models, a DT captures the conditions of the facility and its key indicators. Further, a DT allows testing futuristic prescriptive models and scenarios in the virtual world before implementing them in reality. Metaphorically, and according to this definition, a digital twin for a human being would include the basic data about such person; a model explaining the key characteristics of the human body, such as the respiratory system, the blood circulation system; and models imagining future alternative scenarios for such systems—what happens to the respiratory system if a smoker quits, or what happens to the digestive system if the human being adopts a healthier diet.

Second, what is an intelligent building? There are too many perspectives on that. It is important to discuss a few concepts before providing our own definition. In general, an intelligent building is an advanced form of buildings that integrates the principles of smart buildings (or building automation) and regenerative sustainability. Smartness here refers to a focus on the use of connected hardware, the implementation of a building management system (BMS), and the automation of some of the building performance or operation data analysis. Regenerative sustainability places the well-being of the facility occupant at the centre of sustainable practices. The aim is to make the building a catalyst for higher levels of human-oriented comfort and performance. Because there is a significant overlap between well-being promotion practices and energy savings, regenerative sustainability embeds and achieves the traditional objectives of sustainable design; [74,75]. With that, a regeneratively sustainable building has smart hardware, systematic capture and analysis of data, designed and governed with the objective to advance not only sustainability but, more importantly, the occupant well-being.

A special form of intelligent buildings is referred to as interactively adaptive buildings [73]. Obviously, with well-being at the centre of intelligent buildings, designing dynamic building and data governance systems to adapt to occupant needs is a central theme of intelligent buildings. This adaptability is provided through (1) bidirectional flow of information between the building and its occupants; and (2) active engagement of the occupants in decision-making. Active engagement is not limited to reactive approvals of decisions by occupants. Rather, it empowers the occupants (and operators) to innovate and co-create the decision options themselves, i.e. the occupant or operator is encouraged and supported to develop new operational schemes and, examine their viability through a virtual replica of the building (the digital twin).

With that in mind, our definition of an interactively adaptive building is a data-rich, collaboratively governed open computing environment that utilises smart hardware and data collection, sharing and analytics tools (particularly machine learning) to empower occupants (and other stakeholders such as operators and researchers) to innovate and co-create solutions or operational schemes with the goal of advancing well-being and sustainable practices. The building becomes a platform for collaborative processing of data in order for it to enable stakeholders to generate new knowledge; and to profile users and its own performance to interactively adapt to real-world needs. An interactively adaptive building is a form of algorithmic governance mechanism for facilities that is based on collecting, learning and serving data to occupants and operators with the aim of empowering occupants to define and develop new knowledge and tools to enhance their well-being.

In short, intelligent and sustainable buildings have two main features.

- Algorithmic governance: Intelligent buildings create a data-rich computing environment, where operators or researchers optimise space use and user comfort through automated decisions.
- Co-creation: Intelligent buildings empower occupants to generate knowledge and lead the analysis and co-create solutions. They are a venue for engagement and an incubator for innovation.

9.1.2   Digital twinning

As depicted in Figure 9.1, the two features of intelligent buildings, when combined, give rise to processes of interactive adaptivity, in which the building and its occupants are engaged in informed exchanges about continuous optimisation of building goals and

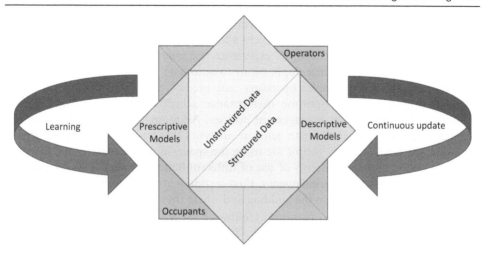

Figure 9.1 Digital twinning as a continuous process of learning and updating.

occupant comfort. As such, we can define Digital Twinning and the role of occupants as follows:

- Digital twinning vs. digital twins: In the proposed definition, digital twins will remain in a perpetual-beta stage. In other words, the process of updating, using, critiquing, and reinventing the digital twin is the tool and outcome of the use of digital twins. That is, the learning from the digital twin to update it is the most crucial task.

- Occupants as stakeholders: Facility users demand more engagement and are technology-savvy. The current socio-technological trends showcase that citizen science is becoming a reality. Citizens possess extensive abilities in information systems. In fact, they are not satisfied with the reactive role. It is no longer enough to provide them with real-time data or consult them in decision-making processes. They possess equally valid knowledge and the most leverage in making the needed change in facility transformation into the green and energy efficient realm. To this end, co-creation advocates that the process of engaging citizens could be more important than the quality of the final product and decision. The concept of empowerment relates to letting citizens lead the innovation and idea generation, and to provide them with the tools and power to create solutions to build new facilities or to better manage existing ones.

In the proposed definition, buildings become a platform for innovation. Stakeholders, including occupants, will be able to use a portal for engagement akin to a social media platform, where they can share ideas and debate alternatives. This level of human-building interaction is at the heart of advanced smart city models (see also Chapter 11). Triangulating diversified longitudinal data and harnessing the knowledge of all stakeholders will create a "building-as-a-platform" environment, where machine learning (ML) tools can be used to conduct descriptive and predictive analysis (Chapter 11). The latter will help model the future states and behaviour of devices or occupants.

Based on similar implementations of digital twinning in other industries, gleaning into the future can considerably improve work processes and decision-making

(Chapter 8). However, in the immediate term, a building-as-a-platform environment will leverage all data across the building life cycle, allowing staff to optimise efficiency, prevent breakdowns, and elevate occupants' experience. Equally important, ML will enable developing business intelligence applications to automate work processes and enhance decision-making. For example, operators can predict equipment failure, prescribe remedial procedures, and optimise maintenance schedules. This can enhance user comfort and optimise asset management policies. Additionally, through the platform, occupants or facility operators can generate analytics that tracks how different groups are using spaces: how many people are using the space, in what capacity, and for what purposes? What are their patterns of use of different building zones? This information could be crucial in case of an emergency. First-respondents will have better information about how many people are in the building and where they are.

The rest of the chapter is organised as follows. First, a brief overview of previous DT implementations is provided in Section 9.2. The overview focuses on the technologies and architectures used when developing DTs. Having discussed DT implementations in the literature, the chapter focuses on the challenges in DT implementation in Section 9.2.1. Next, a brief discussion over the potentials of DT implementation is provided in Section 9.2.2. Given the important role of Machine Learning (ML) in DTs, Section 9.2.3 describes how ML and advanced sensing can push DTs to realise their potentials. Section 9.3 presents a case study of a wide scale DT project to showcase the main considerations, opportunities, and challenges that should be considered in a wide-scale DT implementation. Finally, Section 9.4 summarises and concludes the chapter.

9.2   BACKGROUND

To this date, numerous studies exist on the concept and applications of Digital Twins in the building industry. As discussed, a different definition for DTs is found in each of them. However, regardless of how a DT is defined, all the definitions rely on (1) a model of a physical entity, (2) a live representation of the entity's status (usually in real-time), and (3) a digital brain that analyses all the data and responds to them. In a DT, the physical model should be connected to its corresponding virtual counterpart in a real-time manner so that the virtual model can create living digital simulations that change as the physical counterpart changes [19]. In the building industry, a DT can be achieved by utilising BIM (as the static segment of the DT), IoT-based network systems (for the dynamic representation), and analytics tools (see also Chapters 8, 10, and 11).

In addition, research on DTs aims to push DTs to be able to interpret, predict, and act on events in real-time [201]. As such, they become powerful masterminds that drive performance and innovation. They are intelligent enough to learn from themselves, human experts, and other similar systems. Such an interaction between a real-world building's environment and its digital yet realistically represented model provides many opportunities such as real-time monitoring and analytics, informed decision-making, building efficiency, occupant comfort enhancement, and most importantly, co-creation [41,201,268]. So far, most of the studies focused on the theoretical aspect of DTs, and progressively, work on DT is moving from theorising about its nature into actual implementation [41,437] (Chapter 8).

This section aims to provide a brief overview of the state of the literature on building DTs. To this end, 16 publications have been selected for review from both construction and non-construction journals (e.g., electrical and computer engineering) to include different perspectives. The focus is mostly on technical implementations; however, a couple of prominent reviews and reports are also included. A summary of these papers is provided in Table 9.1. Given the discussion provided in the previous section, the following were analysed for each paper: (1) Applications considered for the DT, (2) Whether the study uses or refers to the use of semantic web technologies and linked data paradigm, (3) the role of people when modelling the DT, (4) Means of visualisation in the DT, (5) Whether the DT works real-time or not, and finally, (6) the type of the data that was used when developing the DT (structured or unstructured).

Literature reveals to us that currently, studies are pushing for standardised databases at the core of a DT. As such, data from different sources are being collected, integrated, and used to provide service. Figure 9.2 shows how DTs are being employed in current practices, from capturing data to providing service. As shown, the architecture of DTs in their current form can be generally viewed as three layers of data acquisition and transmission, data storage and analysis, and services. It should be noted that this is only a simplified view of building DTs with the purpose of describing them as they are currently practiced. There are more detailed architectures of the DT layers, such as that in [234], and discussed in Chapter 8. In addition, this overview also helps framing the challenges and potentials of DTs in the building industry, which are discussed in the following.

9.2.1   The challenges

At the most fundamental level, the data is hardly available. The industry is notorious for the ad-hoc practices in data collection during design and construction phases. In the operations phase, even more challenges exist. A substantial portion of buildings lack a BMS to collect operational data. The majority of these are outdated and lack the ability to verify data integrity. In many cases, engineers have to climb rooftops, crawl under installations, wade into pump rooms, and peer above ceiling tiles to obtain circumstantial information to develop (educated) guesses to improve the energy efficiency of their facilities. Occupant data, a key category of operational data, is hardly collected.

When data is available, significant problems such as data reliability, provenance, and interoperability still exist. At this point, the most frequently mentioned challenge is data exchange and integration [15,234]. There is a significant interoperability gap that still exists in regard to cost, schedule, and energy analysis. It is hard to find reliable and interoperable data to support the engineering aspects of the design of new buildings. It is harder to find such data for the construction phase and even harder for the operations phase. This makes any integrated analysis time-consuming and inefficient. Partially, this is attributed to the fact that BIM does not cover the operations phase of buildings—some even argue that it should not. BIM is also overly focused on visual and material aspects of the engineering facets of the building.

The advancement of the knowledge economy is exasperating the problems. For example, new data streams of significant volume (such as IoT or social media or city-level big data) provide clear benefits for better managing our buildings and linking them to the smart city applications. However, without the ability to triangulate these types of data to location, material, cost, energy, and occupant data it will be hard to put them to use (see also Chapter 10).

Several studies have suggested that the semantic web has the potential to address and solve these challenges. According to these studies, successful implementation of a DT requires correspondence between its various objects and models so that later on, the DT can iterate on a process of continuous learning to discover patterns and

Table 9.1　Summary of the reviewed papers

Articles	Type	Applications	Linked data	People	Visualisation	Real-time	Data
Pargmann et al. [291]	Technical	Monitoring	No	No	AR	Yes	Structured
Ding et al. [106]	Theoretical	Predictive maintenance, quality inspection, automated scheduling, optimisation	No	No	Mobile terminals	Suggested	N.A.
Oliver et al. [96]	Technical/Theoretical	Monitoring, occupant interaction	No	Aims to provide interaction opportunities for people	AR/VR	No	Structured
Tomko and Winter [386]	Report	Prediction, monitoring, optimisation, simulation, active control	No	Suggest to include people as the social aspect of DT	N.A.	Yes	N.A.
Tang et al. [377]	Literature review	BIM-based system data integration and retrieval	Suggested	No	AR/MR/VR is Suggest	Suggested	N.A.
Zheng et al. [437]	Technical	Monitoring	No	No	Desktop platform	Yes	Structured
Khajavi et al. [201]	Technical/Literature review	Monitoring	No	No	Desktop platform	No	Structured
Wu and Liu [419]	Technical	Persuasive technology energy management	No	User feedback	Desktop platform	Yes	Structured
Boje et al. [41]	Literature review	Optimisation, simulation, monitoring, sensing, prediction	Suggested	No	N.A.	Suggested	N.A.
Austin et al. [15]	Technical	Smart city energy management, prediction	Yes	No	Desktop platform	No	Structured
Lu et al. [234]	Technical	Predictive maintenance, maintenance optimisation, monitoring	No	No	Desktop platform	No	Structured
Francisco et al. [139]	Technical	Energy performance benchmarking	No	No	VR	No	Structured
Cheng et al. [67]	Technical	Facility maintenance, monitoring, assessment, prediction, planning	No	N.A.	Desktop platform	Yes	Structured and unstructured
Lin et al. [440]	Technical	Monitoring, safety	No	No	Desktop platform	Yes	Structured
Mohammadi and Taylor [264]	Technical	Urban health knowledge discovery	No	Social media	AR/VR	Yes	Unstructured and structured

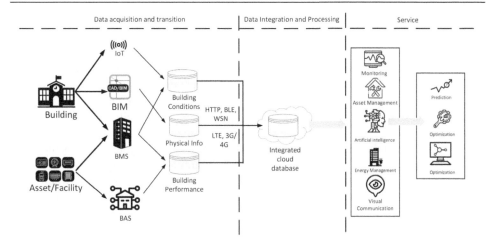

Figure 9.2 Current practices of digital twins from data to service.

cause-effect relationships and semantically refine its models [15]. The solution is then seen in the integration of semantic web technologies and machine learning. With this perspective, Austin et al. [15] developed a semantic modelling and machine learning architectural platform, which was implemented in a case of energy analysis of buildings in Chicago (a city-level DT). However, the potentials of the proposed solution are yet to be explored when dealing with the heterogeneous data sources of a building (building level DT).

Regarding the necessity of using semantic web technologies, there is a whole study in [41] that gives an in-depth explanation of the necessity and emergence of Digital Twins by addressing the limitations of previous technologies (BIM). The paper then recognises data and knowledge connection as the main challenge of DTs, the solutions of which lie in the utilisation of linked data and semantic web technologies. Even in non-DT relevant studies that focus on the integration of BIM and IoT systems, semantic web technology, and linked data is seen as a successful solution [377]. Nevertheless, we do not see many practical studies utilising the semantic web and linked data technologies when designing a DT. As stated in the five stars of open data platform, we need to investigate the bridge between unstructured, structured data and the role of RDF in that [34].

Equally important to finding reliable, interoperable data is the inadequacy of the normative (top-down) research approaches in the domain. BIM was predicated on the belief that a common data model can capture the domain's knowledge and promote interoperability. Real-world applications showed that this is not possible due to the diversity of data types, the context-sensitivity of construction and operations knowledge, and the lack of semantics in the BIM data model (IFC).

In addition to the integration challenge, there is a need for more effective visualisation of information to the many users of DTs. A DT is supposed to be used by many people with different backgrounds and for different purposes. Thus, the same data should not be visualised in the same way to different users. Profiling users and using innovative visualisation technologies is hardly seen in the literature despite the many advances that the construction industry has had in the visualisation domain (particularly the AR/MR/VR domain). Nevertheless, recent studies are starting to investigate the potentials of VR/AR technologies for DT data visualisation [139,291].

Additionally, we need further studies to adequately account for unstructured data. A huge portion of building data is unstructured, in the form of written documents, feedback, social media posts, or comments. Since the data generated by people are not usually in a structured form, perhaps, one of the reasons we do not see people as an element of DTs is the lack of effective solutions for including unstructured data in DTs.

Except in a handful of studies, we do not see people as a part of DT models. When people's feedback is not captured, the performance of facilities may not match their needs, and this can have adverse effects on sustainability. Furthermore, people are increasingly demanding more engagement. The Netsavvy and climate-active generation is not satisfied with the comfort levels in their buildings and their energy performance. They are also not satisfied with no access to data or, in the few cases where it is available, the reactive nature of data provided (existing smart building management systems only allow occupants to view energy data). The prevailing socio-technical trend is that customers should have the ability to access and "act" on data. Furthermore, the extent of the data on people is usually limited to their numbers (how many people are where and when), which can be attributed to the tendency to only include structured data. A DT should be an interconnected system of people's input/views and the status of building performance. The accumulation of such types of data is key to enabling the use of machine learning in intelligent buildings. We need to advance the practice of capturing these sets of data. Such an unstructured data will not be easily connected to a DT that is based on strict structural models such as IFC—which is a major challenge to the need for embedding BIM in DT.

We should not build DTs based on simple reactive principles and technologies. We should formulate our definitions and architecture to support both predictive and prescriptive analysis/modelling. The vision of DTs is to replicate the facility data and its current and possible future status in the virtual world to enable collaborative analysis of real-world data and an easier generation and evaluation of potential futures. This dynamic and unstructured architecture will rely on finding and empowering patterns to generate knowledge, and actively participate in the management and design of buildings. While linked data can play a key role (especially with structured data), the true value will be generated by relying on machine learning of unstructured (and, yes, not interoperable) data. This also requires a bidirectional link between buildings and DTs, allowing DTs to impact their counterparts (and exist like an organism capable of changing its environment [386]). Most of the current applications of DTs do not establish a bidirectional link, and the link is only from the physical to the digital. This can be partially attributed to the privacy and security issues that will arise if a DT (and its users) are given access to control a physical entity. In summary, the following can be considered the main challenges of DTs:

1. Lack of a common interpretation

2. Lack of a practical method for data integration

3. Difficulty to synchronise data in a real-time manner

4. Absence of data and models of people and occupants

5. Limited visualisation capabilities

6. Privacy and security

7. Exclusion of unstructured data

8. Lack of an active role and being either passive, re-active, or predictive

9. Lack of a bi-directional link between the physical and digital counterparts

9.2.2   The potentials

Digital Twinning can offer powerful data solutions that provide asset management efficiencies and ultimately save time, cost, and energy, and improve comfort levels. It can be used for a variety of purposes spanning monitoring, asset maintenance and performance prediction, source of product well-being when selling the product, and a black box in emergencies and disasters. Consequently, digital twinning promises more sustainable buildings (encompassing all the aspects of the environment, economy, and society) and on the long run will enable a better understanding of how cities are functioning as a whole, resulting in the establishment of smart cities (see Chapter 11) [416].

Given the characteristics of Digital Twinning, on the fundamental levels, they facilitate current processes of building management practices (e.g., asset monitoring). More smart applications include predictive maintenance for preventing asset failure, improving energy management, or space management [51,340,421]. Moreover, Digital Twinning can enable persuasive technology as a tool to impact and inform occupants. Persuasive technology refers to design principles of persuasion and control of product automation technology to ensure a change in user behaviour (change in ideas or practice) [419]. A persuasive technology requires day-to-day objects to effectively communicate with users. To do this, users require a sensing device that can store and collect perceived data through network communications for use in a system with interactive visual interfaces [419]. All these requirements fit very well with the capabilities of a DT. In more complex forms, Digital Twinning can even be equipped with socio-psychological concepts to improve effectiveness [70]. Other common Digital Twinning applications include safety (e.g., evacuation or rescue in fire incidents) [66], and structural health monitoring [184].

From an economic perspective, applying data analytics in buildings will help improve their performance in various ways. Increasingly, reports are being published on how much savings can building analytics systems achieve [211]. As a result, there is an awareness that data-driven advanced modelling and analysis is needed to make the right decisions, and there is a push from industrial needs for DT projects. The energy efficiency economy is creating a demand for advanced data analytics in buildings. For example, in Toronto, large buildings are required to report electricity, water, and gas use. Furthermore, the success of AI applications in other sectors made it clear to large industrial players that investments in data management are not just good engineering practices, it is an essential tool for market survival.

Digital Twinning will be able to address the data management issues in the building industry by providing unprecedented datasets and enable building-as-a-platform systems. Digital Twinning will enable data triangulation and provide access to diverse and longitudinal structured data (such as IoT and occupancy data) and unstructured data (such as complaints, work orders, and maintenance logs) of buildings as well as contextual data (such as weather and user community profile). Most importantly, they will provide access to (the typically elusive) occupant data and to occupants themselves, innovative stakeholders with unique knowledge profiles who can collect data and generate ideas.

Finally, Digital Twinning provides a chance for a breakthrough in research approaches. For a long time, research in the domain has been dominated with expert-based, top-down, normative thinking in the form of seeking to create common data models. The size and diversity of the data that will be provided afford us an unparalleled opportunity to explore approaches that rely on machine learning. This will allow us to explore the use of data-driven methods for creating dynamic links between data sets and for using the pattern-based analysis to extract rules bottom-up.

9.2.3   Machine learning and advanced sensing

As discussed, DTs are more than data repositories in that they are capable of capturing the context and dynamics of their physical counterparts thanks to soft sensing and machine learning algorithms. In other words, DTs do not only capture and store raw data (e.g., room CO_2 levels), but they sense events (e.g., use the CO_2 level data to understand room occupancy numbers), discover patterns (e.g., occupancy patterns), and predict the future scenarios (e.g., occupancy load in the near future). Some researchers refer to such an advanced level of sensing as "Soft Sensing" which is a critical component of DTs [163]. The concept refers to using analytics or algorithms for combining several measurements (simplified data) to suggest (i.e., sense) that an event has occurred or is about to occur. There are two major differences between traditional sensing and soft sensing. First, it is software that does the "sensing", not a hardware/sensor/IoT (Chapters 3 or 10). Second, and more importantly, it reports/predicts events or higher-order data that cannot be otherwise detected. For instance, a hardware can be used to identify the presence of an occupant in an area. But soft sensing combines the occupant presence with other data such as weather conditions and work environment conditions and predicts the possibility of an accident. In other words, soft sensing helps capture abstract knowledge that cannot be collected physically. The use of soft sensing is essential to any effort that aims to engage occupants in the operations of intelligent buildings.

Currently, soft sensing, in the building industry, is predominantly used for improving the energy management of buildings by considering the presence of occupants in the algorithmic design of HVAC operations. However, other industries have advanced the soft sensing applications to understand the behaviour and even intentions of users and then use these insights to predict the next moves of users. An accurate estimation of what users are likely to do will provide user-centred intelligence for homes and cities, which can be used for providing high-value services.

According to Melfi et al. [259], knowledge about occupants can be captured in different resolutions, which can be assessed in three dimensions. Table 9.2 (inspired from [259]) shows the three dimensions of occupant info. In the first level, we can identify whether an occupant is present or not, and in the second level, we can understand an approximate number of occupants (rather than just presence). The third level would be identifying the occupants in the building. The identification does not have to be detailed (as it can violate privacy); rather, any level of knowledge about occupants can be categorised as level three (e.g., the distinction between students and teachers). In the fourth level, we can understand the activities of occupants in terms of what they are doing (e.g., resting or working). Higher resolutions require more complex implementations of

Table 9.2 Occupancy resolution [259]

Temporal resolution	Occupant resolution	Spatial resolution
Days	Occupancy	Building
Hours	Count	Floor
Minutes	Identity	Room
Seconds	Activity	Exact position

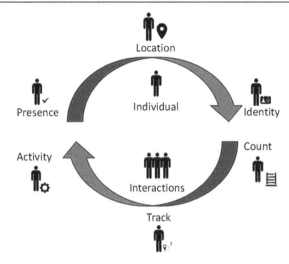

Figure 9.3 Spatio-temporal occupant properties.

soft sensing algorithms. For instance, simple hardware (sensor) can identify whether an occupant is present in a room or not (level 1 resolution). But no existing sensor can tell us whether an occupant is anxiously working or happily resting in a room (level 4).

Of course, this classification is neither comprehensive nor necessarily true. For instance, it might be argued that identification of a user might be more difficult than recognising the activity of a user. Also, there are other aspects of occupants that are not included here. For instance, one might be interested in the physical (blood pressure) or even mental (stress) state of a user. In terms of spatial resolution, the details can be more than just locating users and cover tracking a user movement from one room to another. Therefore, different classifications in different studies are seen for soft sensing. For instance, Labeodan et al. [217] suggested a different spatio-temporal scheme for occupancy measurements as shown in Figure 9.3.

We can capture a wide range of data from occupants such as their infrared radiations to CO_2 levels. According to [354], Passive Infrared (PIR) sensors are the most common methods in traditional approaches of occupancy detection. These sensors measure the infrared radiations, the changes in which is a sign of whether occupants are present or not. These systems are associated with many uncertainties. As a result, energy managers have to take conservative measures, which diminish the energy-saving potentials. Another approach is to use ultrasonic sensors to detect human movements. Similar to PIRs, these sensors are also limited. For instance, they might detect a blowing curtain as a human motion. Limitations of these sensors are further discussed in [354].

As a result, some studies began to explore the CO_2 generation rate of occupants for detecting and counting them. However, CO_2 concentration is affected by many factors (e.g., the air permeability of the envelope) and cannot be used as the sole feature to detect occupants. Therefore, it is used as a complementary feature along with other sensors. For instance, the study in [13] used motion detectors and CO_2 sensors to count the occupants of a room. These traditional methods require a collection of sensors (wired or wireless) to work accurately [108]; and, therefore, are both expensive and non-robust (i.e., if one sensor fails, the whole system becomes either suboptimal or nonfunctional). In response to the limitations of traditional systems (e.g., low resolution, false detections), newer methods shifted from designing detector hardware for gathering physical data to designing smart software for interpreting existing data, i.e., Soft-Sensing.

There are two clear advantages here [354]. First, these, so-called, soft-sensing systems use data that are already collected by building control systems (for instance, BMS) or are available but not collected. As a result, they do not need robust expensive systems merely for collecting data (e.g., infrared radiations). The second advantage is that due to the collection of a variety of data, the aggregated accuracy is far better than traditional systems. The data that is collected through soft-sensing systems has a wide range from elevator usage to devices connected to Wi-Fi endpoints. Soft-sensing systems can also use sensors but are essentially different from traditional systems. The sensors capture information such as temperature, relative humidity (RH), and light intensity and are wirelessly connected to a network (such as Zigbee) [259] (see also Chapter 10).

Of course, data collection is not the only difference between soft-sensing and traditional sensing approaches. The difference mainly lies in the analytics process (see also Chapter 11), where machine learning algorithms are employed to make sense of the available data in an environment. Significant achievements have been made in soft-sensing and energy conservation methods. For instance, Dong and Lam [109] developed a complex environmental sensor network and captured motion, light, sound, CO_2, and RH for counting occupants in a room. At the core of the system, a Gaussian Mixture model-based Hidden Markov Model (HMM) was used to analyse the data. The learning system managed to achieve an outstanding accuracy of 83% on occupancy numbers, which could potentially result in 18.5% energy savings (the saving is obtained from simulation). As discussed above, the data used for soft-sensing does not necessarily require specific sensors. Thanks to the analytical capabilities of ML algorithms, already existing data can be used for accurate detections. For instance, the study in [146] used area access badges, Wi-Fi access points, calendar, and instant messaging clients and achieved a 90% accuracy in occupant detection.

Currently, a soft-sensing application in the building industry is mainly for energy management by optimising HVAC control systems [55] or lighting [216]. However, if we aim to engage occupants as a part of building management or any bigger system (such as a Digital Twin), we need to be able to make higher-level inferences to be able to communicate with the occupant. We need to be able to detect their behaviour, not just their presence. Once their intentions and behaviours are detected, the building can be considered intelligent enough to communicate with occupants and thus engage them in the process of building management. Further, the occupants can have access to a wider yet more specific range of information. For instance, once a system is intelligent enough to identify the profile of the occupant being communicated with, it can employ recommender systems to suggest appropriate information and analytical tools based on the profile of the user and the activity he/she is doing. At this point, the system can be intelligent enough to detect and predict abnormalities and failures. For instance, Forkan et al. [136] developed a system based on HMM to detect anomalies in the daily routines of users. Such systems are called context-aware smart systems and are widely pursued in smart home studies [160].

Usually, the process of predicting user behaviour has two stages. In the first stage, activity types are recognised, and profiles for each activity are created. These processes are usually unsupervised using methods such as clustering algorithms (k-means in particular). In the second stage, the activity profiles and a set of ML or deep learning algorithms are used to predict user behaviours based on a set of events [160]. For instance, Bourobou et al. [53] used k-pattern clustering algorithms to construct user activity models and then used ANNs to predict user activities. Soft-sensing can also be used to understand the intentions of users. For instance, a user picking up a cup indicates the intention of drinking. HMM is particularly useful for this matter [160]. An interesting line of research in this domain is the integration of semantic web technologies and ML methods. Ontologies can be used as initial sources of activity models, which ML algorithms can use to detect user behaviours. As activity data increases, new patterns of activities are discovered, and the model is updated [64].

Although soft-sensing approaches have opened many doors in research, there are a couple of limitations as well. Fundamentally, implementing such advanced sensing systems requires system integration and live access to data which is not desirable for many data managers. Privacy is a major downside of soft-sensing. Capturing occupant data and analysing it in servers located elsewhere might be considered a privacy invasion by many. This is particularly important if the data is occupant images [339]. Edge computing can be a possible solution to the local analysis [55]. But it does come with its own limitations as it is only using local files (see Chapter 10).

One of the main limitations of soft-sensing approaches is the scale of implementation. According to [354], most of the studies of occupancy analysis are proof-of-concepts implemented in short periods of time in a limited number of spaces. Wide-scale implementations will introduce a new set of challenges, especially on the data management side. With massive amounts of information being collected through long periods of time from a large number of systems, managing the data will become a critical challenge. It is currently difficult and infeasible to advance current approaches from a proof-of-concept to real-life systems used in the daily management of buildings. There are still investigations to be made on the potentials of the linked data paradigm and machine learning solutions to address these issues.

9.3   THE UNIVERSITY OF TORONTO INTELLIGENT BUILDINGS DIGITAL TWIN PROJECT

This section presents the University of Toronto (UofT) Intelligent Buildings Digital Twin (IBDT) project to showcase the main considerations, opportunities, and challenges that should be considered in a wide-scale DT implementation. IBDT aims to create a digital twin of UofT 150 buildings throughout their life cycle with the following main objectives. The centre will:

1.   Provide easy access to data for researchers in related fields;
2.   Act as a testbed for the implementation of advanced data management tools;
3.   Examine and showcase the feasibility of artificial intelligence (AI) in the industry;
4.   Serve the needs for the facility management at UofT; and
5.   Empower building occupants to participate in the management of their facilities.

IBDT is situated at the nexus of operational optimisation and research work. The project is implemented in three phases to serve two layers of action: 1- UofT Facilities and Services, and 2- Building Stakeholders. Currently (March 2021), the first phase is completed and the project is progressing through its second phase. Figure 9.4 shows the overall domain of action of the project, and Figure 9.5 shows the main objectives and tasks of each phase of the project.

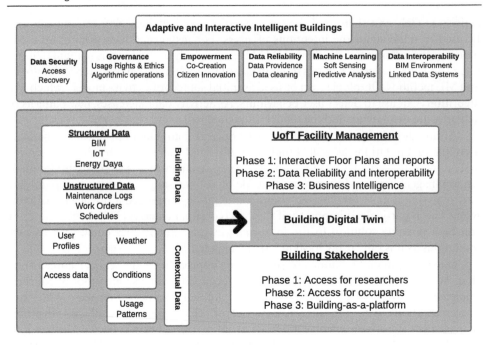

Figure 9.4 Domains of action of the project.

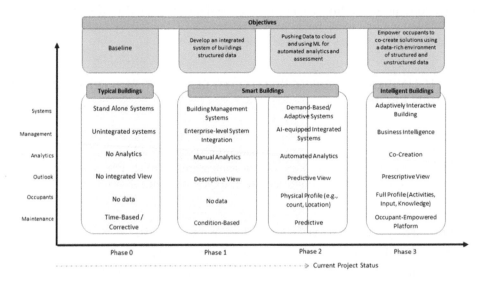

Figure 9.5 The intelligent buildings digital twin Project.

9.3.1   Phases I and 2

As shown in Figure 9.5, the first two phases of the project are focused on transforming the buildings from typical non-automated buildings into smart ones. The project is committed for cost-effective, sustainable, and efficient solutions for all UofT facilities. As such, the design of all building systems should present the best use of technologies. Motivated by the values behind integrated technologies, a system architecture is designed

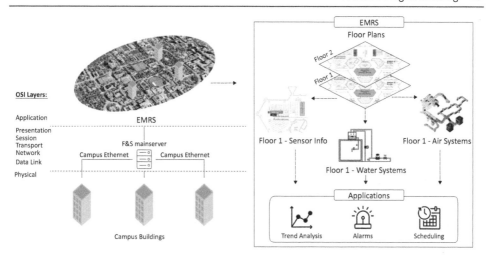

Figure 9.6   Using EMRS for managing campus as a single building.

to seamlessly integrate business and building systems on an enterprise network. As shown in Figure 9.6, the proposed architecture treats the whole campus as a single building. A secured and dedicated CAT 6 fibre Ethernet intranet is used throughout the campus to connect buildings at a main distribution switch. All campus-wide approved IP-based devices are connected to this Ethernet physical layer, and all the systems are hosted and maintained on the main server of the U of T Facility and Services (F&S) [282]. In the proposed architecture, buildings and their systems (e.g., HVAC systems) are considered "Mechanical Rooms" which are connected to each other through the Ethernet network on an enterprise server (see also Chapter 10). The architecture is composed of building-level controls integrated into a campus-wide Enterprise Management and Reporting System (EMRS), which acts as the application layer. This campus wide architecture is developed to centralise the process of managing the following systems: (1) Building Automation Systems (BAS), (2) Energy Metering systems, (3) Security and Access Control Systems, (4) CCTV and Digital Video, (5) Fire and Life Safety Systems.

In the first phase, UoT buildings are equipped with Building Automation Systems (BAS) with complete Direct Digital Control (DDC) solutions for automated control of HVAC and Lighting systems (HVAC+L). All BAS connect to the campus LAN data link layer to form a single campus backbone. The devices and objects from any integrated BAS on campus are compiled and displayed in the EMRS platform. The EMRS is a front-end application layer with access and oversight to all BAS. The EMRS is at the application layer and provides the following services: (1) Data sharing, (2) Alarm and Event Management, (3) Scheduling, (4) Trending, (5) Device Management, and (6) Network Management. BACnet/IP communication protocols are used for the integration of all BAS on the campus.

The specific architecture of BAS used in each building is different. The details of such systems are out of the scope of this chapter, however, generally the architectures are similar to Figure 9.7 (for reference, see also Chapters 10 and 11).

Each building has several field Input/Output (I/O) devices such as sensors and actuators in different locations (rooms and spaces). These field devices (FD) are connected to a set of controllers in a Field-Bus network system. In other words, FDs use the serial Field-BUS protocol (RS485) to communicate to the IP-based building

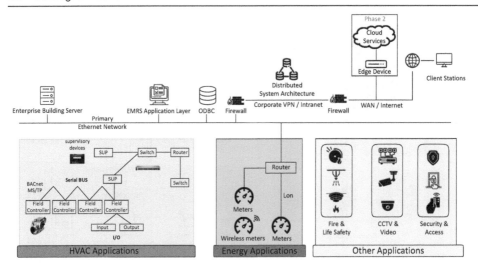

Figure 9.7 Campus building systems architecture.

controllers. The field controllers are connected to each other and to a supervisory device (SUP) using BACnet communication protocols. Through a supervisory (SUP) device, the controllers are then connected to routers and then to the main Ethernet cable. An important factor in the design of the architecture is that buildings should operate independently, and the central system mainly acts as a source of data collection and analysis. As such, the BAS control strategies will reside at the controller level and will communicate with the LAN network using MS/TP or IP data link protocols.

Another major source of target data in this phase is energy metering data. Electricity, heat, steam, and water usage are examples of this sort of data. In this phase, the readings from each metering device are also connected to the main Ethernet. The LonWorks communication protocol is a common protocol used for utility control networks and metering devices. A router is needed to provide network connectivity between the Ethernet LAN and LonWorks. The details of other systems are excluded for the purposes of this chapter.

The architecture is equipped with Enterprise System Integration tools and offers interfacing options to Enterprise Resource Planning (ERP) systems for database synchronisation. Furthermore, Open Database Connectivity (ODBC) support allows read-only access to the EMRS. Additionally, the Distributed System Architecture (DSA) allows multiple/geographically distributed servers, each performing a different application (e.g., HVAC, Life Safety), operate as a single system. Thus, DSA provides a centralised monitoring and scalability. The system also provides Intranet and Internet connectivity for clients to access from anywhere in the world.

9.3.1.1 IBDT for UofT facilities and services

In the first phase, EMRS provides a descriptive view of real-time conditions of building systems and enables condition-based maintenance, as opposed to a typical building, where maintenance is done either to correct a failure (corrective) or on a time-based system. The system serves facility operators by providing them access to interactive floor plans. This allows operators to visualise building data, study conditions, conduct inquiries, track trends, and generate reports. The platform enables the development of spatial intelligence solutions by linking BAS data to facility layouts and floor plans. Instead of accessing data at the device level, this digital twin will allow the analysis of zone-level events and conditions. This has direct uses in addressing the localised needs of occupants, enhancing energy efficiency, and optimising space use.

The second phase of the project aims to provide higher-level controls for the whole building and, at the same time, help in orchestrating BAS and IoT devices. As shown in Figure 9.6, an Edge gateway will be utilised to provide centralised management of all the devices (see also Chapter 10). This phase aims to push analytics from manual to automated by adding ML abilities to the system. As such, maintenance can be done based on predictive analysis (predictive maintenance) which has higher values than condition-based maintenance [347]. Most importantly, this phase aims to improve system interoperability by allowing systems to interact with each other and adapt to the physical conditions. For instance, using Edge-AI technologies [221], the system will integrate physical profiles of occupants, such as their numbers and locations, with HVAC management systems to implement a demand-based management. This will allow HVAC systems to adapt based on the presence of occupants.

Phase 2 will also expand both the structured and unstructured data used in the DT. The platform currently uses floor plans and layouts to visualise a building (a non-BIM DT). In phase 2, BIM models of buildings will be added to the common data environment (CDE) (see also Chapter 6). In addition, schedules of activities and room reservations will also be added to be used for profiling buildings. As for the unstructured data inspection and maintenance logs, work orders, and service requests made by occupants will be added to the platform.

9.3.1.2 IBDT for building stakeholders

On the second layer of action, the project aims to promote advanced research and innovation by researchers and building occupants. In the first phase, IBDT enables UofT researchers to access building data to conduct simulations and analyses (mainly for energy and comfort modelling). Access to data has been a major challenge to researchers in the field. Researchers can use this large, longitudinal, and real-time data, to explore new horizons in the design and analysis of intelligent buildings. Interaction with operators and occupants can also be another unique feature for research work. Using the platform, researchers are able to address real-world problems, draw on the expertise of operators, and have a conducive environment to test the applicability of their models. In the second phase, the project aims to provide a portal for engagement for occupants. Using the portal, occupants can share ideas and debate alternatives. This will advance researchers' abilities to investigate new models for the socio-technical nature of buildings.

9.3.2 Phase 3

In the third phase, business intelligence applications will be developed and served to facility operators. Insights gained will enable operators to improve facility monitoring, explore system inter-dependencies, foster linkage and traceability across teams and ecosystems, and enable a better response to real-time conditions. This can help in simulating and predicting future states (prescriptive view), which can help in the efficient management of incidents and in improving and recalibrating preventive maintenance policies. The key objective of this phase is to encourage and enable occupants to lead decision-making and innovate new products and services. So, in the third phase, a pilot of the building-as-a-platform will be available to occupants. It will include basic analytics services to empower them to share ideas, debate options, study what-if scenarios and negotiate action plans. The vision of this phase is to have buildings that are able to interact with occupants and adapt to their needs.

9.3.3   Project outcomes

In addition to providing effective and secure access to stakeholders (operators, researchers, and occupants), the design of the project allows to conduct research on digital twin architectures and management, develop, and document a best practices map in the use of data warehousing and edge computing in the building sector. The objective here is to conduct applied research into the management aspects of these technologies. Specifically, analysing the needs and suitability of the technologies to different usages within the building industry means to theorise the decision-making criteria and their assessments, the contribution and role of IoT-level analysis as it relates to edge-level analysis.

Of particular interest is advancing occupants' relationship to and use of data. While the technology side of data warehousing is well established, the human and process sides are still in their infancy in the building industry. The project will link the research on data governance to the design of digital twins to examine new technology architectures that promote human-building interactions. To this end, the project will have access to usage data: what patterns of access are associated with which user profile, under which building conditions? Equally important, the system has access to the semantics of occupants' comments. This is an input from the key stakeholders, who have first-hand experiences of and insights into intelligent buildings. These data sets can be used to investigate two broad questions: (1) what is the nature of the complexities of the actor-concept networks? Addressing this question helps understand best practices for profiling users and their usage patterns. (2) What are the characteristics of the moments of truth: what triggers input from users (ideas, complaints, etc.)? Answering this question helps identify the modes of data delivery and operation in intelligent buildings that would make it valuable for real-time usage.

9.4   SUMMARY AND CONCLUSION

This chapter begins with an argument that DTs are not interoperable BIM-based data repositories. It is discussed that full interoperability is not achievable. It is infeasible to accommodate everything within IFC. Given the complexity of intelligent building data, even linked data systems are challenging to develop and infeasible to maintain. Business-wise, DTs are meant for the use of a wide spectrum of users who are business-oriented as much as engineering-oriented. However, standards like IFC are developed by engineers for engineers.

DT users need to harness a building's data and understand what is/will be going on in the building, and use it to support efficient decision-making. The required knowledge for such analytics is contained in unstructured data, which makes standardised data models even more infeasible. DTs are meant to be a virtual twin of a building, not its data repository. They should be able to describe what is going on in a building and predict potential future scenarios. As a result, in addition to data, DTs must encompass two types of models, descriptive models and prescriptive models. Descriptive models will be developed based on pre-defined theories and will use data to understand ongoing conditions. The results will be used in prescriptive models to virtually experiment with potential scenarios before implementation. This makes Digital Twinning a continuous process of using, critiquing, learning, updating, and reinventing.

Digital Twinning will engage the building, its operators, and its occupants in the continuous optimisation of building goals and occupant comfort. As such, intelligent buildings should be equipped with algorithmic governance, creating a data-rich computing environment. This will allow operators to optimise building performance through automated decisions. This will also empower occupants to lead the analysis, generate knowledge, and co-create solutions. The proposed approach will create a "building-as-a-platform" environment, which allows stakeholders (including occupants) to use a portal for engagement (akin to a social media platform), where they can share ideas and debate alternatives.

To materialise the discussion, the chapter presents the University of Toronto's Digital Twin project. Technical discussions on how a wide-scale digital twin project is being implemented are provided. The long-term objective of the project is to develop a pilot of the building-as-a-platform for occupants. The platform will include basic analytics services to empower occupants to negotiate action plans, debate options, study what-if scenarios, and share ideas.

Digital Twinning will address the building industry data management issues by providing unprecedented datasets of longitudinal structured data, unstructured building data, contextual data, and most importantly, occupant data. Additionally, Digital Twinning will revolutionise research approaches from expert-based, top-down, normative thinking to data-driven, bottom-up analysis. Instead of creating common data models, researchers can focus on creating dynamic links between data sets and use pattern-based analysis to extract rules bottom-up.

NOTE

1 Colin J. Parris. Minds + Machines: Meet A Digital Twin. https://www.youtube.com/watch?v=2dCz3oL2rTw

Chapter 10

IoT and edge computing in the construction site

Aaron Costin and Janise McNair

CONTENTS

Technological advancements such as robotics, automation, sensors, as well as the advanced use of wireless laptops, tablets and smart phones, have provided the construction industry with new and improved techniques to increase productivity, quality, and safety on the construction site. Recently, the Internet of Things (IoT) has increased the value and benefits of these technologies by integrating the ability of even more devices to connect and share data over the Internet. In order to leverage such systems, it is important to understand the capabilities and constraints of these systems in the construction environment. Connectivity, bandwidth, cybersecurity, and quality of service (QoS) are a few of the many challenges that will be faced with the dynamic and chaotic nature of the construction environment. This chapter provides an overview of the IoT system architecture on the construction site, considering state-of-the-art technology and next generation IoT infrastructures, including software-defined networks (SDN), edge computing, cloud-based services, and machine learning techniques. Recent innovations and applications will be reviewed to identify the strengths and weaknesses

DOI: 10.1201/9781003204381-12

of IoT-enabled construction sites, identify research gaps, and recommend best practices for the next generation of intelligent construction sites.

10.1   INTRODUCTION

The construction industry's adoption of information and communication technologies (ICT), namely automation, robotics, and building information modelling (BIM), has been slowly, but surely, increasing over the past decades. Recently, there has been an increased enthusiasm for the use of digital ecosystems and smart technologies, such as Cloud Computing and Artificial Intelligence (AI), as the lower costs, greater availability, and tangible benefits have been major driving forces in part by the 4th Industrial Revolution (i.e., Industry 4.0). This enthusiasm has led such a paradigm shift in greater research and adoption, that the industry is witnessing a new era called Construction 4.0, which is an idea *based on a confluence of trends and technologies (both digital and physical) that promise to reshape the way built environment assets are designed and constructed* [341]. Construction 4.0 has a foundation based on Cyber-Physical Systems (CPS), which is the interaction between the physical world and the virtual (cyber). Essentially, CPS involves systems that focus specifically on the physical-virtual data loop. This feedback loop is an important aspect that brings the "smartness" to devices, since Big Data analytics, machine learning (ML), and artificial intelligence (AI) techniques can be utilised to learn user habits.

One major aspect of CPS is the use of physical objects, more aptly "things" that are embedded with sensors, software, and mechanics for the purpose of connecting and exchanging data with other physical objects over the Internet. This large network of connected devices is now common referred to the "Internet of Things" (IoT). The significance is that IoT devices can be embedded into a variety of objects, such as tools, equipment, environmental sensors, medical devices, and even clothing. These IoT-enabled devices are considered "smart objects" since they now possess rich computing and analytical potential. Significantly, these connected IoT devices can enable a production and transmission of information that was unimaginable, ultimately providing new insights that can lead construction companies to goals such as cost reductions, productivity gains, safer operations, new business opportunities.

So, what makes a device "smart"? Does plugging a lamp into a Wi-Fi-enabled outlet and being able to control it make it a "smart lamp"? No, not fully since this is automation and convenience. However, if the outlet can monitor energy consumption and could be integrated with systems to learn a user's habits, which in turn could have time constraints which would optimise the use in order to reduce energy, only then could the system be labelled "smart". Still, the lamp, itself, is not a "smart object", but rather the Wi-Fi-enabled outlet is. Deploying an emergent technology to a specific application is not enough to call it "IoT". There are many great off-the-shelf products and research projects that claim to be under the terminology "IoT" simply because they either utilised a nifty technology or BIM-based solutions. Such technologies and applications can be integrated into an IoT system, but they alone are not "IoT". Continuing with the previous example, a lamp could be made "smart" and be considered "IoT" if it had embedded the mechanisms and sensors needed for electronic computations, which will be discussed further.

This chapter dives deeper into these nuances and provides an overview of the IoT system architecture on the construction site, considering state-of-the-art technology and

Table 10.1 Terms and acronyms

Acronym	Term
AI	Artificial Intelligence
BIM	Building Information Modelling
CPS	Cyber-Physical System
IC	Industrialised Construction
ICT	Information Communication Technologies
IEEE	Institute of Electrical and Electronics Engineers
IPv#	Internet Protocol version number
IoT	Internet of Things
LLP	Low-Power and Lossy Networks
ML	Machine Learning
QoS	Quality of Service
RFID	Radio Frequency Identification
WSN	Wireless Sensor Network
URI	Uniform Resource Identifier

next generation IoT infrastructures. Table 10.1 displays the terms and acronyms used in this chapter.

10.2   CONSTRUCTION INDUSTRY: PUSH-PULL TO CONSTRUCTION 4.0 AND IoT

The construction industry has great impacts on the social, economic, and environmental well-being of society since construction is the foundation of the built environment. The construction of a project involves the planning, designing, financing of the project in order to have successful construction, which is ultimately determined if the project is performed safely, on time, and under budget. The main performance metric of construction is productivity, which is defined as *the relationship of the amount of output produced by a given system during a given period of time, and the quantity of input resources consumed to create, or produce outputs over the same period of time* [422]. Construction input consists of labour (physical work), materials, equipment, energy, capital, and data; while the output is the final goods and services (i.e. the finished building). Therefore, construction aims to provide the highest quality building at the most economical cost by minimising the costs of the input. Hence, any increase in the inputs could negatively affect productivity. The construction industry is one of the largest markets in any nation and makes up a substantial portion of developed countries' economy; however, the industry has been plagued with inefficiencies and outdated practices that have been keeping the productivity rate increases near zero over the past few decades. Productivity in manufacturing has nearly doubled over the same time period, due to improved technologies and automation [62].

This lag in productivity growth in construction (see also Chapter 1) can be attributed to two major factors: the lack of technology innovation optional and the shortage of skilled labour. The lack of technology and innovation has been plaguing the construction industry, mainly due to the resistance to adoption and lack of investments [84]. In addition to the institutional barriers and resistance to change, other factors include technology implementation barriers, opportunity costs, lack of knowledge of the technology, and legal implications [84]. Although the trend is to become more digital, all contract documents and submittals are still paper-based. Additionally, most of the technology adoption took place in the engineering and design phases, such as Computer-Aided Design (CAD) and Building Information Modelling (BIM), and the back-end office task (document control, payroll, etc.) where there is no skilled construction labour. The increased use of technology for up-front planning has been reducing actual construction work. For example, robotic total stations can eliminate the work of three workers, but need a digital plan which is created in the office. The second major factor is the lack of training, mainly due to skills gap between skilled and unskilled labour. Skilled labour positions are important since they have the experience and certification to perform various tasks that unskilled labour cannot legally nor safely due. However, there is a strong resistance by companies to properly train workers mainly because of the reluctance of the added training costs, high turnover of trained workers (i.e. lost investment), and lack of employee acceptance to formal training programmes [150]).

Other attributes to the lack of productivity growth include the rising costs of labour and the costs of safety incidents. The rising cost of labour, especially skilled labour, has a direct impact on the final unit costs of a project, which could affect the final economics of the company since the cost of input has been increasing with minimal output to compensate. Profit margins are fairly low in construction, ranging about 5% for building construction and upwards of 10% for heavy civil. Thus, any increase in costs greatly affects the competitiveness. Skilled labour serves a critical role in society by being the bedrock of a healthy middle class and balanced economy. Furthermore, the lack of productivity due to the rising costs of skilled labour is exasperated by an increasing labour shortage. This shortage greatly impacts the negative productivity trend since contractors will have to spend more time and money training unskilled labour to perform the tasks of skilled labour. Furthermore, the costs of safety incidents directly affect productivity and the budget. In construction, safety and productivity go hand-in-hand since safety has a direct impact on production as well as the profitability of a project. The costs of safety incidents can be detrimental to a company's profits. Not only is the direct incident costly (medical bills, increased insurance premiums, workers compensation, legal claims, fines, etc.), but the indirect costs can be staggering. Indirect costs include the loss of time due to the project being shut down, increasing safety training and education programmes, decrease in worker morale, and potential of lost revenue from not being selected on future projects due to safety incident rates.

The push forward from the dire need of improved methods paired with a pull of the technology advancement in other industries has created a perfect opportunity for the construction industry to capitalise on.

10.3   IoT SYSTEM FRAMEWORK

Despite the name, the "Internet of Things" is not just a large quantity of things (e.g. devices) connected to the Internet. IoT represents the ability to have a multitude of heterogeneous devices communicate with each other without physical connections. These devices, which are equipped with a form of sensor or actuator, a tiny microprocessor, a communication device, and a power source, have come to be known as "smart objects" [399]. The physical devices and technology are a heterogeneous combination of devices, including RFID, wireless, and infrared communication devices, as well as sensors for electrical activity, proximity, motion, etc. These devices represent the ability to have a multitude of heterogeneous devices communicate with each without

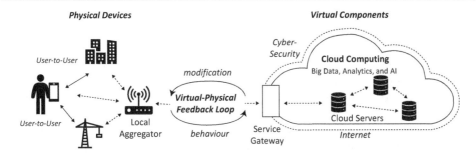

Figure 10.1 Abstraction of the IoT system for construction. (Image from [87].)

physical connections and can produce real-time data that can be used for analytics and optimisation. A basic abstraction on how the IoT components connect to one another is displayed in Figure 10.1 (adapted from [85]), which represents the data flow of smart objects through the IoT framework. The IoT framework is composed of four basic components:

1. Smart objects
2. Local aggregators
3. Service gateways
4. Cloud networks

Most notably, these systems include objects and sensors that can communicate over local aggregators, such as wireless local area networking (LAN) (e.g. Wi-Fi), Bluetooth, or Long-Term Evolution (LTE) to connect to the Internet to achieve various tasks. As the IoT devices represent the physical world, the Cloud that the data are sent into and processed is known as the virtual world. The virtual components are the Cloud servers that host the various algorithms for Big Data, Analytics, and Artificial Intelligence (AI). The intent is to utilise the IoT physical devices to perform feedback control loops, where the devices provide and receive information from a control system that executes a certain application [47]. This feedback loop is an important aspect that brings the "smartness" to devices, since Big Data, analytics, and artificial intelligence (AI) techniques can be utilised to learn user habits.

Significantly, feedback is needed to adjust functionality based on human behaviour and preference. A significant component of this network are the human and social aspects, which are critical for the virtual-physical feedback system to be able to adjust functionality based on human behaviour and preference. This feedback loop is essential to producing and assessing information, such as how the system can adapt to user preferences and behaviours, how human behaviours can be modified to adapt to changing environments, and how the system can be modified to achieve desired results (e.g. maximising worker safety). Thus, this is a continuous feedback loop that modifies both human behaviour and system optimisation.

10.3.1 Edge computing

A major barrier commonly faced for construction projects is the low access to high speed internet and computing. Furthermore, end users mostly have limited storage capacities and finite processing capabilities, thus running computing-intensive applications, such as BIM, in a resource-limited area, namely the contractor trailer. This is clearly a major challenge.

To combat this, one effective method is the use of edge computing, which is a new computing method that brings computation and data storage closer to the location where it is needed. Many Internet users are familiar with cloud computing, the delivery of computing services – including servers, storage, databases, networking, software, analytics, and intelligence – over the Internet ("the cloud"). Cloud computing includes familiar software as a service (SaaS) instances, such as Google Apps, Twitter, Facebook, and Instagram, supported by processing engines like Google File System located deep in the Internet architecture.

On the other hand, much of the IoT-created data will be stored, processed, analysed, and acted upon close to, or at the edge of, the network with very short response times for large quantities of private or proprietary data. Cloud computing is not equipped to support these local, busy, private applications. Data is increasingly produced at the edge of the network, therefore, it would be more efficient to also process the data at the edge [355]. Shi et al. [355] define edge computing as the enabling technologies that allow computation to be performed at the edge of the network, on downstream data on behalf of cloud services and upstream data on behalf of IoT services. Figure 10.2 gives a schematic overview of this landscape of computing on cloud, edge and devices to be used for the construction site.

A smartphone can be the edge device between things close to a person and cloud, while a gateway in a smart home is the edge between home things and cloud. Smart homes and buildings are well-suited for edge network processing. Many wireless sensors, robots, cameras, proximity sensors, health and security indicators and other devices are deployed

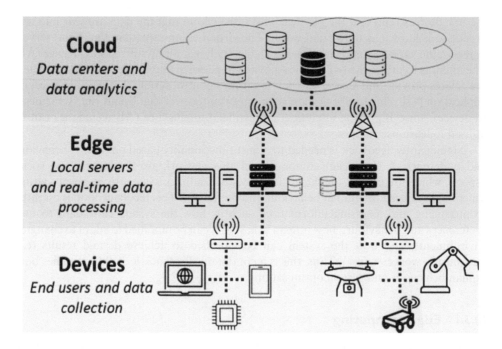

Figure 10.2 Edge computing.

and reporting large amounts of data that should be protected and private, remaining near the construction site. With an edge gateway running a specialised edge operating system (edgeOS) on site, the IoT system can be connected and managed easily and the data can be processed locally to release the delay encountered over the backbone Internet infrastructure. Many proposed and present-day construction site solutions utilise web or cloud-based approaches, but new solutions must embrace the more efficient edge computing approach [205].

10.3.2 IoT and edge computing enabled intelligent job site

Recent innovations such as location-aware computing and Machine Learning algorithms offer the significant potential of improving and supporting important decision-making tasks in the field. Furthermore, these devices can remotely control robotic autonomous machines such as unmanned aerial vehicles (UAVs), computer numerical control (CNC) machines, 3D printers. Significantly, these decision-making data can be utilised for standards, regulation, and education that can ultimately address the various issues previously discussed. Construction is a dynamic environment with new hazards being generated continually based on the current task, relative position of equipment, vehicles and workers, current environmental factors, etc. Therefore, integration of data from localisation systems with other Internet-enabled sources (schedule, weather report, daily progress reports, etc.) would provide additional and valuable information about the real-time safety risks accounting for the dynamic hazards in addition to predefined ones. Since these information streams require robust data transfer packets, having an edge node on site enables the ability to process them. These techniques are being deployed to increase productivity, quality, and safety on the construction job site (Figure 10.3).

Recent literature reviews have been conducted for IoT applications in construction [377]; methods for interoperability [85]; and construction safety sensors and technology [87,384]. Specifically relating to IoT, a recent literature review has found that key drivers for adoption include interoperability, data privacy and security,

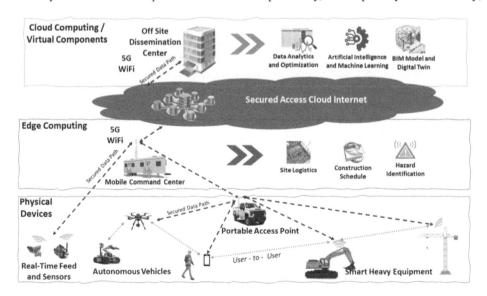

Figure 10.3 IoT and edge computing enabled intelligent construction sites.

flexible governance structures, and proper business planning and models [148]. These technologies are being utilised to ultimately increase the efficiency, safety, and quality of construction projects. Data streams that have been received through IoT devices and sensor networks could be integrated with Building Information Modelling (BIM) tools and make a paradigm shift in construction efficiency improvement [377].

Thibaud et al. [384] state that the use of IoT in construction is sky rocketing and emphasised that the IoT can significantly help in alleviating the high-risk concerns in construction industry. Construction sites could be classified as high-risk work environments. Kanan et al. [196] utilise an IoT architecture and propose an autonomous system for hazard recognition and sending safety notifications to site labourers. The validation of system results shows that this new approach could be a cheap and effective system that has low operational costs that supports the feasibility of their further utilisation. Louis and Dunston [230] provide a practical and sensor-agnostic implementation of operation-level decision-making by utilising IoT networks along with advancements in modelling and simulation tools.

AI is making a strong emergence in the construction industry, although its level of adoption is the lowest compared to other sectors [38]. For example, Smartvid.io[1] is developing an automated speech and image recognition software that utilises AI for construction monitoring and control. A user is able to define a "watchlist" of items for safety, quality and productivity, and the algorithms sift through a mass quantity of images and videos to identify problems and take swift action. What used to take multiple safety and project managers to ensure safety and productivity, can all be done automatically in real-time. Volvo Construction Equipment is developing an advanced warning system that sends a warning message to the operator to reduce the risk of accidents by utilising AI algorithms to detect and decipher specific objects using several computer vision methods.

10.4   IoT CONNECTIVITY AND REQUIREMENTS

All these connected devices produce and transmit an extensive amount of data, requiring the robust collection and analytical mechanisms to process the data. Additionally, with these data being transmitted from the various heterogeneous IoT sources, the ability to interconnect, share, and use the data seamlessly and efficiently, known as interoperability, becomes a major challenge. Interoperability is the ability of systems to mutually operate with each other and make use of data produced by the various systems without effort on the part of the end user of the systems [85]. *The IoT is a very complex heterogeneous network, enabling a seamless integration of these things is a huge challenge* [425]. IoT systems that exchange information need two things to be fully interoperable: (1) systematic interoperability, which is the ability of heterogeneous systems to connect with each other to enable a data exchange, and (2) semantic interoperability, which is the ability for each system to understand the meaning of the data that is being exchanged [85].

The standard protocol for global Internet communication, called the Internet Protocol (IP), accommodates wide area network (WAN) operations by transferring datagrams in the form of IP packets across one or more communication networks [224]. The key feature of the Internet is the ability to connect many hosts of various origins and types using a standard set of networking protocols, namely IPv4 and IPv6. An Internet connection is based on packet switching principles which divide data into manageable

Table 10.2 Layers of Internet of Things protocols and technologies

Function	IoT protocols
Networking	6LowPAN (lightweight IPv6), RPL, LoRaWAN
Identification	EPC, uCode, IPv6, URIs
Communications	Xbee, Wi-Fi, Bluetooth, LPWAN
Discovery	Physical Web, mDNS, DNS-SD
Data protocols	MQTT, CoAP, AMQP, Websocket
Device management	TR-069, OMA-DM
Semantic	JSON-LD, Web of Things Model
Multi-layer frameworks	Alljoyn, IoTivity, Weave, Homekit

packages of information that can be routed through the Internet using the standard protocols for transport, addressing, and routing. To accomplish this task, Internet routers handle billions of packets per second, and require multiple tens of thousands of Bytes of information per packet, including 40–80 Bytes of overhead, and computer processing operations at each endpoint for coding, quality and security.

With the advent of miniaturisation and advances in computer processing came "Low-Power and Lossy Networks" (LLNs), i.e., networks typically composed of many embedded devices with limited power, memory, and processing resources interconnected by a variety of links, such as IEEE 802.15.4/Zigbee/Xbee or low-power IEEE 802.11/Wi-Fi. Application areas for LLNs are industrial monitoring, building automation (heating, ventilation, and air conditioning (HVAC), lighting, access control, fire), connected home, health care, environmental monitoring, urban sensor networks, energy management, assets tracking, and refrigeration [48]. IoT serves the LLNs by adapting the key functions of the Internet – interoperability; heterogeneous device connections; universal addressing schemes; and reconfigurable routing – to LLN nodes without the high power consumption, high cost, and large router requirements. Table 10.2 shows a sample set of IoT protocols and solutions for the different levels of the IoT platform. We will go through some of these levels in the next subsections.

10.4.1 Networking: 6LowPAN, RPL, and LoRaWAN

In IoT, Internet Protocol versions 4 and 6 (IPv4 and IPv6) are replaced by IPv6 over Low-power Wireless Personal Area Networks (6LoWPAN),which provides for reduced packet sizes – from multiple tens of thousands of Bytes to less than 200 Bytes – through encapsulation and header compression mechanisms. 6LoWPAN also replaces operation over Ethernet, a high power consumption and large packet network, in favour of the low power IEEE802.15.4/Zigbee/Xbee or IEEE802.11/Wi-Fi-based networks.

Instead of the complex IP routing protocols to carry information along designated paths from source to one or more destinations, IoT uses the IETF Routing Protocol for LLNs (RPL) [417]. RPL is capable of building routes quickly and transmitting routing information among small nodes with minimum overhead. It supports routers, or nodes, that are interconnected by lossy links, typically supporting only low data rates, that are usually unstable with relatively low packet delivery rates. RPL also supports networks of up to thousands of nodes, wherein the traffic patterns are not simply point-to-point, but in many cases point-to-multipoint or multipoint-to-point. Furthermore, such networks may potentially comprise up to thousands of nodes. These characteristics offer unique challenges to a routing solution. RPL objectives include minimising energy consumption, minimising latency, or satisfying quality constraints.

Low-Power Wide-Area Networks (LPWANs) are wireless telecommunication networks that are designed to transmit small data packets over long-range distances using the unlicensed spectrum. For many industries, including supply chain, agriculture, healthcare, energy, and urban planning, LPWAN networks are a much better fit than IEEE802.15.2/Bluetooth or IEEE802.11/Wi-Fi. Long Range Wide Area Network (LoRaWAN) is a proprietary radio modulation technique based on chirp spread spectrum (CSS) technology that must be used to connect to the LoRaWAN LPWAN network. Demonstrations of LoRaWAN have shown reasonable performance at communication distances up to about 400 miles (600 km) and a single LoRa base station can connect to sensors more than 15–30 miles (22.5–45 km) away in rural areas.

10.4.2 Identification

For unique identification, the Internet Protocol (IP) uses an IP address scheme that identifies the point of attachment to the network using the many possible IP addresses for the devices (2^{32} for IP version 4 and 2^{128} for IP version 6). In IoT networks, the IP address may create unnecessary overhead in the data packet. Furthermore, often in IoT networks, a particular device does not have to be contacted, but rather a region of a room, or location of an alarm, or a group of sensors.

Novel techniques are available for the identification of IoT nodes or regions. The Electronic Product Code (EPC) is designed by GS1 EPCglobal as a universally unique product code much like the Universal Product Code (UPC), and the European Article Number (EAN) codes, which are barcode symbols for tracking trade items in stores. The EPC contains an item reference and serial number to recognise each product. It can be used with RFID tags to support supply chain management and users can retrieve information about the corresponding items using the Object Name System (ONS) or other discovery service on the Internet [164].

A Uniform Resource Identifier (URI) is a compact sequence of characters that identifies an abstract or physical resource. A URI can be further classified as a locator (URL), a name (URN), or both. A more commonly known version of a URI is a Uniform Resource Locator (URL), which in addition to identifying a resource, provides a means of locating the resource by describing its primary access mechanism (e.g., its network "location"). A ucode is an identifier to be stored in many types of tags (RFID tags, optical code, infrared markers, and even sound sources), specified by the Ubiquitous ID Centre. Its use is to identify objects and places where existing standards do not fit the application needs. Applications that use ucode take advantage of the Internet extensively.

10.4.3 Communications

As mentioned previously, there are several options available for communications over the wireless link for IoT networks, ranging from low power, short-range links (Xbee) to very long range wireless links (LoRaWAN). IEEE 802.15 is a working group of the Institute of Electrical and Electronics Engineers (IEEE) "IEEE 802" standards committee which specifies wireless personal area network (WPAN) standards. Two of the standards are regularly employed with IoT networks, IEEE802.15.4 (Zigbee) and IEEE802.15.1 (Bluetooth). IEEE 802.15.4 is the standard for low-rate WPAN communications. The design goal is low power, simple design, low cost, low-to-medium bit rates, and moderate delays without too stringent quality requirements. It only defines the communications, i.e., the Physical layer (PHY) and Medium Access Control Layer (MAC) of the network. The higher-layer Zigbee (Xbee) protocols are left to industry and the individual applications designed by users.

IEEE 802.15.1 is a standard for replacing short-range wired communications, such as printer cables, headphone cables, RS232 links and voice cables with a wireless option. Depending on configuration, 5–7 end devices can be active at once and end devices can only transmit when polled by the controller device. A larger network can be formed by a node that can control one Bluetooth piconet and be an endnode of another piconet at the same time, creating a scatternet. IEEE 802.11 is part of the IEEE 802 set of wireless local area network (WLAN) protocols that specifies communications (MAC and PHY) protocols for various frequencies, including but not limited to 2.4, 5, 6, and 60 GHz frequency bands. Wi-Fi is the highest power alternative used in IoT networks. It can operate in an ad-hoc configuration or with an access point/controller. The individual standards in use now are 802.11a, 802.11b, 802.11g, and 802.11n. The newest standard, 802.11ac, is the newest and fastest standard; it promises to exceed 1 Gbps of throughput. However, Wi-Fi can potentially increase transmission power consumption up to 3–6 times that of the 802.15 protocols.

10.4.4   Discovery

Multicast Domain Name System (mDNS) is a DNS-like service discovery protocol to resolve host names to IP addresses in a local network without using any unicast DNS server [441]. The protocol operates on IP multicast UDP packets through which a node in the local network enquires the names of all other nodes. DNS Service Discovery (DNS-SD) uses standard DNS messages to discover services in an IoT network. Based on mDNS, DNS-SD is used to resolve services available in a network in two steps:

1.   in the first step, host names of the service providers are resolved, and

2.   in the second step, IP addresses are paired with the host names using mDNS.

It is important to identify host names as IP addresses can change in the network. Like in mDNS, the protocol keeps the host names constant in the network despite the event that devices or nodes have dynamic IP addresses. Since the host name always remains constant in the network, it is possible to identify the devices uniquely and reliably within the local IoT network.

Neither mDNS nor DNS-SD requires any additional infrastructure (DNS Name Server in the network) or manual configuration or administration of connected IoT devices. Google has developed a discovery protocol called "Physical Web". The Physical Web is an open approach to enable quick and seamless interactions with physical objects and locations. It enables you to see a list of URLs being broadcast by nearby objects. Google's goal is instant node interaction: *Our core premise is that you should be able to walk up to any 'smart' physical object (e.g. a vending machine, a poster, a toy, a bus stop, a rental car) and interact with it without first downloading an app. The user experience of smart objects should be much like links in a web browser: i.e., just tap and use*[2].

10.4.5   Data protocols

The higher layers of the Internet protocols (see Table 10.2) provide many important services, including access to users, distributed information services, independence from differences in data representation, and control structures for communication between applications, as well as establishes, manages, and terminates connections between cooperating applications [370].

Message Queuing Telemetry Transport (MQTT) is a many-to-many communication protocol for passing messages between multiple clients through a central broker. It can use the Transmission Control Protocol (TCP) for its transport layer, which is widely characterised as "reliable, ordered and error-checked". MQTT uses a "publish/subscribe" model and requires a central MQTT broker to manage and route messages among an MQTT network's nodes. A central MQTT broker manages and routes messages among an MQTT network's nodes. Others subscribe to messages (content, type, subject). Generally, a broker subscribes to all messages and then manages information flow to nodes.

The Constrained Application Protocol (CoAP) is a specialised web transfer protocol for use with constrained nodes and networks. It enables web services and allows integration to the web and HTTP architectures. The Advanced Message Queuing Protocol (AMQP) 1.0 is a standardised framing and transfer protocol for asynchronously, securely, and reliably transferring messages between two parties [442]. The technical standardisation forum for the AMQP protocol and extension specifications is OASIS, and it has achieved formal approval as an international standard as ISO/IEC 19494:2014 [262]. The protocol can be used for symmetric peer-to-peer communication, for interaction with message brokers that support queues and publish/subscribe entities, as Azure Service Bus does. It can also be used for interaction with messaging infrastructure where the interaction patterns are different from regular queues, as is the case with Azure Event Hubs.

The WebSocket API is an advanced technology that makes it possible to open a two-way interactive communication session between the user's browser and a server. With this API, you can send messages to a server and receive event-driven responses without having to poll the server for a reply.

10.5   IoT SYSTEM NETWORK MANAGEMENT

IoT system network traffic will come from heterogeneous devices and construction network structures that experience different types of events, requiring more complex networking and control protocols. Managing these evolving networks using traditional network management schemes can increase the cost of network operation and maintenance and leave significant vulnerabilities in fault tolerance and cybersecurity.

Software-Defined Networking (SDN) is a networking paradigm in which the forwarding hardware is decoupled from control decisions. The network intelligence is logically centralised in software-based controllers (the control plane), and network devices become simple packet forwarding devices (the data plane) that can be programmed through an open interface. SDNs help to assemble new services and infrastructure quickly to meet dynamically changing environment objectives.

Furthermore, the software implementation of the control plane and the built-in data collection mechanisms are excellent tools to implement Machine Learning (ML) network control applications. Extracting knowledge from data collection to understand and predict state of the ACRES network will be crucial to implementing data and cybersecurity management. Two significant challenges are proposed to be addressed: real-time SDN machine learning systems, and real-time IoT system management through a distributed SDN (D-SDN) architecture [443]. The challenges of the distributed, real-time approach include coordination between SDN controllers; synchronisation, timeliness of responses across the network, which includes latency and jitter in the communication and implementation of the management response; and dynamic failover in the case of failure or compromise of one of the D-SDN controllers. Building a trust relationship between applications and controllers is also an open issue. To address these needs, we will develop a novel distributed, cross-layer controlled SDN with an ML capability for real-time response at local IoT nodes.

The SDN framework is a viable solution for IoT network management due to its high scalability to improve the reliability of sensing and sharing within wireless sensor networks. However, we must solve the network coverage problem which states that sensor network node transmissions will be impacted by node channel disturbances and energy depletion [111]. Clustering the sensor network under virtual switches has divided the sensor network reasonably and reduced the regional network load. Using multiple switches enables an SDN framework based on multiple controllers within the WSN.

This introduces several main challenges, such as (1) in-band control (opposed to SDN implementation in wired networks that can leverage from outband control); (2) higher communication latency; (3) smaller link layer frames; and (4) limited energy supply. Another challenge due to WSN characteristics is the possible disruption and delay in communication, low energy supply and reduced data frame length [63]. A specific solution for such scenario is a software-defined clustered sensor network (SDCSN), where multiple controllers can interconnect SDN domains through border controllers, which may lead to a secure model for WSN ad-hoc networks. Other clustered SDN architectures deploy a distributed security solution where the flow traffic between the sensor nodes can be controlled in a collaborative manner by the SDN cluster head [284]. For mobile sensors, a cloud-based architecture can be used to develop a movement strategy determined by the controller to maintain the barrier coverage for various real environments such as 3D terrain and obstruction, as well as sensing a new environment and reporting it to the cloud. The controller learns the new environment and generates a corresponding new strategy. This design reduces to a similar design to new policy generation in SDNs [207].

Machine learning can be used for a plethora of applications within the SDN controller environment. Traditional supervised machine learning methods generally assume that every possible class and the distribution of possible samples for each of these classes are appropriately characterised by training data [420]. For implementation in a live construction environment however, it is infeasible to assume that all possible behaviours can be identified and characterised in training data prior to implementation of a system. Techniques must take into account the different classes of network traffic created from the diversity of the nodes connected, and the dynamic surrounding environment that provides a multitude of scenarios. Furthermore, malicious attacks and their associated behaviours on the communication infrastructure may need to be re-imagined and implemented in novel ways to avoid stopping construction activities for each attack. Thus, a system is needed that can adapt to changes in communication behaviour through application-specific QoS and novel cybersecurity-specific SDN controller architectures.

10.6 GAPS

This section reviews the most pressing gaps revealed in literature for the use of IoT in the construction sector.

10.6.1 Lack of evaluation and performance metrics for IoT technology in the construction environment

Construction is a very hazardous and dynamic environment, and many sensors and technology created for indoors (e.g. Smart Homes) may not be suitable for ruggedness of outdoors such as weather, impacts, and debris. For example, Korman and Zulps [209] state that *there are still some technical challenges that are being addressed, such as battery life, administration of wearables on a daily basis, and creating robust analytics.* Furthermore, the dynamic nature (machinery, heavy equipment, conductive materials) provides potential barriers to the performances of various wireless network technology (e.g. Wi-Fi) [86]. Therefore, there is a need to create a framework and open database for assessing and recording the characteristics and performance metrics of IoT devices specific to the construction environment.

10.6.2 No valuation assessment tools of the implementation of IoT-based solutions in construction

Costs have been the traditional driving force for making decisions in building construction. Even when considering other aspects, such as safety, sustainability, and productivity, costs are typically the driving metric in making the final decision. As the use of IoT increases, the need for an impact assessment tool is critical for industry stakeholders to make smart decisions regarding deploying IoT-based solutions. This is reinforced by a recent case study by [230]: *the IoT concept and associated technologies present a wealth of opportunities for construction industry advances in process controls and automation, but the explosion of data must be met with domain-specific methodologies designed and tested to assure that project managers can make both faster and smarter decisions.* There is a need to define a valuation assessment framework that can empirically and statistically compare the IoT solutions in the construction environment.

10.6.3 No standards for IoT deployment in construction

Most of the IoT case studies were single use (either a single technology or a single application). Because IoT spans multiple industries with many manufacturers and differs broadly in application scenarios and user requirements, large-scale commercial deployment of related IoT services could be very challenging. It is stated that *large-scale service deployment of new technologies needs to be framed within a set of standards* [16]. There is a need for a set of guidelines and best practices, specifically for construction, to be established and accepted into industry standards.

10.6.4 No "one-fits-all" IoT solution for construction

Literature has revealed an abundant number of technology applications. However, most deployments have been either a single-use application and or a single technology due to many challenges that are currently faced on the construction site (e.g. scalability, QoS). Thus, there has yet to be a comprehensive, IoT solution that can incorporate a multitude of heterogeneous solutions. This last gap may be truly unattainable, as there are countless scenarios and considerations needing to be addressed in the construction environment. Needless to say, this "solution" may not necessarily be a single application, but rather a framework or methodology that can be implemented across the entire domain. Still, such a framework has yet to be developed and validated.

10.7   OUTLOOK AND CONCLUSION

If successfully implemented, automation and construction will ultimately shape the future of work in construction. Therefore, it is imperative that the current work force adapt and create a symbiotic relationship with the emerging technology: where the automation and technology could improve productivity while the worker learns the skill sets needed to use and optimise the technology, ultimately ensuring job security. As the many challenges faced by the construction industry continue to grow, technology will play an integral role in supporting the current workforce. The augmentation of technology will continue to educate and inspire the next generation of workers. Technologies can increase worker efficiency and can even preserve jobs by helping companies improve efficiencies, reduce costs, and increase value. As seen in other industries, demand for labour will increase the need for technologies and automation. It is important to not let machines pass up human involvement, but rather take advantage of emergence to enhance the workers, the construction industry, and society as a whole.

In order to advance the frontiers of science, technology, and education in the construction industry, while fostering a lifelong and pervasive learning with technology, it is imperative to advance the human-technology partnership by having the workers utilise and learn from the emerging technological advancements. *Combining automation with labor, we can do less with more, allowing construction firms to keep up with competitive construction schedules.*[3] Technology will play an integral part within the construction industry, and it is vital to augment human behaviour to utilise these technologies, ultimately giving companies the advantages of automation while resolving the shortage of labourers and the growing demands of commercial construction and future work.

During the IoT introduction process, the following issues still need to be explored and solved: the required human resource and technology, the security of networks and communication, the integration between the different systems, and the speed and the huge demands of the network. In the future, these IoT technologies will continue to impact and shape the construction industry, not only bringing new challenges and opportunities, but also promoting the industry towards facilitating the objective of Construction 4.0.

NOTES

1 https://www.smartvid.io/.
2 http://google.Github.Io/physical-web/.
3 https://constructionblog.autodesk.com/ai-automation-construction/.

Chapter 11

Smart cities and buildings

Hendro Wicaksono, Baris Yuce, Kris McGlinn, and Ozum Calli

CONTENTS

Smart buildings function within the wider context of the smart city, which itself must function within the wider energy and transport (smart) grids. It is essential therefore that smart buildings be integrated into this wider context. This requires intelligent approaches for managing and coordinating the diverse range of processes and technologies involved and a move towards a "digital infrastructure" which can transform how these smart environments operate and can be monitored but, more importantly, can circumvent the constraints of physical infrastructure through the capacity of data centres or the capacity of available communication pipes. This chapter explores the concept of the smart city,

DOI: 10.1201/9781003204381-13

and the role that smart buildings, smart energy grids and smart transportation takes within, with a particular emphasis on the state of art with respect to the integration of data across these different domains, from the micro to the macro (see also Chapter 3), from building sensors to smart grids. It explores different data analytics approaches, and it does this with reference to specific use cases, focusing on techniques in the main application areas along with relevant implemented examples, while highlighting some of the key challenges currently faced and outlining future pathways for the sector.

11.1 INTRODUCTION

The world's population has reached about 7.3 billion and it is expected that by 2100 the population will be around 11.2 billion (according to UN, 2015). In addition, nearly 52% of this population (about 3.8 billion) live in urban areas. Moreover, 33% of the population lives in urban clusters which makes total urban living around 85% (nearly 6.2 billion) of the overall world population[1]. With this growth in the urban environment, the structural problems and urban level resource management have become a challenging issue for cities and mega-cities and can have negative effects on billions of people's lives. As a result, over the past decade, there has been growing interest in the concept of smart (or intelligent) cities, due to the enormous potential for improving the environmental and socio-economic impact cities have both at national and international levels of this urban growth [144,186,424].

Through the application of new management models, based on ICT systems, Artificial Intelligence (AI) technologies, other digital technologies, sensory-based infrastructures, cloud platforms, digital twin applications to monitor, control and manage the population growth migration, climate change, energy, water, education, health, transportation, housing and urban life issues [334], smart cities set out to address the complexity generated for local authorities' governmental skills and capabilities. This new approach has been labelled as smart city or "smarter cities" and ultimately, is intended to improve the quality of life of all residents [61].

This chapter begins in Section 11.2 by exploring the concept of the smart city and the key components which make up the smart city, with particular attention to the use of IT technologies to enable seamless integration of physical components with a digital infrastructure through contemporary computing and networking technologies and advanced data management tools. This includes a discussion of the features of a smart city, smart buildings and homes, smart energy systems and grids, and smart mobility and transport. Then, Section 11.3 elaborates on the processes of digitisation and digitalisation that occur in the transformation towards smart cities.

Section 11.4 then gives more detail regarding data infrastructure and open data initiatives in smart cities. This section reviews existing approaches and results for five use case cities: Bonn, Dublin, Toronto, Singapore, and Tokyo. The chapter then finally gives more detail on the use of linked data in smart cities (Section 11.5) and on data analytics approaches (Section 11.6).

11.2 SMART CITIES

When considering the architecture of the smart city, there are several typical main components. These are smart buildings (offices, public buildings, homes), smart factories, smart grid, telematics, and telecare [182]. In the next sections (Sections 11.2.1–11.2.4), we explore each of these components, with specific reference to their integration into the smart city.

II.2.1 Features of a smart city

In smart cities, one of the key enablers is the Internet of Things (IoT), and as such, a smart city cannot exist without IoT infrastructure (see also Chapters 3 and 10). Together with IoT, a smart city should address the following features to demonstrate smartness [228].

- Smart people: Smart people are those who are educated for high-level thinking and have the necessary ICT skills [334]. The people of the smart city should use their "smart thinking" in the form of a mindset of environmental protection, sustainability and a healthy lifestyle, which includes aspects such as recycling, water and energy saving [152].

- Smart economy: Smart economy concerns ICT-based economy and other Industry 4.0 technologies. This component's characteristics are diverse employment opportunities with labour market flexibility, diversification promoting entrepreneurship and innovation.

- Smart governance: Smart governance is an integration of new technologies into traditional urban governance processes using such platforms as e-governance, e-management, or e-democracy, [193].

- Smart living: Smart living encompasses the development of the smart buildings for education, tourism, healthcare, and public safety platforms to improve the quality of life of citizens.

- Smart environment: Smart environment feature covers the concepts of safe, green and sustainable environments, and developing new ICT tools and platforms to manage these topics in the smart city.

- Smart mobility: Smart mobility concerns the sustainable movement of people, vehicles and other mobility transport technologies to improve the logistics and movements using new ICT tools in transport systems. Transport plays a significant role in the smart city, and the next section examines this.

In this chapter, we are not going to present the smart economy directly. Some economic aspects of the smart solution will be highlighted as smart economy applications are directly related to the economic applications of the smart solutions that are utilising Industry 4.0 technologies.

II.2.2 Smart buildings and homes

It is estimated that in Western countries (e.g. Europe, the United States) people spend up to 90% of their time indoors [204,337]. Furthermore, buildings account for approx. 39% in the US [28] total energy consumption and 50% in the EU [429] total energy consumption, when taking into the whole Building Life Cycle (BLC), which includes the life stages of a building from design, through construction, operation, and on to eventual demolition and recycling [49]. With growing urban populations, buildings and the management of people's needs as they use and move between them, play an important role within the smart city. This section discusses what we mean by a smart building, how these buildings' "smartness" is enabled and how this plays into the wider context of the smart city.

II.2.2.I   Definitions and terms

Within the literature there are many definitions for smart buildings, and often the term smart and intelligent are used interchangeably. The term "intelligent building" is older than that of smart building, having been first coined back in the early 1980s, and more commonly seen in the literature [99,133,134,365,418]. Building intelligence has historically been closely linked to the growth of information technologies (IT). By the nineties, there was a growing realisation that existing definitions placed an unbalanced focus on technologies and began to include both consideration of technologies and the needs of users [365,418]. This shift towards a human-centric view continued into the next century, and expanded also to include the environment [72,175]. Modern intelligent buildings bring together an understanding of not only the needs and goals of the users of the building, but also how the building fits into the wider context of its environment. This furthermore includes the capabilities to "sense" these different environments and therefore provide services that have a view of the user's context which matches the user's view of their own context [145,175].

In the past decade, the terms "smartness" and "smart building" have become more popular to describe intelligent buildings, being adopted by, for example, the European Union in relation to the Energy Performance of Buildings Directive (EPBD) [123]. Like intelligence, there remain though several different views on what a smart building is [58,91,361,364] ranging from large in scope, i.e. to *integrate and account for intelligence, enterprise, control, and materials and construction as an entire building system, with adaptability, not reactivity, at the core, in order to meet the drivers for building progression: energy and efficiency, longevity, and comfort and satisfaction* [58], to more specific, for example with an energy focus, where smart means being able to manage the amount of renewable energy sources in the building and the smart grid, through advanced control systems, smart metres, energy storage, and demand-side flexibility, while also being able to react to the users' and occupants' needs and is able to diagnose faults in building operations [91]. These definitions can also be applied to the full range of building types, e.g. a hospital, an office, or a home, although the latter are often referred to directly as smart homes.

II.2.2.2   An interconnected smart environment

Ultimately though, the concept of the smart/intelligent building aligns closely with that of the "smart environment", which integrates an environment with "sensors, actuators, displays, and computational elements" [79]. To be smart, a smart building must be enabled to collect data about the current state of the building, have analysis done on this data and then, either:

1.   Provides analysis and feedback to an appropriate stakeholder, who must enact some change or reconfiguration of the building's systems, or

2.   An automated control system acts upon the available data to reconfigure the building automatically.

This requires integration of many interconnected components, and challenges remain to build, configure and maintain smart buildings, related to security and privacy, data acquisition, storage and processing, feasibility and evidence of return on investment, as well as lack of coordination between IoT developers and the building industry [192].

The design and construction of the smart building must also be smart, involving the use of new technologies which enable the optimisation of delivery of new building projects, to the management of construction sites, construction technologies and communication and data management [11]. This and its operation must all be considered within the larger context of the smart city, its telematics, telecare and energy grid. The smart building must therefore be integrated seamlessly into this larger context, and here digitisation and IoT play a significant role [182].

II.2.2.3 Smart building architectures

Buildings are designed for operation and the majority of the life cycle of the building, and hence its interplay with the city is during this stage. From this perspective, conceptually, different smart building architectures have been developed. These can generally be categorised into different layers: for example, the sensor (and actuator) layer, interoperability layer, computational layer, and application layer (Figure 11.1). Collectively, these monitor environment states, perform statistical and algorithmic analysis, and provide feedback and control mechanisms to users [179,198,257].

As an alternative architecture in the smart home domain, a smart building can be divided into four components [288]:

1. The sensor and actuation infrastructure
2. Middleware
3. Processing engine
4. User interaction interface

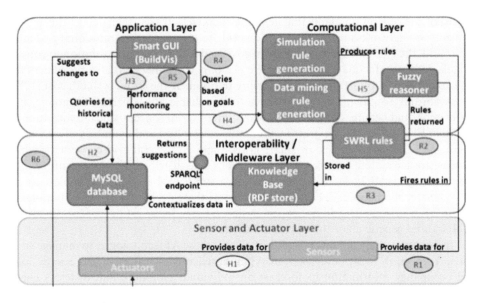

Figure II.1 The architecture of the proposed ontology, ANN-genetic algorithm, and fuzzy logic based smart energy management solution [179].

In this architecture, the sensor and actuation infrastructure handles all interaction between the digital infrastructure and the physical environment. The middleware integrates the infrastructure through a common interface. The processing engine conducts some "processing" on the collected data to learn about the environment and building user's activities, so as to improve the building's energy consumption. Data on the environment includes measurements relating to temperature, CO_2, humidity, etc. User activities include scheduled activities (office work, meeting, lunch, etc.) and interrupt activities (toilet break, drink, exercise) [257]. A user interaction interface then supports interaction with end users; sending them notifications to stimulate behaviour, gather feedback and commands from them. Other architectural configurations for buildings and smart homes can also be found in [292,306] which define the similarities and differences in the sensing, data management, and reasoning and human interaction layers.

Similar layered architectures have been proposed for the smart city by [233,366]. The former divides the smart city into four layers: sensor level, integration level, intelligent level, and application level; while the latter relies on five layers: Layer 1 – sensor networks, applications, services and tools; Layer 2 – sensing, acquisition, and collection of data; Layer 3 – data processing and monitoring; Layer 4 – data integration and management; Layer 5 – secure deployment and quality control. Careful consideration and alignment of the architectural patterns should ease the integration process for future smart cities and smart buildings.

II.2.3   Smart energy systems and grids

As mentioned previously, energy plays a big role in buildings, and as such, the management of energy use between buildings and the energy grid plays a significant role in the smart city. Smart grids are one of the most essential infrastructures for smart city applications. A smart grid is a new concept for upgrading traditional energy systems into a smarter grid that receives the energy from suppliers and delivers it to the consumers using secure, smart, and reliable ICT systems. In a nutshell, the traditional grids consist of the generation plants that generate energy/electricity, then deliver through the transmission networks and the distribution grid. To transform this traditional grid into a smart grid, the grid management has been governed with the smarter solution to conduct reliable decisions. In smart grids, there are not only consumers but also prosumers that play key roles to support the grids' reliability to optimise the entire system's cost [127].

According to [200], smart grids have to be reliable and efficient, economic, high performing and secure systems. An overall smart grid concept and architecture are shown in Figure 11.2. Smart grids facilitate the connection and operation of generators of all sizes and technologies. In smart grids, there are several grid elements available such as storage systems, micro generation, renewable distributed energy resources, and distributed grids [103]. Since the decisions in a smart grid are highly complex, several AI system technologies have been used to support these decisions. The most popular AI technologies in a smart grid are expert systems, fuzzy logic, artificial neural networks (ANNs), and genetic algorithm-based solutions.

In literature, there are several studies utilising these AI technologies to manage smart grids. Yuce et al. [428] developed an ANN-based model using current energy loads and occupancy information to forecast an individual building's and district's energy consumption. In the proposed model, a set of parallel feed-forward ANNs is used for the prediction process. The training and testing stages of parallel ANNs use a dataset of a group of six buildings (ANN1, ANN2, ANN3, ANN4, ANN5, ANN6). The inputs of

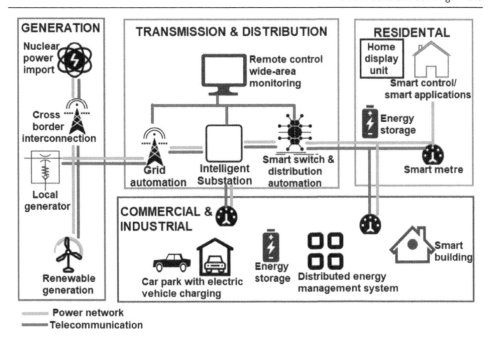

Figure II.2 A generalised architecture for a smart grid diagram [6].

each ANN are determined using Principal Component Analysis (PCA) and Multiple Regression Analysis (MRA) methods. Comparison between the actual and predicted electricity consumption for the winter season are shown in Figure 11.3. The results show that peak demand can be predicted successfully and used to forecast and provide demand-side flexibility to the aggregators for district energy management systems with high efficiency.

Similarly, Skolthanarat et al. [362] also proposed an ANN-based predicting system for the load demand. They preferred the radial basis neural networks model and least numbers of hidden neurons are utilised to prevent the generalisation and overfitting phenomenon. The radial basis neural networks with the proposed algorithm performed excellent predictions of the load demand.

However, accurate prediction is not the only problem; optimising the smart grid is another problem. Hence, Muralitharan et al. [267] proposed a multi-objective evolutionary algorithm, to minimise the energy consumption cost and the waiting time for appliance execution. The results show that if the consumers exceed the threshold limit, they must temporarily add stops for the scheduled running electrical appliances to maintain the energy cost.

Moreover, Yuce et al. [429] presented an Artificial Neural Network/Genetic Algorithm (ANN–GA) based smart appliance scheduling approach to optimise energy systems in domestic buildings. Using this approach, the overall cost of grid energy bills is reduced using maximum use of renewable sources (PV and wind turbine) in peak hours, after forecasting and optimising entire systems for a week while reducing the use of grid energy. An unoptimised schedule and 40% grid cost reduced schedule is shown in Figures 11.4 and 11.5, with D1 = washing machine, D2 = dishwasher, D3 = tumble dryer, D4 = iron, D5 = cooker, D6 = microwave, D7 = vacuum cleaner, D8 = phone charger and D9 = car charger.

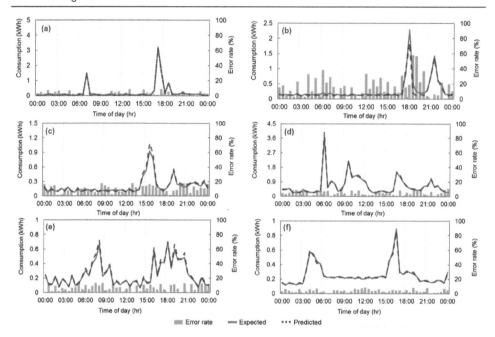

Figure II.3 The actual and predicted electricity consumption for six buildings: (a) ANN1, (b) ANN2, (c) ANN3, (d) ANN4, (e) ANN5, (f) ANN6 for a typical day in winter [428].

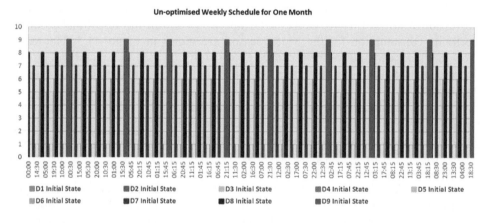

Figure II.4 The schedule without the optimised solution [429].

II.2.4   Smart mobility and transport

Smart transportation is a final essential feature of the smart city environment that concerns any ICT developments for new transportation systems. As populations grow and residents' needs for more advanced transportation systems increase, new and more effective transportation systems are needed. The main challenges in smart transportation are related to transport ICT technologies, road optimisation routes and development of user-friendly transport networks using carbon-neutral and low-carbon emissions systems. Hence in literature several studies have focused on tackling these challenges.

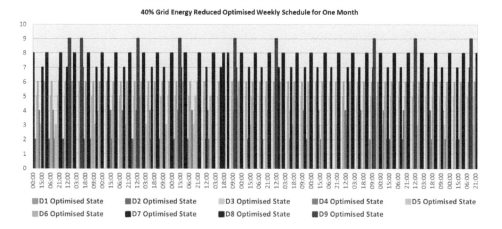

Figure II.5 The schedule for 40% less grid energy usage [429].

Babar and Arif [18] have presented a model for a smart transportation system using Big Data analytics by leveraging GPS and sensor data, collected in an Apache-Spark system. They utilised this to monitor the vehicle and generate decisions for vehicle schedule optimisation, facility surveillance and speed control problems using a real-time approach. This enables real-time processing and provides friendly communication in an IoT-based smart transportation environment.

Another key application field for the smart transportation systems is the transportation network management problem. In this regard, Wang et al. [407] developed a three-layer transport management system to support smart urban mobility on bus transportation.

In their solution, the first layer focuses on the data collection process that utilises a big data analytics approach using Hadoop to compute the bus travelling time and passenger demands. In the second layer, they have developed a network analysis approach to identify the passenger transit patterns and bus delays using a causal relationship analysis. In the last layer, they generate the outcome of the analysis and visualisation of the sensory data, mainly environmental data. The proposed solution is based on a multi-step nearest neighbour search algorithm and is tested in the city of Fortaleza in Brazil. In their study, 600 trips, 1,000 m and 174 regions were investigated. To identify a region's mobility demand, average passenger boarding volume per stop and modified region PageRank measurements are utilised. For testing, the results of 10% highest regions based on metrics are selected and normalised between 0 and 1 and colour-coded. To evaluate and visualise the proposed solution, bus departure time, bus average speed and passenger numbers are utilised [407], where the user can see and decide if they need to use certain type of bus and manage their routes.

A similar methodology has been proposed by Jan et al. [189], who have utilised a four layers solution that consists of data collection and acquisition, network, data processing, and application for traffic congestion and parking management problems. The proposed solution is for real-time application for the driver's road traffic management issues. However, to make the solution more adaptive and intelligent, several AI-based solutions have also been discussed such as fuzzy based traffic management systems to deal with uncertainty in smart traffic systems.

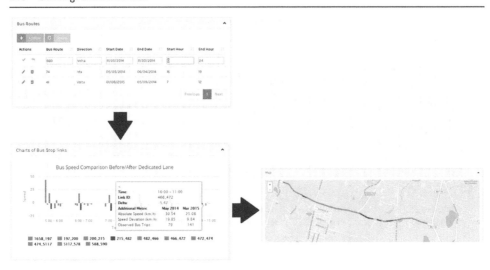

Figure II.6   The results for the traffic lights with a fixed cycle [430].

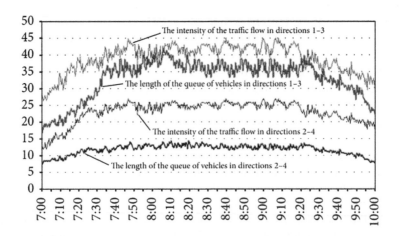

Figure II.7   The results for the traffic lights with using the AFLTCS-based method [430].

Yusupbekov et al. [430] have proposed an adaptive fuzzy logic traffic control system (AFLTCS) to manage uncertain information in heavy traffic streams. The aim of their study is to generate an effective solution to manage the traffic lights for the crossroads using fuzzy systems to minimise the total delay at the crossroads. The authors have tested the proposed solution on a busy road in Tashkent city, Uzbekistan. They have compared their results for the peak hours (from 07:00 to 10:00) for the fixed cycle lights management (Figure 11.6) and AFLTCS-based traffic lights (Figure 11.7). The figures present the results of the simulation of the road intersection that is managed by traffic lights with fixed cycles and with AFLTCS-based traffic lights.

These solutions demonstrate how integrated, smart solutions which make use of a range of data sources can be used to optimise transport in a modern smart city. Digitisation plays a large part in enabling this integration. In the next section we look at digitisation and the important role data and data management play, as a key technology to enable in smart cities.

II.3 DIGITAL TRANSFORMATION IN SMART CITIES

A city can be considered as a complex and dynamic socio-technical system that is built by the city's environment, urban society, the urban economy and the institutions in place [132]. A city is a multi-dimensional system and can be viewed from different perspectives. People can look at a city as a political system that defines who has the responsibilities to decide about what matters. A city is also an economic system since it generates wealth and jobs, involves economic interactions, and provides conditions for workings. To function as a socio-technical system, typically, a city offers services to all stakeholders, such as housing, public transportation, hospitals, schools, energy and water supplies, shopping centres, and business centres. The level of services in cities are more advanced comparing to other smaller settlements such as towns and villages.

Digitisation is a process to convert an analogue object or information into a digital representation [56]. Digitisation processes enable digitalisation. Digitalisation means the utilisation of digitisation processes using digital technologies, such as Internet of Things, cloud and big data, in an organisation to improve the performance of an organisation and the scope of its activities [194].

A city can be considered as a large-scale organisation consisting of different physical infrastructures and human stakeholders who conduct private and business activities involving interactions among them. A human settlement having more than 100 thousand inhabitants can be categorised as a city [281]. Digitalisation in the context of smart cities leverages digitisation in different services to improve the quality of the services. An example of digitalisation of a service in the energy sector is the introduction of IoT by using sensor devices to collect power consumption data and cloud services to display the development of power consumption and forecast of future power consumption (see Sections 11.2.2–11.2.3). Another example is in mobility service through digital ticketing and real-time arrival and departure time predictions (Section 11.2.4).

Smart cities introduce digitalisation using ICT to facilitate and improve the interactions among city infrastructures and people in the cities through digital services [132]. A smart city is a system built by four main components, i.e. [208,219]:

1. Technologies including sensor devices, network infrastructure, communication interfaces and control centres to facilitate the digitalisation and interconnections among people and infrastructures;
2. Tasks indicating services provided to the people, such as city administration, healthcare, public safety, education and environment protection;
3. Structures describing hierarchies, formal structures and communication methods;
4. People living in the city.

Digital transformation transforms a city to a smart city through digitalisation of services based on digital strategies and policies developed in city governance framework. Digital transformation utilises technologies to transform tasks and is managed through city governance to give positive impacts of the whole people living in the city including both citizens and stakeholders. It captures and creates values through the development of smart services and business models created around them [194,401]. The roles of digitisation, digitalisation, and digital transformation in the context of smart city development are summarised in Figure 11.8. The smartness of a city can be measured by the level of digital transformation in six dimensions, i.e. economy, people, governance,

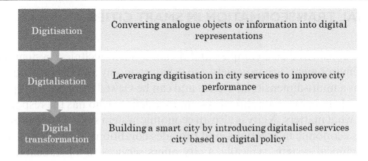

Figure II.8 Digitisation, digitalisation, and digital transformation in smart cities.

mobility, environment and living. The set of six smart city dimensions is defined as smart city wheels in [149].

II.4   DATA INFRASTRUCTURE AND OPEN DATA INITIATIVES IN SMART CITIES

The introduction of digital technologies, such as IoT, in the city to collect data from and control the physical infrastructures and interconnected smart services accessible from the smart phones, generates a large amount of data. The Internet of Things comprises a huge number of sensing and control devices that are interconnected to exchange data. The amount of the exchanged data is large. They are transferred every second. The data generated in the Internet of Things have heterogeneous formats. It is also difficult to assure the quality due to the uncertainties on the data, for instance inconsistency, incompleteness, and ambiguities. This uncertainty is called veracity [321]. Due to these characteristics, the data is qualified to be categorised as big data [214]. Additionally, the interconnected smart services, such as smart energy services and mobility services, do not only generate data but also consume them to create values.

To allow value creation from data collected through digitisation and the integration of IoT in the city infrastructure, a data infrastructure is required in the smart city context. The data infrastructure manages the data storage, provides features and tools for analytics, and assures data security, privacy, and quality. The data infrastructure enables the delivery of smart and interconnected services to citizen and city stakeholders such as policymakers, industry, and investors. The data infrastructure is depicted in Figure 11.9.

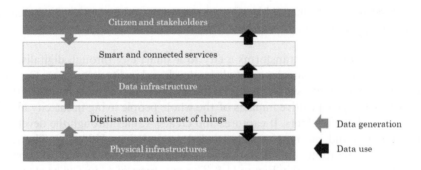

Figure II.9 Data infrastructure for smart city.

Cities around the developed world countries, such as Barcelona, Chicago, Manchester, Amsterdam, and Helsinki, have been developing open data initiatives to give stronger impacts to the smartness of the city and at the same time as the results of data that are continuously generated by different services in the smart city context. The open data initiatives have potential impacts on accelerating open innovation economy that relies on the availability of data sets [283]. Those impacts can be assessed, monitored, and improved by using Randomised Controlled Trials (RCT) that combine theoretical models and experiments. It allows a contextualisation and characterisation of the impacts of open data initiatives on cities, particularly on sustainable development [272]. In developing world countries, the design and implementation of open data initiatives require a conceptualised framework that uses similar contextual attributes that can be adapted in different environments [105]. The following sections give examples of open data infrastructure in five different cities to support the smart city initiatives.

II.4.1   Use case: Bonn

The Open Data portal of the city of Bonn[2] was launched on May 28, 2014, providing data from the administration of the city and other related institution within the city to the public whether it would be private or public entities for computation and data processing as open data. The portal has a separate catalog known as the Open Government Data (OGD) Cockpit which is responsible for providing information on the IT interfaces and data tips in relation to the city. This portal is introduced as a standard for data that serves for the public in Bonn. The city simultaneously is in the process of developing guidelines for more implementation of such Open Data for the general public use. In addition to the main goal of providing the vast amount of data to the public, this portal serves as a waiver of political influences in the use of big data with the incorporation of third parties making the project transparent between the city and the public.

The implementation of this portal would allow the visualisation of databases, information transparency to the public, new ideas to prosper that create applications and services based on data and information flow. The Open Data portal of the city covers and offers data pertaining to several Smart City dimensions. The inclusion of smart governance is witnessed as the Bonn council and its related information system is run with the IT interface OParl[3]. The Smart Economy is evident with the availability of the budget plans and business reports in the portal. The implementation of the IT interface Open 311 that deals with the online citizen management portrays the annexation of the Smart people dimension. Lastly, the city map of the federal city of Bonn with several layers is available in the portal promoting the Smart Mobility characteristic of a smart city.

Further data sets such as population, data, education and science data, infrastructure, building and living, environment and climate, etc. are available in this data portal and fall under at least one of the six dimensions of a smart city (see Section 11.2.1).

II.4.2   Use case: Dublin

Smart Dublin is a project that focuses on enhancing the quality of life in the Dublin region with the help of combining the knowledge and experience from technology providers, academia, and citizens[4]. The teams are comprised of highly skilled project partners such as the likes of Google, Microsoft, Intel, and Deloitte. The project also includes research centres, higher education and state agencies that improve the standard and the quality of the output. The main objective of Smart Dublin is to future-proof the city of Dublin through the means of trialling and scaling innovations to many local trials. These local trials are categorised into different entities such as environment, people, mobility, economy, government and living conditions.

The Smart governance aspect of the smart city dimensions is covered by this project as they make data related to Government and participation available. A major portion of the Smart People characteristic is covered by this project as they work on data characterised as Art Culture and Heritage, Population & Communities and Recreation and Amenities. The Smart Mobility dimension of the city is improved by processing Transport & Infrastructure and Planning & Land Use data. However, it is crucial to note the Planning and Land use would also play a role in improving the Smart Environment along with Environment & Energy data. The Economy and Innovation data present in this project contribute to improve the Smart Economy aspect of the Smart city dimensions. Finally, Public Health and Safety data will improve the Smart living condition of the city.

II.4.3   Use case: Toronto

An Open Data Portal was launched by the city of Toronto in 2009 to provide open data to the community[5]. Ever since, the Toronto's digital and civic department has been able to use this to its advantage in terms of civic decision-making as policymakers and navigate and address socio-political barriers. After noticing the success of the usage of this Open Data Portal, it evolved further with the inclusion of developers, policymakers, academics, and civic advocates.

The common goal of this community is to make data available to the public which can be processed and published which would incite modern designs and innovations in the city. The key team members in the Open Data team are from the Common Components, Digital Communications, Technical infrastructure Services and Digital Technology Services and Strategic Communications. Data pertaining to all six of the smart city characteristics is available in Toronto's Open Data Portal. The Smart governance aspect of the smart city is evident in data that is available under the topic of City government, Municipal Licensing and Standards. The data related to economic development and culture, Financial Planning as well as Accounting Services highlights the Smart Economy aspect. The inclusion of Smart Mobility is seen through the availability of Transportation services, locations and mapping, city planning data. The Smart Environment aspect of the city is emphasised by data available under the topic of Climate change, Parks, Forestry and Recreation, Solid Waste Management services. The data from the City Clerk's office, City Manager's office directs towards the inclusion of the Smart People aspect in a smart city. Lastly, the Smart Living characteristic can be seen from the availability of the Public Safety, Community Service, Poverty reduction data.

It is important to note that there are several more data sets available in the open data portal that fall under at least one of the dimensions of a smart city.

II.4.4   Use case: Singapore

The Data.sov.sg one-stop portal[6] was made available to the public in 2011. Initially, the portal was commenced with 70 datasets and currently there have been 100 apps that have been developed through this portal. Interestingly, the portal is now being made available to the public in a better readable and accessible format. Instead of just providing the repositories, the data is being transformed to active charts and articles.

There are four main goals that are set to be achieved from the portal. Primarily, the portal aims to provide access to data related to the government to the general public. Through this, it intends to communicate relevant information to the public through visualisations. The data is promoted for app developers which generate value. Finally, the increase of research and analysis is furthered by making these data available to the public.

The Smart Economy aspect is emphasised by making data available under the topic of Economy and Finance while the data related Smart Mobility can be accessed under Infrastructure and Transport. Education and Health data are available from this one-stop portal which emphasises the Smart Living characteristic within a Smart City. Smart Environment related data are made available under the topic of Environment. Finally, the Society data from the portal highlights the Smart People aspect that is found within a smart city.

II.4.5   Use case: Tokyo

The Tokyo Open Data catalog[7] is an open data portal that provides information related to the city and is available for the public use. It is mainly used to research and understand the city further and develop smartphone applications to understand the city of Tokyo in a user-friendly manner.

The data pertaining to the Smart governance dimension of a smart city is made available under the topic of Tokyo Metropolitan Government Bureau. Similarly, the Smart Economy and Smart Environment data can be obtained under the themes Industrial Job creation and Environment, respectively. Town Planning and Transport and information data reflects the Smart Mobility characteristic. Finally, Smart Living dimension within a smart city is covered by making Medical/Welfare and Life data available to the public.

II.5   LINKED DATA FOR SMART CITIES

Smart cities are typically implemented through a set of smart and interconnected services addressing various public sectors. The service requires data located in various systems and locations. To enable interoperability among the systems that generate and consume the data, a flexible and extensible data model is needed. linked data is an approach that allows the interconnection of information on semantic level. The semantics of the information is described through ontologies [326]. Both open and closed data can be linked to permit the implementation of integrated services to the citizen, such as transportation, energy, and water services.

Additionally, better decision-making and utilisation of resources can be achieved through employing data analytics on top of the linked data, as will be discussed in Section 11.6. Therefore, it will improve the smartness of cities [190].

II.5.1   Implementation of linked data in smart city projects

Linked data can facilitate the interconnections among smart city services from multiple domains. The city of Catania, for example, uses linked data to link geolocation data and data from public transportation, urban fault reporting, road maintenance, and waste management services [78]. The city of Catania uses the GeoSPARQL and BBC Programmes ontology[8] to describe the semantics of some parts of its linked data.

Linked data is also applied to link the data collected through the IoT infrastructure in a city on the semantic level. The semantics of the collected data are defined by matching them with previously defined ontologies. The applications on top of the IoT and linked data infrastructure aim to improve the lives of elderly citizens through Ambient Assisted Cities [266]. The concept is implemented in the City4Age project that uses Madrid, Athens, Singapore, Montpellier, Lecce, and Birmingham as the test beds [289]. linked data has also been employed to semantically set relationships among spatial features such as built and natural environment to allow better understanding of physical phenomena of geospatial features [319].

Several communities develop ontology catalogs that contain various ontologies from multiple domains. The ontologies can be used as semantic data models to design and develop smart city applications. Linked Open Vocabularies (LOV), LOVs for IoT (LOV4IoT), OpenSensingCity (OSC), and READY4SmartCities are examples of ontology catalogs. A study conducted by Gyrard et al. [162] found that LOV4IoT contains ontologies that promote the implementation of Internet of Things thus it is relevant for designing smart city applications.

To introduce and apply linked data in a project, a methodology is required that consists of multiple steps involving both domain and ontology experts. For a project in the Architecture, Engineering, and Construction (AEC) domain, the methodology starts with defining use cases, identifying data requirements, map the data requirements to existing ontologies, development of ontologies if necessary, defining links, and publishing the data [255]. Identification and definition of use cases in the linked data application domain are essential to ensure that the linked data fulfil the requirements of the project or application. Usually, there are common use cases shared across projects in a certain domain. Therefore, it is necessary to have a use case repository containing those common use cases. Each use case describes amongst others the involved stakeholders, required data domain, and available ontologies or datasets [256]. Thus, projects having similar use cases can reuse and link those ontologies and datasets. The methodology has been adopted in the smart cities, particularly in a smart energy system that considers multiple use cases involving citizens, industry, and municipalities, and also requires data from heterogeneous sources [387,415].

II.5.2   Linked data for smart energy system

Linked data has been used to link data sources required to develop an integrated smart energy service provided to citizens, business stakeholders, and policymakers [415]. This approach has been implemented in a smart energy system to link and integrate open data, such as weather and geospatial data, metering data from sensors, e.g. power consumption, temperature and humidity, and data from simulation tools such as EnergyPlus [413]. The system has been prototypically validated in three different European cities, i.e. Cambridge, Seville, and Lizanello. Figure 11.10 shows an overview of the data infrastructure. The district profile employs linked data to link and route data from different sources to the services provided to users. The services are integrated into a web-based platform.

The platform provides apps to citizens to help them to improve their power consumption pattern, their awareness regarding energy efficiency measures, and their communication with stakeholders, such as utility companies, to implement the measures. Policymakers can use the platform to monitor the district energy efficiency and to support decision-making on the introduction of renewable energy sources in the district.

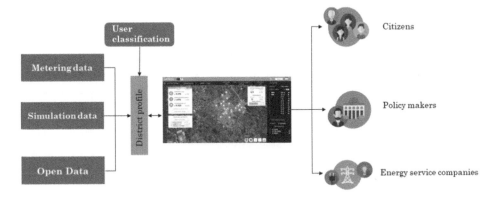

Figure II.I0 Overview of the smart energy system data infrastructure.

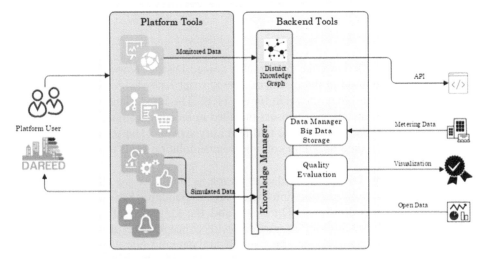

Figure II.II DAREED platform architecture.

The platform also helps energy service companies to have more transparency of their customer behaviours, to define new tariffs, and to improve the communication with their customers [414].

As depicted in Figure 11.11, the platform consists of two main components i.e. platform and backend tools. The platform tools are web-based applications integrated in the platform. It has a user access management layer that controls the access to the different tools. Only particular user roles have access to a certain tool. For example, the tool to simulate the impacts of introduction of renewable power generators can only be accessed by users such as municipalities or building operators, not the citizens.

The tools in the platform facilitate the complete cycle of decision-making in smart cities from only providing necessary information, create and improve awareness of energy efficiency among different city actors, conducting simulations to assess different energy efficiency measures, and implementing the measures.

The backend tools are responsible to collect, store, present, and ensure the quality of data coming from various data sources. As shown in Figure 11.11, the Knowledge Manager plays the central role. It employs a knowledge graph which is based on linked

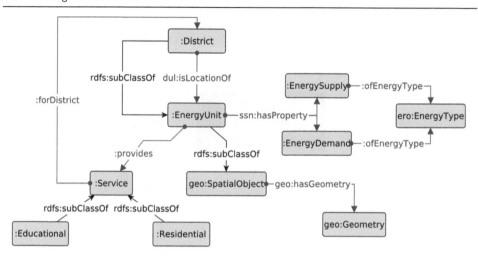

Figure II.I2 DAREED ontology [387].

data technology. The Knowledge Graph, also named District Knowledge Graph, links various data sources used in the platform. The District Knowledge Graph is described semantically by the ontology shown in Figure 11.12. The Data Manager in the backend stores the data from metering of sensors installed in built environment in the city, such as power consumption, temperature, and humidity sensors. The sensor metering data are linked with open data and data coming from simulation tools through the District Knowledge Graph.

The schema of District Knowledge Graph is described in the ontology. The ontology reuses and links existing ontologies, such as Semantic Sensor Network (SSN), GeoSPARQL, Energy Resources Ontology, and the Ontology of units of Measure (OM) [415]. The SSN Ontology is extended with concepts to express energy balance from district to building levels including supply and demand sides [387]. This approach allows the integration of energy behaviour data outputted from simulation tools like EnergyPlus.

The interactions between the platform tools and District Knowledge Graph are managed by the Knowledge Manager. The Knowledge Manager translates the JSON requests from the platform tools into SPARQL queries. As illustrated in Figure 11.13, the Knowledge Manager has two possible modes to query the data requested by the platform tools. The first mode is to query the District Knowledge Graph using SPARQL. The queried data are data coming from the linked open data cloud (RDF format) that have been linked with the District Knowledge Graph. The second mode is to perform SPARQL queries directly to the District Knowledge Graph and additionally retrieve data from the Data Manager through JSON query. The second mode is chosen when the platform tools request metering data along with data from the District Knowledge Graph.

II.6   DATA ANALYTICS APPROACHES

Data analytics is an approach consisting of a set of techniques to integrate data from a wide variety of different sources, identify patterns, correlations, and other useful information, drawing inferences, and finally making predictions [158,286]. Data analytics enables more accurate decision-making. Thomas Davenport classified data analytics

- SPARQL query on Knowledge Graph

- SPARQL query on Knowledge Graph and additional data retrieval

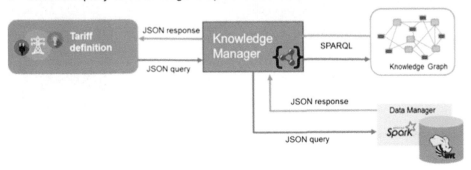

Figure II.13 Interaction with linked data in DAREED.

into four types based on the questions that can be answered through it. The first type, descriptive analytics, focuses on the question "What has happened?". The second, diagnostic analytics, answers "Why it has happened?". The third type, which is predictive analytics, addresses the question "What will happen?". Finally, prescriptive analytics gives answers to the question "How to make the best of it?" [94,183]. The following sections give an overview on the four types of data analytics and how they can be applied to smart cities.

II.6.1 Descriptive analytics

Since decision-making often draws from information and experiences from the past, descriptive analytics is an essential approach in a smart city project to understand the states of the city in the past. Various techniques can be applied for descriptive analytics such as dashboards, data plots, statistical summaries and plots, reports, and data visualisation in maps.

Using visualisations, complex scenarios applying IoT in various smart city services ranging from building management, healthcare, and transportation can be more understood by city actors [191]. A visualisation system of free parking places in the city that is accessible through smartphones and tablets help citizens to plan their trips to the city centre. The system reads data generated by sensors in a Wireless Sensor Network (WSN) installed in city parking places [155]. An application of descriptive analytics in smart cities is also a system that monitors people who violate traffic rules, for example bikers who do not wear helmets [360]. Traffic congestion monitoring, crime and safety reports and trends identification, and monitoring city pipeline networks are examples of projects that implement descriptive analytics [69,202,376].

In the smart energy service mentioned in previous sections, energy consumption data of buildings located in the city are visualised in a map. When a building is selected, the platform sends a data request to the Knowledge Manager. The Knowledge Manager retrieves the metering data from the Data Manager and links them with the District Knowledge Graph containing the link to geolocation data. The retrieved data are then sent back to the platform for display. Figure 11.14 shows the geo-visualisation of the city of Seville, Spain, provided by the District Monitoring Tool of the platform. Figure 11.15 gives an example of the visualisation of energy data of a selected building.

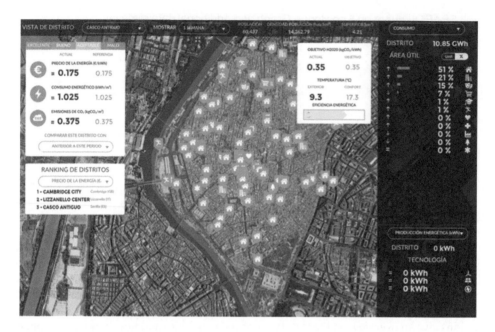

Figure 11.14 Example of descriptive analytics through geo visualisation of energy data in DAREED.

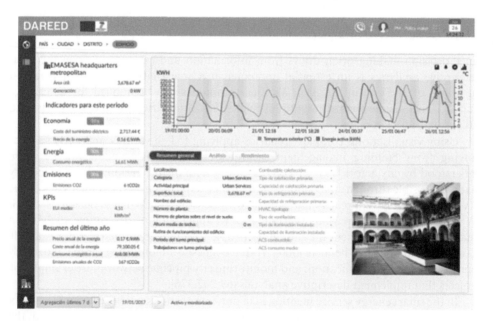

Figure 11.15 Example of building energy data visualisation.

II.6.2 Diagnostics analytics

In smart cities, the diagnosis of the faults, or identification of any security risks, is essential to achieve a reliable and robust system. Hence, the diagnostic analytics play a very essential role. In literature, diagnostic analytics is utilised to identify problems that have happened or are happening in a mechanical, electrical or ICT system. It also helps to identify the causes for desired systems' performances which also covers to understand the impact of the input factors and operational policies for the systems. Hence it is highly important for smart city applications. For instance, Elshaafi et al. [119] proposed a trustworthy management system to quantify the security of component services and their compositions for cyberattacks and to help smart systems prevent the cyberattacks.

The proposed solution is providing a secure and trustworthy service by quantifying the security levels. To manage a cyberattack, they have used a surface metric originated from changes in the value of a security and utilising an aspect's weight change approach. An example of a cyberattack to a system is presented in Elshaafi et al. [119], which presents a resource attack surface that has a score of 1 (i.e., secure resource) when the access right (and other aspects) level score is high (≤ 1), in the regions where it has an average weight above 1.

Another diagnostic analytics study has been conducted for fault detection problems [314]. They have proposed an integrated and hybrid artificial Intelligence (AI) solution for the building management systems (BMS) to increase the comfort and to reduce the operation costs. In the proposed solution, a fault detection solution is developed using an artificial neural network (ANN) to evaluate the current and historical information. The proposed solution aims to detect the abnormalities of energy consumption, optimising the use of different resources after the fault detection process as shown in Figure 11.16. The proposed solution is tested on 7 halls, 154 offices, 12 data room, and 22 power rooms' electricity consumptions, and successfully detected anomalies in energy consumption. The results of the detection process is shown in Figure 11.17.

II.6.3 Predictive analytics

Predictive analytics focuses on predicting or forecasting future events or trends based on the patterns and correlations of data collected from the past. Statistical analysis and machine learning are typically employed to carry out predictive analytics. For example, a machine learning technique, multiple kernel regression (MKr), that extends support

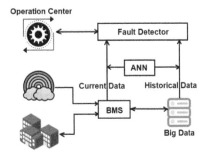

Figure II.16 The proposed anomaly detection solution to detect abnormal energy usage [3I4].

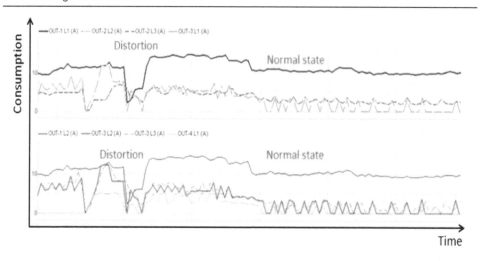

Figure II.I7 Electricity fault detection process [3l4].

vector regression (SVR), is used to predict water demand in a city by considering water consumption and climate data such as temperature, wind speed, precipitation, and air pressure [173]. A decision tree based technique is also used to detect anomalies and predict energy consumption in buildings [179,412]. Zero-shot machine learning is proposed to estimate the flow of the urban transportation network [227].

In the healthcare domain, machine learning and deep reinforcement learning techniques have been developed to diagnose heart disease automatically [390], predict cardiovascular patient risk after percutaneous coronary intervention (PCI) [431], as well as identify and diagnose breast cancer [258,353,388]. These prediction capabilities can improve the health quality of citizens living in a city and allow healthcare service providers and policymakers to plan and manage healthcare facilities efficiently.

In the energy sector of smart cities, machine learning approaches, such as Deep Neural Network (DNN), Rpart and Random Forest have been implemented to predict energy consumption of city public spaces [433]. Time series analysis, such as auto-regressive integrated moving average (ARIMA) and seasonal auto-regressive integrated moving average (SARIMA), are employed to forecast building power consumption. The power consumption of most public buildings follows a daily and weekly seasonal pattern. Forecast of time series containing a multiple seasonality can be done using methods such as Trigonometric Seasonal Box-Cox Transformation with ARMA residuals Trend and Seasonal Components (TBATS) and Artificial Neural Network (ANN) [197]. Figure 11.18 shows an example of power consumption forecast of a building in the city of Seville displayed in the smart energy platform.

II.6.4 Prescriptive analytics overview

In smart city applications, several data analytics models have been proposed. One of them is prescriptive analytics which is associated with both descriptive and predictive analytics. A descriptive analytics approach focuses on what has happened, while predictive analytics focuses on what might happen. The main purpose of the prescriptive analytics receives data from both descriptive and predictive sources and searches for the best solution among choices [163]. The approach utilises the outcome of the optimisation

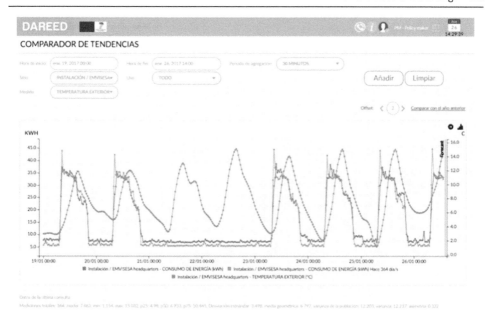

Figure II.18   Power consumption forecast in DAREED.

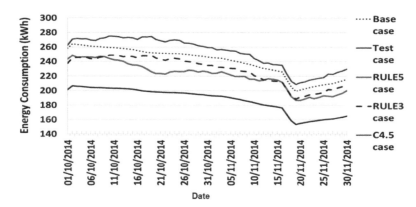

Figure II.19   Energy savings with the proposed solution for two-month [179].

and simulation systems to suggest possible options [100]. Hence, it is a reliable approach to utilise in smart city applications to solve complex problems.

In literature, several approaches have been used to tackle these problems. Howell et al. [179] proposed a cloud-based smart building energy management solution to tackle the energy smart infrastructure feature of the smart city. The proposed methodology uses a semantic middleware which is integrated to a sensory based network and empowered with intelligent data analytic tools like genetic algorithms, ANNs, and fuzzy logic and linked with a user-friendly and smart web-based user interface. The solution was applied in a public building for the period October–November, and managed to reduce energy consumption around 21% as shown in Figure 11.19.

Similarly, Fotopoulou et al. [137] have proposed an innovative energy-aware system to support the design and development of user specific energy management and awareness services. The solution aims to support occupants' energy consumption and

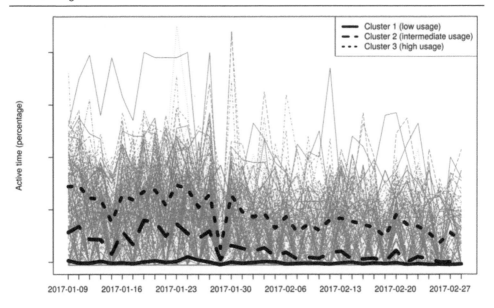

Figure II.20 The clustering solutions for three level of cluster groups: cluster I (solid lines), low energy
consumption; cluster 2 (dashed lines), intermediate energy consumption; and cluster 3
(dotted lines), high energy consumption [137].

efficiency awareness behaviours to make a positive impact on total energy cost. The
proposed solution aims to use a classification and rule-based solution to extract user
behaviour in their daily energy consumption patterns. They have applied this solution
in HVAC systems to extract clusters based on the k-means algorithm. The algorithm
extracts the HVAC energy consumptions based on three clusters as in low, intermediate,
and high energy usage zones. This clustering process is shown in Figure 11.20.

II.7   CONCLUSION

This chapter introduced the concept of the smart city and its relationship to smart
buildings. It presented several of the main features of the smart city, in particular, the
smart building, the smart energy and grids, and the smart mobility and transport. It
provided some overviews of the different architectures of smart cities, and how they relate
to smart building architectures. Next, it explored how digital transformation can support
the move towards smarter cities, and the different steps that must be taken to move toward
complete digital representations of cities and the services they enable with reference to
five use cases across five cities: Bonn, Dublin, Toronto, Singapore, and Tokyo.

Next, the use of linked data to support digitisation of cities and energy systems
was presented, with reference to specific examples. Finally, an overview of different
approaches to data analytics (descriptive, diagnostic, predictive, and prescriptive) were
discussed, and how these approaches are essential for supporting improved decision-
making across all the components that make up a smart city. Collectively, the application
of digitisation and data analytics are core aspects of the smart city, and can be applied
to support the move towards the realisation of smart, sustainable cities which meet the
needs of both individuals, and society at large.

II.8 ACKNOWLEDGEMENT

We thank TUBITAK 2219 programme to fund Dr. Ozum Calli to conduct her post-Doc study at the University of Exeter. We also thank to European Commission who funded some parts of this chapter through DAREED project (Grant agreement ID: 609082)

NOTES

1 Hannah Ritchie and Max Roser. Urbanization - Our World in Data, 2018. https://ourworldindata.
   org/urbanization
2 https://opendata.bonn.de/.
3 https://oparl.org/.
4 https://smartdublin.ie/.
5 https://www.toronto.ca/city-government/data-research-maps/open-data/.
6 https://data.gov.sg/.
7 https://portal.data.metro.tokyo.lg.jp/.
8 https://www.bbc.co.uk/ontologies/po.

Bibliography

1. N. Abbasabadi and M. Ashayeri. Urban energy use modeling methods and tools: A review and an outlook. *Building and Environment*, 161:106270, 2019.

2. Joseph Abhayaratna, Linda van den Brink, Nicholas Car, Rob Atkinson, Timo Homburg, Frans Knibbe, Kris McGlinn, Anna Wagner, Mathias Bonduel, Mads Holten Rasmussen, and Florian Thiery. OGC Benefits of Representing Spatial Data Using Semantic and Graph Technologies. Technical Report 19-078r1, Open Geospatial Consortium, June 2020.

3. K. Afsari, C. M. Eastman, and D. R. Shelden. Cloud-based BIM data transmission: current status and challenge. In *Proceedings of the International Symposium on Automation and Robotics in Construction*, volume 33, pages 1073–1080. IAARC Publications, 2016.

4. International Energy Agency. Energy Performance Certification of Buildings. A policy tool to improve energy efficiency, IEA Report, 2010.

5. Abiola Akanmu and Chimay J. Anumba. Cyber-physical systems integration of building information models and the physical construction. *Engineering, Construction and Architectural Management*, 22(5):516–535, September 2015.

6. Abdullah Hamed Al-Badi, Razzaqul Ahshan, Nasser Hosseinzadeh, Reza Ghorbani, and Eklas Hossain. Survey of Smart Grid Concepts and Technological Demonstrations Worldwide Emphasizing on the Oman Perspective. *Applied System Innovation*, 3(1), 2020.

7. E. Alreshidi, M. Mourshed, and Y. Rezgui. Requirements for cloud-based BIM governance solutions to facilitate team collaboration in construction projects. *Requirements engineering*, 23(1):1–31, 2018.

8. Grigor Angjeliu, Dario Coronelli, and Giuliana Cardani. Development of the simulation model for Digital Twin applications in historical masonry buildings: The integration between numerical and experimental reality. *Computers & Structures*, 238:106282, October 2020.

9. P. T. Anh Mai, J. K. Nurminen, and M. Di Francesco. Cloud Databases for Internet-of-Things Data. In *IEEE International Conference on Internet of Things*, pages 117–124, Taipei, Taiwan, 2014. IEEE.

10. B. Anthony Jnr, S. Abbas Petersen, D. Ahlers, and J. Krogstie. API deployment for big data management towards sustainable energy prosumption in smart cities-a layered architecture perspective. *International Journal of Sustainable Energy*, 39(3):263–289, 2020.

11. Rasa Apanaviciene, Andrius Vanagas, and Paris Fokaides. Smart Building Integration into a Smart City (SBISC): Development of a New Evaluation Framework. *Energies*, 13:2190, May 2020.

12. Dörthe Arndt, Ben De Meester, Anastasia Dimou, Ruben Verborgh, and Erik Mannens. Using rule-based reasoning for RDF validation. In *International Joint Conference on Rules and Reasoning*, pages 22–36. Springer, 2017.

13. Abhay Arora, Manar Amayri, Venkataramana Badarla, Stéphane Ploix, and Sanghamitra Bandyopadhyay. Occupancy estimation using non intrusive sensors in energy efficient buildings. In *14th Conference of International Building Performance Simulation Association, Hyderabad, India, Dec. 7–9*, 2015.

14. Mohammad Sadegh Aslanpour, Mostafa Ghobaei-Arani, and Adel Nadjaran Toosi. Auto-scaling web applications in clouds: A cost-aware approach. *Journal of Network and Computer Applications*, 95:26–41, 2017.

15. Mark Austin, Parastoo Delgoshaei, Maria Coelho, and Mohammad Heidarinejad. Architecting smart city digital twins: combined semantic model and machine learning approach. *Journal of Management in Engineering*, 36(4):04020026, 2020.

16. Ibukun Awolusi, Chukwuma Nnaji, Eric Marks, and Matthew Hallowell. Enhancing construction safety monitoring through the application of internet of things and wearable sensing devices: A review. In *ASCE International Conference on Computing in Civil Engineering 2019*, pages 530–538, 2019.

17. Salman Azhar, Abid Nadeem, Johnny Y. N. Mok, and Brian H. Y. Leung. Building Information Modelling (BIM): A new paradigm for visual interactive modeling and simulation for construction project. In *First International Conference on Construction in Developing Countries (ICCIDC-I)*, volume 1, pages 435–446, Karachi, Pakistan, 2008.

18. Muhammad Babar and Fahim Arif. Real-time data processing scheme using big data analytics in internet of things based smart transportation environment. *Journal of Ambient Intelligence and Humanized Computing*, 10(10):4167–4177, October 2019.

19. Gianluca Bacchiega. Creating an embedded digital twin: monitor, understand and predict device health failure. *Inn4mech-Mechatronics and Industry*, 4, 2018.

20. Y. N. Bahar, C. Pere, J. Landrieu, and C. Nicolle. A thermal simulation tool for building and its interoperability through the building information modeling (BIM) platform. *Buildings*, 3(2):380–398, 2013.

21. Bharathan Balaji, Arka Bhattacharya, Gabriel Fierro, Jingkun Gao, Joshua Gluck, Dezhi Hong, Aslak Johansen, Jason Koh, Joern Ploennigs, Yuvraj Agarwal, Mario Berges, David Culler, Rajesh Gupta, Mikkel Baun Kjærgaard, Mani Srivastava, and Kamin Whitehouse. Brick: Towards a unified metadata schema for buildings. In *Proceedings of the 3rd ACM International Conference on Systems for Energy-Efficient Built Environments*, BuildSys '16, page 41–50, New York, NY, USA, 2016. Association for Computing Machinery.

22. Raphael Barbau, Sylvere Krima, Sudarsan Rachuri, Anantha Narayanan, Xenia Fiorentini, Sebti Foufou, and Ram D. Sriram. OntoSTEP: Enriching product model data using ontologies. *CAD Computer Aided Design*, 44(6):575–590, 2012.

23. Filipe Barbosa, Jonathan Woetzel, Jan Mischke, Maria Joao Ribeirinho, Mukund Sridhar, Matthew Parsons, Nick Bertram, and Stephanie Brown. Reinventing construction through a productivity revolution. *McKinsey Global Institute*, 2017.

24. Sergey Yevgenievich Barykin, Andrey Aleksandrovich Bochkarev, Olga Vladimirovna Kalinina, and Vladimir Konstantinovich Yadykin. Concept for a supply chain digital twin. *International Journal of Mathematical, Engineering and Management Sciences*, 5(6):1498–1515, 2020.

25. Srijita Basu, Arjun Bardhan, Koyal Gupta, Payel Saha, Mahasweta Pal, Manjima Bose, Kaushik Basu, Saunak Chaudhury, and Pritika Sarkar. Cloud computing security challenges & solutions – A survey. In *2018 IEEE 8th Annual Computing and Communication Workshop and Conference (CCWC)*, pages 347–356. IEEE, 2018.

26. Michael Batty. Digital twins. *Environment and Planning B: Urban Analytics and City Science*, 45(5):817–820, September 2018.

27. Kirk Beaty, Andrzej Kochut, and Hidayatullah Shaikh. Desktop to cloud transformation planning. In *2009 IEEE International Symposium on Parallel & Distributed Processing*, pages 1–8. IEEE, 2009.

28. Burcin Becerik-Gerber, Mohsin K. Siddiqui, Ioannis Brilakis, Omar El-Anwar, Nora El-Gohary, Tarek Mahfouz, Gauri M. Jog, Shuai Li, and Amr A. Kandil. Civil Engineering Grand Challenges: Opportunities for Data Sensing, Information Analysis, and Knowledge Discovery. *Journal of Computing in Civil Engineering*, 28(4):04014013, July 2014.

29. J. Beetz. *Facilitating distributed collaboration in the AEC / FM sector using semantic web Technologies*. PhD thesis, Eindhoven University of Technology, 2009.

30. Jakob Beetz, Léon van Berlo, Ruben de Laat, and Pim van den Helm. BIMserver. org – An open source IFC model server. In *Proceedings of the CIB W78 conference*, pages 1–8, 2010.

31. Jakob Beetz, Jos Van Leeuwen, and Bauke de Vries. IfcOWL: a case of transforming EXPRESS schemas into ontologies. *Artificial Intelligence for Engineering Design, Analysis and Manufacturing*, 23(1):89–101, 2009.

32. Jakob Beetz, Jos P. van Leeuwen, and Bauke de Vries. An ontology web language notation of the industry foundation classes. In *Proceedings of the 22nd CIB W78 Conference on Information Technology in Construction*, pages 193–198, 2005.

33. Mangesh Bendre, Bofan Sun, Ding Zhang, Xinyan Zhou, Kevin ChenChuan Chang, and Aditya Parameswaran. Dataspread: Unifying databases and spreadsheets. In *Proceedings of the VLDB Endowment International Conference on Very Large Data Bases*, volume 8, page 2000. NIH Public Access, 2015.

34. Tim Berners-Lee. linked data – Design Issues, 2006. https://www.w3.org/DesignIssues/LinkedData.html.

35. Tim Berners-Lee, James Hendler, and Ora Lassila. The semantic web. *Scientific american*, 284(5):34–43, 2001.

36. David Bernstein. Containers and cloud: From lxc to docker to kubernetes. *IEEE Cloud Computing*, 1(3):81–84, 2014.

37. M. Bew and M. Richards. BIM Maturity Model. In *Construct IT Autumn 2008 Members' Meeting. Brighton, UK*, 2008.

38. J. L. Blanco, S. Fuchs, M. Parsons, and M. J. Ribeirinho. Artificial intelligence: Construction technology's next frontier. Technical report, McKinsey & Company, 2018.

39. Calin Boje. Developing the Crowd Simulation Scenario (CSS) ontology supporting building evacuation design. In *Proceedings of the 7th Linked Data in Architecture and Construction Workshop*, pages 50–63, 2019.

40. Calin Boje, Veronika Bolshakova, Annie Guerriero, Sylvain Kubicki, and Gilles Halin. Semantics for linking data from 4D BIM to digital collaborative support. *Frontiers of Engineering Management*, 2020.

41. Calin Boje, Annie Guerriero, Sylvain Kubicki, and Yacine Rezgui. Towards a semantic Construction Digital Twin: Directions for future research. *Automation in Construction*, 114:103179, June 2020.

42. A. Bolton, L. Butler, I. Dabson, M. Enzer, M. Evans, T. Fenemore, and F. Harradence. The Gemini Principles. Technical report, University of Cambridge, Centre for Digital Built Britain, 2018.

43. Mathias Bonduel, Anna Wagner, Pieter Pauwels, Maarten Vergauwen, and Ralf Klein. Including widespread geometry formats in semantic graphs using RDF literals. In *Proceedings of the European Conference on Computing in Construction (EC3 2019)*, pages 341–350, Chania, Greece, 2019.

44. Mathias Bonduel, Anna Wagner, Pieter Pauwels, Maarten Vergauwen, and Ralf Klein. Including Widespread Geometry Schemas Into Linked Data Based BIM Applied to Built Heritage. *Smart Infrastructure and Construction*, 172(1):34–51, March 2020.

45. Peter Bonsma, Iveta Bonsma, Tsvetelina Zayakova, Andre van Delft, Rizal Sebastian, and Michel Böhms. Open standard CMO for parametric modelling based on semantic web. In *eWork and eBusiness in Architecture, Engineering and Construction*, pages 923–928, 2014.

46. Peter Bonsma, Iveta Bonsma, Anna Elisabetta Ziri, Ernesto Iadanza, Federica Maietti, Marco Medici, Federico Ferrari, Rizal Sebastian, Sander Bruinenberg, and Pedro Martín Lerones. Handling huge and complex 3D geometries with semantic web technology. In *IOP Conference Series: Materials Science and Engineering*, volume 364, 2018.

47. Borja Bordel, Ramón Alcarria, Tomás Robles, and Diego Martín. Cyber-physical systems: Extending pervasive sensing from control theory to the Internet of Things. *Pervasive and mobile computing*, 40:156–184, 2017.

48. Carsten Bormann, Mehmet Ersue, and Ari Keranen. Terminology for constrained-node networks, RFC 7228. Technical Report IETF RFC 7228, Internet Engineering Task Force (IETF), 2014.

49. André Borrmann, Markus König, Christian Koch, and Jakob Beetz. *Building Information Modeling: Technology Foundations and Industry Practice*. Springer International Publishing, first edition, 2018.

50. A. Faye Borthick, Gary P. Schneider, and Therese R. Viscelli. Analyzing data for decision making: Integrating spreadsheet modeling and database querying. *Issues in Accounting Education*, 32(1):59–66, 2017.

51. Lorenzo Bottaccioli, Alessandro Aliberti, Francesca Ugliotti, Edoardo Patti, Anna Osello, Enrico Macii, and Andrea Acquaviva. Building energy modelling and monitoring by integration of IoT devices and building information models. In *41st Annual Computer Software and Applications Conference*, volume 1, pages 914–922, Turin, Italy, 2017. IEEE.

52. Mike Botts, George Percivall, Carl Reed, and John Davidson. OGC Sensor Web Enablement: Overview and High Level Architecture. In Silvia Nittel, Alexandros Labrinidis, and Anthony Stefanidis, editors, *GeoSensor Networks: Second International Conference, GSN 2006, Boston, MA, USA, October 1-3, 2006, Revised Selected and Invited Papers*, pages 175–190. Springer Berlin Heidelberg, Berlin, Heidelberg, 2008.

53. Serge Thomas Mickala Bourobou and Younghwan Yoo. User activity recognition in smart homes using pattern clustering applied to temporal ANN algorithm. *Sensors*, 15(5):11953–11971, 2015.

54. M. K. Bracht, A. P. Melo, and R. Lamberts. A metamodel for building information modeling-building energy modeling integration in early design stage. *Automation in Construction*, 121:103422, 2021.

55. Larry J. Brackney, Anthony R. Florita, Alex C. Swindler, Luigi Gentile Polese, and George A. Brunemann. Design and performance of an image processing occupancy sensor. In *Proceedings: The Second International Conference on Building Energy and Environment 2012987 Topic 10. Intelligent buildings and advanced control techniques*. Citeseer, 2012.

56. J. Scott Brennen and Daniel Kreiss. Digtalization. In Robert T. Craig Klaus Bruhn Jensen, editor, *The International Encyclopedia of Communication Theory and Philosophy*, pages 556–566. John Wiley & Sons, Chichester, 2016.

57. F. G. Brundu, E. Patti, A. Osello, M. Del Giudice, N. Rapetti, A. Krylovskiy, M. Jahn, V. Verda, E. Guelpa, L. Rietto, and A. Acquaviva. IoT software infrastructure for energy management and simulation in smart cities. *IEEE Transactions on Industrial Informatics*, 13(2):832–840, 2016.

58. A. H. Buckman, M. Mayfield, and Stephen B. M. Beck. What is a smart building? *Smart and Sustainable Built Environment*, 3(2):92–109, September 2014.

59. buildingSMART International. Enabling an Ecosystem of Digital Twins. Technical report, BuildingSmart International, 2020.

60. Raf Buyle, Laurens De Vocht, Mathias Van Compernolle, Dieter De Paepe, Ruben Verborgh, Ziggy Vanlishout, Björn De Vidts, Peter Mechant, and Erik Mannens. OSLO: Open standards for linked organizations. In *Proceedings of the international conference on electronic governance and open society: Challenges in Eurasia*, pages 126–134, 2016.

61. Nelio Cacho, Frederico Lopes, Everton Cavalcante, and Irani Santos. A smart city initiative: The case of Natal. In *2016 IEEE International Smart Cities Conference (ISC2)*, pages 1–7, 2016.

62. Sriram Changali, Mohammad Azam, and Mark van Nieuwland. The construction productivity imperative. Technical report, McKinsey & Company, 2015.

63. Claude Chaudet and Yoram Haddad. Wireless software defined networks: Challenges and opportunities. In *Microwaves, Communications, An-tennas and Electronics Systems (COMCAS)*, Tel Aviv, Israel, 2013. IEEE.

64. Liming Chen, Chris Nugent, and George Okeyo. An ontology-based hybrid approach to activity modeling for smart homes. *IEEE Transactions on human-machine systems*, 44(1):92–105, 2013.

65. Peter Pin-Shan Chen. The Entity-Relationship Model – toward a Unified View of Data. *ACM Transactions on Database Systems*, 1(1):9–36, March 1976.

66. Xiu-Shan Chen, Chi-Chang Liu, and I.-Chen Wu. A BIM-based visualization and warning system for fire rescue. *Advanced Engineering Informatics*, 37:42–53, 2018.

67. Jack CP Cheng, Weiwei Chen, Keyu Chen, and Qian Wang. Data-driven predictive maintenance planning framework for MEP components based on BIM and IoT using machine learning algorithms. *Automation in Construction*, 112:103087, 2020.

68. Zoé Chevallier, Béatrice Finance, and Benjamin Cohen Boulakia. A reference architecture for smart building digital twin. In *Proceedings of the International Workshop on Semantic Digital Twins, CEUR Workshop Proceedings*, Heraklion, Greece, volume 2615, pages 1–12, June 3, 2020.

69. Cristian Chilipirea, Andreea-Cristina Petre, Loredana-Marsilia Groza, Ciprian Dobre, and Florin Pop. An integrated architecture for future studies in data processing for smart cities. *Microprocessors and Microsystems*, 52:335–342, 2017.

70. Ming-Chuan Chiu, Tsai-Chi Kuo, and Hsin-Ting Liao. Design for sustainable behavior strategies: Impact of persuasive technology on energy usage. *Journal of Cleaner Production*, 248:119214, 2020.

71. H. Cho, D. Cabrera, S. Sardy, R. Kilchherr, S. Yilmaz, and M. K. Patel. Evaluation of performance of energy efficient hybrid ventilation system and analysis of occupants' behavior to control windows. *Building and Environment*, 188:107434, 2021.

72. Derek J. Clements-Croome. *Intelligent Buildings: Design Management and Operation*. Thomas Telford Publishing, first edition, 2004.

73. Raymond J. Cole, John Robinson, Zosia Brown, and Meg O'Shea. Re-contextualizing the notion of comfort. *Building Research & Information*, 36(4):323–336, 2008.

74. Sylvia Coleman and John B. Robinson. Introducing the qualitative performance gap: stories about a sustainable building. *Building Research & Information*, 46(5):485–500, 2018.

75. Sylvia Coleman, Marianne F. Touchie, John B. Robinson, and Terri Peters. Rethinking performance gaps: A regenerative sustainability approach to built environment performance assessment. *Sustainability*, 10(12):4829, 2018.

76. Autonomic Computing et al. An architectural blueprint for autonomic computing. *IBM White Paper*, 31(2006):1–6, 2006.

77. P. Conejos Fuertes, F. Martínez Alzamora, M. Hervás Carot, and J. C. Alonso Campos. Building and exploiting a Digital Twin for the management of drinking water distribution networks. *Urban Water Journal*, 17(8):704–713, September 2020.

78. Sergio Consoli, Valentina Presutti, Diego Reforgiato Recupero, Andrea G. Nuzzolese, Silvio Peroni, Misael Mongiovi', and Aldo Gangemi. Producing Linked Data for Smart Cities: The Case of Catania. *Big Data Research*, 7:1–15, 2017.

79. Diane J. Cook and Sajal K. Das. *Smart Environments: Technology, Protocols and Applications*. John Wiley & Sons, Inc., September 2004.

80. Steven Anson Coons. An outline of the requirements for a computer-aided design system. In *Proceedings of the May 21-23, 1963, spring joint computer conference*, pages 299–304, 1963.

81. Vincenzo Corrado, Ilaria Ballarini, Leandro Madrazo, and German Nemirovskij. Data structuring for the ontological modelling of urban energy systems: The experience of the SEMANCO project. *Sustainable Cities and Society*, 14:223–235, 2015.

82. G. Costa, A. Sicilia, G. N. Lilis, D. V. Rovas, and J. Izkara. A comprehensive ontologies-based framework to support the retrofitting design of energy-efficient districts. In *Proceedings of the 11th European Conference on Product and Process Modelling*, volume 2016, pages 673–681. CRC Press, 2016.

83. G. Costa, A. Sicilia, X. Oregi, J. Pedrero, and L. Mabe. A catalogue of energy conservation measures (ECM) and a tool for their application in energy simulation models. *Journal of Building Engineering*, 29:101102, 2020.

84. Aaron Costin, Alireza Adibfar, Hanjin Hu, and Stuart S. Chen. Building Information Modeling (BIM) for transportation infrastructure – Literature review, applications, challenges, and recommendations. *Automation in Construction*, 94:257–281, 2018.

85. Aaron Costin and Charles Eastman. Need for interoperability to enable seamless information exchanges in smart and sustainable urban systems. *Journal of Computing in Civil Engineering*, 33(3):04019008, 2019.

86. Aaron Costin, Jochen Teizer, and B. Schoner. RFID and BIM-enabled worker location tracking to support real-time building protocol and data visualization. *Journal of Information Technology in Construction*, 20:495–517, 2015.

87. Aaron Costin, Andrew Wehle, and Alireza Adibfar. Leading indicators – a conceptual IoT-based framework to produce active leading indicators for construction safety. *Safety*, 5(4):86, 2019.

88. Council of European Union. Energy Efficiency Directive (2012/27/EU), 2018.

89. Drury B. Crawley, Linda K. Lawrie, Frederick C. Winkelmann, Walter F. Buhl, Y. Joe Huang, Curtis O. Pedersen, Richard K. Strand, Richard J. Liesen, Daniel E. Fisher, Michael J. Witte, and J. Glazer. EnergyPlus: creating a new-generation building energy simulation program. *Energy and buildings*, 33(4):319–331, 2001.

90. J. Cuenca, F. Larrinaga, and E. Curry. DABGEO: A reusable and usable global energy ontology for the energy domain. *Journal of Web Semantics*, 61-62:100550, 2020.

91. Joud Al Dakheel, Claudio Del Pero, Niccolò Aste, and Fabrizio Leonforte. Smart buildings features and key performance indicators: A review. *Sustainable Cities and Society*, 61:102328, October 2020.

92. L. Daniele, F. den Hartog, and J. Roes. Created in close interaction with the industry: the smart appliances reference (SAREF) ontology. In Roberta Cuel and Robert Young, editors, *Formal Ontologies Meet Industry*, pages 100–112, Cham, 2015. Springer International Publishing.

93. Souripriya Das, Seema Sundara, and Richard Cyganiak. R2RML: RDB to RDF Mapping Language. W3C Recommendation, W3C, 2012.

94. Thomas Davenport. Make Better Decisions. *Harvard Business Review*, 87:117–120, November 2009.

95. Juan Manuel Davila Delgado, Liam J. Butler, Ioannis Brilakis, Mohammed Z. E. B. Elshafie, and Campbell R. Middleton. Structural Performance Monitoring Using a Dynamic Data-Driven BIM Environment. *Journal of Computing in Civil Engineering*, 32(3):04018009, May 2018.

96. Oliver Dawkins, Adam Dennett, and Andrew Hudson-Smith. Living with a Digital Twin: Operational management and engagement using IoT and Mixed Realities at UCL's Here East Campus on the Queen Elizabeth Olympic Park. In *Proceedings of the 26th annual GIScience Research UK conference: GISRUK 2018*, 2018.

97. Christophe Debruyne, Éamonn Clinton, Lorraine McNerney, Atul Nautiyal, and Declan O'Sullivan. Serving Ireland's geospatial information as linked data. In *Proceedings of the ISWC 2016 Posters & Demonstrations Track, CEUR Workshop Proceedings*, Kobe, Japan, volume 1690, October 19, 2016.

98. Christophe Debruyne, Kris McGlinn, Lorraine McNerney, and Declan O'Sullivan. A Lightweight Approach to Explore, Enrich and Use Data with a Geospatial Dimension with

semantic web Technologies. In *Proceedings of the Fourth International ACM Workshop on Managing and Mining Enriched Geo-Spatial Data*, pages 1–6, New York, NY, USA, 2017. ACM.

99. T. Derek and J. Clements-Croome. What do we mean by intelligent buildings? *Automation in Construction*, 6(5):395–400, 1997.

100. Prachi S. Deshpande, Subhash C. Sharma, and Sateesh K. Peddoju. *Security and Data Storage Aspect in Cloud Computing*, volume 52 of *Studies in Big Data*. Springer, Singapore, 2019.

101. R. Di Giulio, G. Bizzarri, B. Turillazzi, L. Marzi, and C. Quentin. The BIM-GIS model for EEBS integrated in healthcare districts: An Italian case study. In *Sustainable Places 2015*, pages 111–119. Sigma Orionis, 2015.

102. Dominic DiFranzo, Alvaro Graves, John S. Erickson, Li Ding, James Michaelis, Timothy Lebo, Evan Patton, Gregory Todd Williams, Xian Li, Jin Guang Zheng, Johanna Flores, Deborah L. McGuinness, and Jim Hendler. The web is my back-end: Creating mashups with linked open government data. In *Linking government data*, pages 205–219. Springer, 2011.

103. G. Dileep. A survey on smart grid technologies and applications. *Renewable Energy*, 146:2589–2625, 2020.

104. Anastasia Dimou, Miel Vander Sande, Pieter Colpaert, Ruben Verborgh, Erik Mannens, and Rik Van de Walle. RML: a generic language for integrated RDF mappings of heterogeneous data. In Christian Bizer, Tom Heath, Sören Auer, and Tim Berners-Lee, editors, *Proceedings of the 7th Workshop on Linked Data on the Web*, volume 1184 of *CEUR Workshop Proceedings*, 2014.

105. Wafeequa Dinah, Pheladi Tracy Lefika, and Bwalya Kelvin Joseph. The Role of Open Data in Smart Cities: Exploring Status in Resource-Constrained Countries. In Manuel Pedro Rodríguez Bolívar, Kelvin Joseph Bwalya, and Christopher G. Reddick, editors, *Governance Models for Creating Public Value in Open Data Initiatives*, pages 105–121. Springer International Publishing, Cham, 2019.

106. Kai Ding, He Shi, Jizhuang Hui, Yongjian Liu, Bin Zhu, Fuqiang Zhang, and Wei Cao. Smart steel bridge construction enabled by bim and internet of things in industry 4.0: A framework. In *2018 IEEE 15th International Conference on Networking, Sensing and Control (ICNSC)*, pages 1–5. IEEE, 2018.

107. N. T. Dinh and Y. H. Kim. Restful architecture of wireless sensor network for building management system. *KSII Transactions on Internet and Information Systems*, 6:46–63, 2012.

108. Robert H. Dodier, Gregor P. Henze, Dale K. Tiller, and Xin Guo. Building occupancy detection through sensor belief networks. *Energy and buildings*, 38(9):1033–1043, 2006.

109. Bing Dong and Khee Poh Lam. Building energy and comfort management through occupant behaviour pattern detection based on a large-scale environmental sensor network. *Journal of Building Performance Simulation*, 4(4):359–369, 2011.

110. Raimundo F. Dos Santos Jr. and Chang-Tien Lu. Geography Markup Language (GML). In Shashi Shekhar and Hui Xiong, editors, *Encyclopedia of GIS*, pages 364–368. Springer, Boston, MA, USA, 2008.

111. Ying Duan, Wenfeng Li, Xiawen Fu, Yun Luo, and Lin Yang. Methodology for Reliability of WSN Based on Software Defined Network in Adaptive Industrial Environment. *IEEE/CAA Journal of Automatica Sinica*, 5(1):74–82, 2018.

112. Frederico Durao, Jose Fernando S. Carvalho, Anderson Fonseka, and Vinicius Cardoso Garcia. A systematic review on cloud computing. *The Journal of Supercomputing*, 68(3):1321–1346, 2014.

113. Charles Eastman. The use of computers instead of drawings in building design. *AIA Journal*, 63(3):46–50, 1975.

114. Charles Eastman. General purpose building description systems. *Computer-Aided Design*, 8(1):17–26, 1976.

115. Charles M. Eastman. *Building product models: computer environments, supporting design and construction.* CRC press, 1999.

116. Charles M. Eastman, Paul Teicholz, Rafael Sacks, and Kathleen Liston. *BIM handbook: A guide to building information modeling for owners, managers, designers, engineers and contractors.* John Wiley & Sons Inc., New York, NY, USA, 2011.

117. Abdulmotaleb El Saddik. Digital twins: The convergence of multimedia technologies. *IEEE multimedia*, 25(2):87–92, 2018.

118. M. Elagiry, N. Charbel, P. Bourreau, E. Di Angelis, and A. Costa. IFC to building energy performance simulation: a systematic review of the main adopted tools and approaches. In *8th Conference of IBPSA Germany and Austria*, pages 527–534, 2020.

119. Hisain Elshaafi, Jimmy McGibney, and Dmitri Botvich. Attack surface-based security metric framework for service selection and composition. *International Journal of Autonomous and Adaptive Communications Systems*, 10(1):88–113, 2017.

120. Douglas C. Engelbart. Augmenting human intellect: A conceptual framework. Technical report, Stanford Research Institute, Menlo Park, CA, 1962.

121. I. Esnaola-Gonzalez, J. Bermúdez, I. Fernandez, and A. Arnaiz. EEPSA as a core ontology for energy efficiency and thermal comfort in buildings. *Semantic Web*, 1:1–33, 2016.

122. I. Esnaola-Gonzalez, J. Bermúdez, I. Fernández, and A. Arnaiz. Two Ontology Design Patterns toward Energy Efficiency in Buildings. In *Proceedings of the 9th Workshop on Ontology Design and Patterns (WOP 2018), CEUR Workshop Proceedings*, volume 2195, pages 14–28, 2018.

123. European Commission. Energy performance of buildings directive | Energy.

124. Simon Evans, Cristina Savian, Allan Burns, and Chris Cooper. Digital twins for the built environment. Technical report, The Institution of Engineering and Technology, 2019.

125. M. Fadhel, E. Sekerinski, and S. Yao. A comparison of time series databases for storing water quality data. In *Interactive Mobile Communication, Technologies and Learning*, pages 302–313. Springer, Cham, 2018.

126. H. Fahad Alomirah and H. M. Moda. Assessment of Indoor Air Quality and Users Perception of a Renovated Office Building in Manchester. *International Journal of Environmental Research and Public Health*, 17(6):1–9, 2020.

127. X. Fang, S. Misra, G. Xue, and D. Yang. Smart Grid – The New and Improved Power Grid: A Survey. *IEEE Communications Surveys Tutorials*, 14(4):944–980, 2012.

128. M. T. Farias, A. Roxin, and C. Nicolle. A rule based system for semantical enrichment of building information exchange. In Theodore Patkos, Adam Wyner, and Adrian Giurca, editors, *Proceedings of the RuleML 2014 Challenge and the RuleML 2014 Doctoral Consortium*, volume 1211 of *CEUR Workshop Proceedings*, pages 2–9, Prague, Czech Republic, 2014.

129. Christoph Fehling, Frank Leymann, Ralph Retter, Walter Schupeck, and Peter Arbitter. *Cloud computing patterns: fundamentals to design, build, and manage cloud applications.* Springer, 2014.

130. Roy Fielding, Jim Gettys, Jeffrey Mogul, Henrik Frystyk, and Tim Berners-Lee. RFC2068: Hypertext Transfer Protocol–HTTP/1.1. Technical Report RFC2068, Internet Engineering Task Force, 1997.

131. Michael Filzmoser, Iva Kovacic, and Dragos-Cristian Vasilescu. Development of BIM-supported integrated design processes for teaching and practice. *Engineering Project Organization Journal*, 6(2-4):129–141, 2016.

132. Matthias Finger and Mohamad Razaghi. Conceptualizing "Smart Cities". *Informatik-Spektrum*, 40:6–13, 2017.

133. Marion R. Finley, Ancilla Karakura, and Raphaël Nbogni. Survey of Intelligent Building Concepts. *IEEE Communications Magazine*, 29(4):18–23, 1991.

134. Barry M. Flax. Intelligent Buildings. *IEEE Communications Magazine*, 29(4):24–27, April 1991.

135. David N. Ford and Charles M. Wolf. Smart Cities with Digital Twin Systems for Disaster Management. *Journal of Management in Engineering*, 36(4):1–10, 2020.

136. Abdur Rahim Mohammad Forkan, Ibrahim Khalil, Zahir Tari, Sebti Foufou, and Abdelaziz Bouras. A context-aware approach for long-term behavioural change detection and abnormality prediction in ambient assisted living. *Pattern Recognition*, 48(3):628–641, 2015.

137. Eleni Fotopoulou, Anastasios Zafeiropoulos, Fernando Terroso-Sáenz, Umutcan Şimşek, Aurora González-Vidal, George Tsiolis, Panagiotis Gouvas, Paris Liapis, Anna Fensel, and Antonio Skarmeta. Providing personalized energy management and awareness services for energy efficiency in smart buildings. *Sensors*, 17(9), September 2017.

138. Martin Fowler. *UML Distilled: A Brief Guide to the Standard Object Modeling Language*. The Addison-Wesley Object Technology Series. Addison-Wesley Professional, third edition, 2003.

139. Abigail Francisco, Neda Mohammadi, and John E. Taylor. Smart City Digital Twin–Enabled Energy Management: Toward Real-Time Urban Building Energy Benchmarking. *Journal of Management in Engineering*, 36(2):04019045, 2020.

140. Aidan Fuller, Zhong Fan, Charles Day, and Chris Barlow. Digital Twin: Enabling Technologies, Challenges and Open Research. *IEEE Access*, 8:108952–108971, June 2020.

141. S. Gandhi and J. R. Jupp. Characteristics of Green BIM: process and information management requirements. In Bernard, Alain and Rivest, Louis and Dutta, Debasish, editor, *International Conference on Product Lifecycle Management*, pages 596–605, Berlin Heidelberg, 2013. Springer.

142. H. Gao, C. Koch, and Y. Wu. Building information modelling based building energy modelling: A review. *Applied Energy*, 238:320–343, 2019.

143. S. Geissler, A. Androutsopoulos, A. G. Charalambides, C. Jareno Escudero, O. M. Jensen, O. Kyriacou, and H. Petran. ENERFUND-Identifying and rating deep renovation opportunities. In *IOP Conference Series: Earth and Environmental Science*, volume 323, page 012174. IOP Publishing, 2019.

144. Amirhosein Ghaffarianhoseini, H. Alwaer, Ali Ghaffarianhoseini, Derek Clements-Croome, Kaamran Raahemifar, and John Tookey. Intelligent or smart cities and buildings: a critical exposition and a way forward. *Intelligent Buildings International*, 10(2):122–129, December 2017.

145. Amirhosein Ghaffarianhoseini, Umberto Berardi, Husam AlWaer, Seongju Chang, Edward Halawa, Ali Ghaffarianhoseini, and Derek Clements-Croome. What is an intelligent building? Analysis of recent interpretations from an international perspective. *Architectural Science Review*, 59(5):338–357, September 2016.

146. Sunil Kumar Ghai, Lakshmi V. Thanayankizil, Deva P. Seetharam, and Dipanjan Chakraborty. Occupancy detection in commercial buildings using opportunistic context sources. In *2012 IEEE international conference on pervasive computing and communications workshops*, pages 463–466. IEEE, 2012.

147. Ali Gharaei, Jinzhi Lu, Oliver Stoll, Xiaochen Zheng, Shaun West, and Dimitris Kiritsis. Systems Engineering Approach to Identify Requirements for Digital Twins Development. In *Proceedings of the International Conference on Advances in Production Management Systems*, pages 82–90, March 2020.

148. Arka Ghosh, David John Edwards, and M. Reza Hosseini. Patterns and trends in Internet of Things (IoT) research: future applications in the construction industry. *Engineering, Construction and Architectural Management*, 28(2):457–481, 2020.

149. Rudolf Giffinger and Haindlmaier Gudrun. Smart cities ranking: An effective instrument for the positioning of the cities. *ACE: Architecture, City and Environment*, 4(12):7–26, February 2010.

150. P. Goodrum, Y. Wang, C. Haas, R. Glover, and S. Vaziri. Construction industry craft training in the United States and Canada. Technical report, Construction Industry Institute (CII), University of Texas, Austin, TX, USA, 2007.

151. G. Gourlis and I. Kovacic. Building Information Modelling for analysis of energy efficient industrial buildings – A case study. *Renewable and Sustainable Energy Reviews*, 68:953–963, 2017.

152. Sujata S. Govada, Widemar Spruijt, and Timothy Rodgers. Smart City Concept and Framework. In T. M. Vinod Kumar, editor, *Smart Economy in Smart Cities: International Collaborative Research: Ottawa, St.Louis, Stuttgart, Bologna, Cape Town, Nairobi, Dakar, Lagos, New Delhi, Varanasi, Vijayawada, Kozhikode, Hong Kong*, pages 187–198. Springer, Singapore, 2017.

153. Toni Greif, Nikolai Stein, and Christoph M. Flath. Peeking into the void: Digital twins for construction site logistics. *Computers in Industry*, 121:103264, 2020.

154. Michael Grieves. Digital Twin: Manufacturing Excellence Through Virtual Factory Replication, 2014. White Paper.

155. R. Grodi, D. B. Rawat, and F. Rios-Gutierrez. Smart parking: Parking occupancy monitoring and visualization system for smart cities. In *SoutheastCon 2016*, pages 1–5, 2016.

156. Gerhard Gröger, Thomas H. Kolbe, Claus Nagel, and Karl-Heinz Häfele. OGC City Geography Markup Language (CityGML) Encoding Standard. Technical Report OGC 12-019, Open Geospatial Consortium, April 2012.

157. Thomas R. Gruber. A translation approach to portable ontology specifications. *Knowledge Acquisition*, 5(2):199–220, 1993.

158. Venkat N. Gudivada. Data Analytics: Fundamentals. In Mashrur Chowdhury, Amy Apon, and Kakan Dey, editors, *Data Analytics for Intelligent Transportation Systems*, chapter 2, pages 31–67. Elsevier, 2017.

159. Daqiang Guo, Shiquan Ling, Yiming Rong, and George Q. Huang. A framework for personalized production based on digital twin, blockchain and additive manufacturing in the context of Industry 4.0. In *Proceedings of the 2020 IEEE 16th International Conference on Automation Science and Engineering (CASE)*, pages 1181–1186, August 2020.

160. Kun Guo, Yueming Lu, Hui Gao, and Ruohan Cao. Artificial intelligence-based semantic internet of things in a user-centric smart city. *Sensors*, 18(5):1341, 2018.

161. G. Guyot, M. H. Sherman, and I. S. Walker. Smart ventilation energy and indoor air quality performance in residential buildings: A review. *Energy and Buildings*, 165:416–430, 2018.

162. A. Gyrard, A. Zimmermann, and A. Sheth. Building IoT-Based Applications for Smart Cities: How Can Ontology Catalogs Help? *IEEE Internet of Things Journal*, 5(5):3978–3990, 2018.

163. Hadi Habibzadeh, Andrew Boggio-Dandry, Zhou Qin, Tolga Soyata, Burak Kantarci, and Hussein T. Mouftah. Soft sensing in smart cities: Handling 3Vs using recommender systems, machine intelligence, and data analytics. *IEEE Communications Magazine*, 56(2):78–86, 2018.

164. Hisakazu Hada and Jin Mitsugi. EPC based internet of things architecture. In *Proceedings of the 2011 IEEE International Conference on RFID-Technologies and Applications*, pages 527–532, Sitges, Spain, 2011.

165. A. Haller, K. Janowicz, S. J. Cox, M. Lefrançois, K. Taylor, D. Le Phuoc, and C. Stadler. The modular SSN ontology: A joint W3C and OGC standard specifying the semantics of sensors, observations, sampling, and actuation. *Semantic Web*, 10(1):9–32, 2019.

166. Armin Haller, Krzysztof Janowicz, Simon Cox, Danh Le Phuoc, Kerry Taylor, Maxime Lefrançois, Rob Atkinson, Raúl García-Castro, Joshua Lieberman, and Claus Stadler. Semantic Sensor Network Ontology, W3C Recommendation 19 October 2017.

167. Karl Hammar, Erik Oskar Wallin, Per Karlberg, and David Hälleberg. The realestatecore ontology. In Chiara Ghidini, Olaf Hartig, Maria Maleshkova, Vojtěch Svátek, Isabel

Cruz, Aidan Hogan, Jie Song, Maxime Lefrançois, and Fabien Gandon, editors, *International Semantic Web Conference*, pages 130–145, Cham, 2019. Springer International Publishing.

168. J. W. Hand, D. B. Crawley, M. Donn, and L. K. Lawrie. Improving the data available to simulation programs. In *Proceedings of Building Simulation*, pages 373–380, Montreal, Canada, 2005.

169. V. S. K. V. Harish and A. Kumar. A review on modeling and simulation of building energy systems. *Renewable and Sustainable Energy Reviews*, 56:1272–1292, 2016.

170. K. Eric Harper, Christopher Ganz, and Somayeh Malakuti. Digital Twin Architecture and Standards. *IIC Journal of Innovation*, 12:72–83, November 2019.

171. Dandan He, Zhongfu Li, Chunlin Wu, and Xin Ning. An e-commerce platform for industrialized construction procurement based on BIM and linked data. *Sustainability*, 10(8):2613, 2018.

172. G. R. Hemanth, S. Charles Raja, J. Jeslin Drusila Nesamalar, and J. Senthil Kumar. Cost effective energy consumption in a residential building by implementing demand side management in the presence of different classes of power loads. *Advances in Building Energy Research*, 0:1–26, 2020.

173. M. Herrera, J. Izquierdo, R. Pérez-Garćia, and D. Ayala-Cabrera. On-line Learning of Predictive Kernel Models for Urban Water Demand in a Smart City. *Procedia Engineering*, 70:791–799, 2014. 12th International Conference on Computing and Control for the Water Industry, CCWI2013.

174. John R. Herring. OpenGIS® Implementation Standard for Geographic information – Simple feature access – Part 1: Common architecture. Standard, Open Geospatial Consortium Inc., 2011. OGC 06-103r4.

175. Mervi Himanen. *The Intelligence of Intelligent Buildings. The Feasibility of the Intelligent Building Concept in Office Buildings.* phdthesis, Helsinki University of Technology, Espoo, Finland, 2003.

176. D. C. Hittle and R. Bishop. An improved root-finding procedure for use in calculating transient heat flow through multilayered slabs. *International Journal of Heat and Mass Transfer*, 26(11):1685–1693, 1983.

177. Pascal Hitzler, Sebastian Rudolph, and Markus Krötzsch. Ontologies in OWL. In *Foundations of Semantic Web Technologies*, chapter 4, pages 111–158. Chapman and Hall/CRC, 2009.

178. Aidan Hogan, Eva Blomqvist, Michael Cochez, Claudia d'Amato, Gerard de Melo, Claudio Gutierrez, José Emilio Labra Gayo, Sabrina Kirrane, Sebastian Neumaier, Axel Polleres, Roberto Navigli, Axel-Cyrille Ngonga Ngomo, Sabbir M. Rashid, Anisa Rula, Lukas Schmelzeisen, Juan Sequeda, Steffen Staab, and Antoine Zimmermann. Knowledge Graphs, 2020.

179. S. K. Howell, H. Wicaksono, B. Yuce, K. McGlinn, and Y. Rezgui. User Centered Neuro-Fuzzy Energy Management Through Semantic-Based Optimization. *IEEE Transactions on Cybernetics*, 49(9):3278–3292, 2019.

180. Shaun Howell, Yacine Rezgui, and Thomas Beach. Integrating building and urban semantics to empower smart water solutions. *Automation in Construction*, 81:434–448, 2017.

181. Shaun Howell, Yacine Rezgui, and Thomas Beach. Water utility decision support through the semantic web of things. *Environmental Modelling and Software*, 102:94–114, 2018.

182. Terence K. L. Hui, R. Simon Sherratt, and Daniel Díaz Sánchez. Major requirements for building Smart Homes in Smart Cities based on Internet of Things technologies. *Future Generation Computer Systems*, 76:358–369, 2017.

183. Laden Husamaldin and Nagham Saeed. Big Data Analytics Correlation Taxonomy. *Information*, 11(1), 2020.

184. Stalin P. Ibanez, Theresa Fitz, and Kay Smarsly. A semantic model for wireless sensor networks in cognitive buildings. In *Computing in Civil Engineering 2019: Smart Cities, Sustainability, and Resilience*, pages 234–241. American Society of Civil Engineers, Reston, VA, USA, 2019.

185. International Organization for Standardization. ISO 10303-11: Industrial automation systems and integration – Product data representation and exchange – Part 11: Description methods: The EXPRESS language reference manual, 2004.

186. Elvira Ismagilova, Laurie Hughes, Yogesh K. Dwivedi, and K. Ravi Raman. Smart cities: Advances in research – An information systems perspective. *International Journal of Information Management*, 47:88–100, 2019.

187. ISO. ISO 19650-1:2018 Organization and Digitization of Information about Buildings and Civil Engineering Works, Including Building Information Modelling (BIM) – Information Management Using Building Information Modelling – Part 1: Concepts and Principles. Technical report, ISO, 2018.

188. Pooyan Jamshidi, Aakash Ahmad, and Claus Pahl. Cloud migration research: a systematic review. *IEEE Transactions on Cloud Computing*, 1(2):142–157, 2013.

189. Bilal Jan, H. Farman, M. Khan, M. Talha, and I. Din. Designing a Smart Transportation System: An Internet of Things and Big Data Approach. *IEEE Wireless Communications*, 26:73–79, 2019.

190. Marijn Janssen, Ricardo Matheus, and Anneke Zuiderwijk. Big and Open Linked Data (BOLD) to Create Smart Cities and Citizens: Insights from Smart Energy and Mobility Cases. In Efthimios Tambouris, Marijn Janssen, Hans Jochen Scholl, Maria A. Wimmer, Konstantinos Tarabanis, Mila Gascó, Bram Klievink, Ida Lindgren, and Peter Parycek, editors, *Electronic Government*, pages 79–90, Cham, 2015. Springer International Publishing.

191. W. Ji, J. Xu, H. Qiao, M. Zhou, and B. Liang. Visual IoT: Enabling Internet of Things Visualization in Smart Cities. *IEEE Network*, 33(2):102–110, 2019.

192. Mengda Jia, Ali Komeily, Yueren Wang, and Ravi S. Srinivasan. Adopting Internet of Things for the development of smart buildings: A review of enabling technologies and applications. *Automation in Construction*, 101:111–126, 2019.

193. Huaxiong Jiang, Stan Geertman, and Patrick Witte. Comparing smart governance projects in China: A contextual approach. In S. Geertman, Q. Zhan, A. Allan, and C. Pettit, editors, *Proceedings of the International Conference on Computers in Urban Planning and Urban Management*, Lecture Notes in Geoinformation and Cartography, pages 99–114. Springer, January 2019.

194. Bokolo Anthony Jnr. Managing digital transformation of smart cities through enterprise architecture – a review and research agenda. *Enterprise Information Systems*, 15(3):299–331, 2020.

195. R. Judkoff, D. Wortman, and J. Burch. Empirical validation of building energy-analysis simulation programs: a status report. Technical report, Solar Energy Research Institute, Golden, CO, USA, 1982.

196. Riad Kanan, Obaidallah Elhassan, and Rofaida Bensalem. An IoT-based autonomous system for workers' safety in construction sites with real-time alarming, monitoring, and positioning strategies. *Automation in Construction*, 88:73–86, 2018.

197. Orhan Altuğ Karabiber and George Xydis. Electricity Price Forecasting in the Danish Day-Ahead Market Using the TBATS, ANN and ARIMA Methods. *Energies*, 12(5), 2019.

198. Aqeel H. Kazmi, Michael J. O'Grady, Declan T. Delaney, Antonio G. Ruzzelli, and Gregory M. P. O'Hare. A review of wireless-sensor-network-enabled building energy management systems. *ACM Transactions on Sensor Networks*, 10(4):1–43, June 2014.

199. Anne Kemp. Information management according to BS EN ISO 19650 – Guidance Part 2: Processes for Project Delivery. Technical report, UK BIM Framework, 2019.

200. Ali Keyhani. *Design of Smart Power Grid Renewable Energy Systems*. Wiley, 2011.

201. Siavash H. Khajavi, Naser Hossein Motlagh, Alireza Jaribion, Liss C. Werner, and Jan Holmstrom. Digital Twin: Vision, Benefits, Boundaries, and Creation for Buildings. *IEEE Access*, 7:147406–147419, 2019.

202. Zaheer Khan, Ashiq Anjum, Kamran Soomro, and Muhammad Atif Tahir. Towards cloud based big data analytics for smart future cities. *Journal of Cloud Computing*, 4, 2015.

203. Bohyun Kim. Responsive web design, discoverability, and mobile challenge. *Library technology reports*, 49(6):29–39, 2013.

204. Neil E. Klepeis, William C. Nelson, Wayne R. Ott, John P. Robinson, Andy M. Tsang, Paul Switzer, Joseph V. Behar, Stephen C. Hern, and William H. Engelmann. The National Human Activity Pattern Survey (NHAPS): A resource for assessing exposure to environmental pollutants. *Journal of Exposure Analysis and Environmental Epidemiology*, 11(3):231–252, July 2001.

205. Petar Kochovski and Vlado Stankovski. Supporting smart construction with dependable edge computing infrastructures and applications. *Automation in Construction*, 85:182–192, 2018.

206. Aliia Kodzhoshalieva. Reducing Cognitive Load for a Data-Intensive Web Application User. Master's thesis, Åbo Akademi, Åbo, 2020.

207. Linghe Kong, Siyu Lin, Weiliang Xie, Xiaoyu Qiao, Xi Jin, Peng Zeng, Wanli Ren, and Xiao-Yang Liu. Adaptive barrier coverage using software defined sensor networks. *IEEE Sensors Journal*, 16(20):7364–7372, 2016.

208. H. Kopackova and P. Libalova. Smart city concept as socio-technical system. In *2017 International Conference on Information and Digital Technologies (IDT)*, pages 198–205, 2017.

209. David B. Korman and Albert Zulps. Enhancing construction safety using wearable technology. In *ASSE Professional Development Conference and Exposition*, 2017.

210. M. Koubarakis, K. Kyzirakos, M. Karpathiotakis, Ch. Nikolaou, M. Sioutis, G. Garbis, and K. Bereta. Introduction in stRDF and stSPARQL. Technical report, National and Kapodistrian University of Athens, 2012. Getting Started Tutorial.

211. Hannah Kramer, Guanjing Lin, Jessica Granderson, Claire Curtin, and Eliot Crowe. Synthesis of Year One Outcomes in the Smart Energy Analytics Campaign. Technical report, Building Technology and Urban Systems Division, Lawrence Berkeley National Laboratory, September 2017.

212. E. Krygiel and B. Nies. *Green BIM: successful sustainable design with building information modeling*. John Wiley & Sons, 2008.

213. S. Kubler, J. Robert, A. Hefnawy, K. Främling, C. Cherifi, and A. Bouras. Open IoT ecosystem for sporting event management. *IEEE Access*, 5:7064–7079, 2017.

214. Sachin Kumar, Prayag Tiwari, and Mikhail Zymbler. Internet of Things is a revolutionary approach for future technology enhancement: a review. *JO – Journal of Big Data*, 6(1), 2019.

215. Kostis Kyzirakos, Manos Karpathiotakis, and Manolis Koubarakis. Strabon: A Semantic Geospatial DBMS. In P. Cudré-Mauroux, editor, *The Semantic Web – ISWC 2012*, volume 7649 of *Lecture Notes in Computer Science*, pages 295–311, 2012.

216. Timilehin Labeodan, Christel De Bakker, Alexander Rosemann, and Wim Zeiler. On the application of wireless sensors and actuators network in existing buildings for occupancy detection and occupancy-driven lighting control. *Energy and Buildings*, 127:75–83, 2016.

217. Timilehin Labeodan, Wim Zeiler, Gert Boxem, and Yang Zhao. Occupancy measurement in commercial office buildings for demand-driven control applications – A survey and detection system evaluation. *Energy and Buildings*, 93:303–314, 2015.

218. Krishan Kant Lavania, Yogita Sharma, and Chandresh Bakliwal. A Review on Cloud Computing Model. *International Journal on Recent and Innovation Trends in Computing and Communication*, 1(3):161–163, 2013.

219. H. Leavitt. Applied organizational change in industry: Structural technological and humanistic approaches. In *Handbook of organizations*, pages 1144–1170. Carnegie Institute of Technology, Graduate School of Industrial Administration, 1962.

220. Chiu-Ming Lee, Wei-Liang Kuo, Tzu-Jan Tung, Bo-Kai Huang, Shu-Hsiang Hsu, and Mohamed Al-Hussein. Government open data and sensing data integration framework for smart construction site management. In *Proceedings of the 36th International Symposium on Automation and Robotics in Construction (ISARC)*, 2019.

221. Yen-Lin Lee, Pei-Kuei Tsung, and Max Wu. Techology trend of edge AI. In *2018 International Symposium on VLSI Design, Automation and Test (VLSI-DAT)*, pages 1–2. IEEE, 2018.

222. Maxime Lefrançois. Planned ETSI SAREF Extensions based on the W3C&OGC SOSA/SSN-compatible SEAS Ontology Patterns. In Anna Fensel and Laura Daniele, editors, *1st International Workshop on Semantic Interoperability and Standardization in the IoT (SIS-IoT)*, volume 2063 of *CEUR Workshop Proceedings*, Amsterdam, Netherlands, 2017.

223. Maxime Lefrançois and Antoine Zimmermann. The Unified Code for Units of Measure in RDF: cdt:ucum and other UCUM Datatypes. In Aldo Gangemi, Anna Lisa Gentile, Andrea Giovanni Nuzzolese, Sebastian Rudolph, Maria Maleshkova, Heiko Paulheim, Jeff Z. Pan, and Mehwish Alam, editors, *The semantic web: ESWC 2018 Satellite Events*, volume 11155 of *Lecture Notes in Computer Science*, pages 196–201, Cham, 2018. Springer.

224. Alberto Leon-Garcia and Indra Widjaja. *Communication networks*. McGraw-Hill, Inc., 2003.

225. Pekka Leviäkangas, Seong Mok Paik, and Sungkon Moon. Keeping up with the pace of digitization: The case of the Australian construction industry. *Technology in Society*, 50:33–43, 2017.

226. T. Li, Y. Liu, Y. Tian, S. Shen, and W. Mao. A storage solution for massive IoT data based on NoSQL. In *2012 IEEE International Conference on Green Computing and Communications*, pages 50–57. IEEE, 2012.

227. Zhiyuan Liu, Yang Liu, Qiang Meng, and Qixiu Cheng. A tailored machine learning approach for urban transport network flow estimation. *Transportation Research Part C: Emerging Technologies*, 108:130–150, 2019.

228. M. Lom and O. Pribyl. Modeling of a Smart City's Building Blocks Using Multi-Agent Systems. *Neural Network World*, 27:317–331, 2017.

229. E. Loscos, Haiyan Xie, R. H. E. M. Koppelaar, M. Borràs, D. Martín-Moncunill, E. Coloma, I. Pérez Arnal, R. Alonso, M. Elagiry, Sergio Velazquez, J. Porkka, P. Vicente Legazpi, A. M. Veleiro Blanco, and Daan Oostinga. Digital Twin Definitions for Buildings. Technical report, SPHERE Project, 2019.

230. Joseph Louis and Phillip S. Dunston. Integrating IoT into operational workflows for real-time and automated decision-making in repetitive construction operations. *Automation in Construction*, 94:317–327, 2018.

231. Marco Lovati, Jennifer Adami, and David Moser. Open source tool for a better design of BIPV+ battery system: an applied example. In *Proceedings of the 35th European Photovoltaic Solar Energy Conference and Exhibition*, pages 1641–1646, 2018.

232. Marco Lovati, Mattia Dallapiccola, Jennifer Adami, Paolo Bonato, Xingxing Zhang, and David Moser. Design of a residential photovoltaic system: the impact of the demand profile and the normative framework. *Renewable Energy*, 160:1458–1467, 2020.

233. Hsi-Peng Lu, Chiao-Shan Chen, and Hueiju Yu. Technology roadmap for building a smart city: An exploring study on methodology. *Future Generation Computer Systems*, 97:727–742, 2019.

234. Qiuchen Lu, Ajith Kumar Parlikad, Philip Woodall, Gishan Don Ranasinghe, Xiang Xie, Zhenglin Liang, Eirini Konstantinou, James Heaton, and Jennifer Schooling. Developing a digital twin at building and city levels: case study of West Cambridge campus. *Journal of Management in Engineering*, 36(3):05020004, 2020.

235. G. P. Lydon, S. Caranovic, I. Hischier, and A. Schlueter. Coupled simulation of thermally active building systems to support a digital twin. *Energy and Buildings*, 202:109298, November 2019.

236. Azad Madni, Carla Madni, and Scott Lucero. Leveraging Digital Twin Technology in Model-Based Systems Engineering. *Systems*, 7(1):7, January 2019.

237. L. Madrazo, A. Sicilia, and G. Gamboa. SEMANCO: Semantic tools for carbon reduction in urban planning. In *Proceedings of the 9th European Conference on Product and Process Modelling*. CRC Press, 2012.

238. L. Madrazo, Á. Sicilia, M. Massetti, Fabian López-Plazas, and Eric Ortet. Enhancing energy performance certificates with energy related data to support decision making for building retrofitting. *Thermal Science*, 22(3):957–969, 2018.

239. Pieter-Jan Maenhaut, Hendrik Moens, Veerle Ongenae, and Filip De Turck. Migrating legacy software to the cloud: approach and verification by means of two medical software use cases. *Software: Practice and Experience*, 46(1):31–54, 2016.

240. L. Mahdjoubi, C. A. Brebbia, and R. Laing. *Building Information Modelling (BIM) in design, construction and operations*. WIT Press, 2015.

241. W. Mahnke, S. H. Leitner, and M. Damm. *OPC Unified Architecture*. Springer Berlin Heidelberg, 2009.

242. T. Maile, M. Fischer, and V. Bazjanac. Building energy performance simulation tools-a life-cycle and interoperable perspective. Technical report, Center for Integrated Facility Engineering, 2007.

243. Andrew Malcolm, Jeroen Werbrouck, and Pieter Pauwels. LBD Server: Visualising Building Graphs in Web-Based Environments Using Semantic Graphs and GlTF-Models. In *Formal Methods in Architecture*, pages 287–293. Springer International Publishing, 2021.

244. Essam Mansour, Andrei Vlad Sambra, Sandro Hawke, Maged Zereba, Sarven Capadisli, Abdurrahman Ghanem, Ashraf Aboulnaga, and Tim Berners-Lee. A demonstration of the solid platform for social web applications. In *Proceedings of the 25th International Conference Companion on World Wide Web*, pages 223–226, 2016.

245. V. Marinakis, H. Doukas, J. Tsapelas, S. Mouzakitis, Á. Sicilia, L. Madrazo, and S. Sgouridis. From big data to smart energy services: An application for intelligent energy management. *Future Generation Computer Systems*, 110:572–586, 2020.

246. Gerardo Santillan Martinez, Seppo Sierla, Tommi Karhela, and Valeriy Vyatkin. Automatic Generation of a Simulation-Based Digital Twin of an Industrial Process Plant. In *IECON 2018 – 44th Annual Conference of the IEEE Industrial Electronics Society*, pages 3084–3089. IEEE, October 2018.

247. L. Maturi, S. Giona, D. Moser, R. Lollini, M. Lovati, P. Alonso, I. Weiss, C. Bales, J. M. Vega de Seoane, A. Becker, S. Hallbeck, E. Widlak, D.-J. Bles, V. Zanon, E. Rico, L. Papaiz, A. Perez Carballo, C. de Nacquard, O. Caboni, V. Esposito, U. Westman, and S. Metayer. EnergyMatching Project–Adaptable and Adaptive RES Envelope Solutions to Maximize Energy Harvesting and Optimize EU Building and District Load Matching. In *Proceedings of the 36th European Photovoltaic Solar Energy Conference and Exhibition*, pages 1827–1831, 2019.

248. George McDaniel. *IBM dictionary of computing*. McGraw-Hill, Inc., 1994.

249. Kris McGlinn, Darragh Blake, and Declan O'Sullivan. GViz – An Interactive WebApp to Support GeoSPARQL over Integrated Building Information. In *Companion Proceedings of the 2019 World Wide Web Conference (WWW '19 Companion)*, San Francisco, CA, USA, 2019. ACM.

250. Kris McGlinn, Rob Brennan, Christophe Debruyne, Alan Meehan, Lorraine McNerney, Eamonn Clinton, Philip Kelly, and Declan O'Sullivan. Publishing authoritative geospatial data to support interlinking of building information models. *Automation in Construction*, 124:103534, 2021.

251. Kris McGlinn, Christophe Debruyne, Lorraine McNerney, and Declan O'Sullivan. Integrating Building Information Models with Authoritative Irish Geospatial Information. In *ISWC 2017, the 16th International Semantic Web Conference*, pages 66–74, Vienna, Austria, 2017.

252. Kris McGlinn, Christophe Debruyne, Lorraine McNerney, and Declan O'Sullivan. Integrating Ireland's Geospatial Information to Provide Authoritative Building Information Models. In *Proceedings of the 13th International Conference on Semantic Systems – Semantics2017*, volume 13, pages 57–64, Amsterdam, Netherlands, 2017. ACM Press.

253. Kris McGlinn, Christophe Debruyne, and Declan O'Sullivan. Interlinking Ireland's Authoritative Geospatial Information On Buildings With Central Statistics Office Data. In *Proceedings of the 10th Annual European Forum for Geography & Statistics*, Dublin, Ireland, 2017.

254. Kris McGlinn, Anna Wagner, Pieter Pauwels, Peter Bonsma, Philip Kelly, and Declan O'Sullivan. Interlinking geospatial and building geometry with existing and developing standards on the web. *Automation in Construction*, 103:235–250, 2019.

255. Kris McGlinn, Matthias Weise, and Hendro Wicaksono. Towards a Shared Use Case Repository to support Building Information Modelling in the Energy Efficient Building domain. In *Proceedings of the 33rd CIB W78 Conference 2016*, 2706-6568, Brisbane, Australia, 2016.

256. Kris McGlinn, Hendro Wicaksono, Willie Lawton, Matthias Weise, Nikolaos Kaklanis, Ioanna Petri, and Dimitrios Tzovaras. Identifying Use Cases and Data Requirements for BIM Based Energy Management Processes. In *CIBSE Technical Symposium*, Edinburgh, UK, April 2016.

257. Kris McGlinn, Baris Yuce, Hendro Wicaksono, Shaun Howell, and Yacine Rezgui. Usability evaluation of a web-based tool for supporting holistic building energy management. *Automation in Construction*, 84:154–165, 2017.

258. Scott Mayer McKinney, Marcin Sieniek, Varun Godbole, Jonathan Godwin, Natasha Antropova, Hutan Ashrafian, Trevor Back, Mary Chesus, Greg S. Corrado, Ara Darzi, Mozziyar Etemadi, Florencia Garcia-Vicente, Fiona J. Gilbert, Mark Halling-Brown, Demis Hassabis, Sunny Jansen, Alan Karthikesalingam, Christopher J. Kelly, Dominic King, Joseph R. Ledsam, David Melnick, Hormuz Mostofi, Lily Peng, Joshua Jay Reicher, Bernardino Romera-Paredes, Richard Sidebottom, Mustafa Suleyman, Daniel Tse, Kenneth C. Young, Jeffrey De Fauw, and Shravya Shetty. International evaluation of an AI system for breast cancer screening. *Nature*, 577(7788):89–94, January 2020.

259. Ryan Melfi, Ben Rosenblum, Bruce Nordman, and Ken Christensen. Measuring building occupancy using existing network infrastructure. In *2011 International Green Computing Conference and Workshops*, pages 1–8. IEEE, 2011.

260. Tarcisio Mendes De Farias, Ana-Maria Roxin, and Christophe Nicolle. IfcWoD,Semantically Adapting IFC Model Relations into OWL Properties. In *Proc. of the 32nd CIB W78 Conference 2015*, pages 175–185, Eindhoven, The Netherlands, October 2015.

261. microdesk. Project Website: https://www.microdesk.com/articles/cloud-usage-in-aec-industry/.

262. Microsoft. AMQP 1.0 in Azure Service Bus and Event Hubs protocol guide – Azure Service Bus, 2021.

263. C. Mirarchi, M. N. Lucky, S. Ciuffreda, M. Signorini, S. Lupica Spagnolo, C. Bolognesi, and A. Pavan. An approach for standardization of semantic models for building renovation processes. *ISPRS – International Archives of the Photogrammetry, Remote Sensing and Spatial Information Sciences*, XLIII-B4-2020:69–76, 2020.

264. Neda Mohammadi and John Taylor. Knowledge discovery in smart city digital twins. In *Proceedings of the 53rd Hawaii international conference on system sciences*, 2020.

265. James Moyne, Yassine Qamsane, Efe C. Balta, Ilya Kovalenko, John Faris, Kira Barton, and Dawn M. Tilbury. A Requirements Driven Digital Twin Framework: Specification and Opportunities. *IEEE Access*, 8(June):107781–107801, 2020.

266. Rubén Mulero, Vladimir Urosevic, Aitor Almeida, and Christos Tatsiopoulos. Towards ambient assisted cities using linked data and data analysis. *J. Ambient Intell. Humaniz. Comput.*, 9(5):1573–1591, 2018.

267. K. Muralitharan, R. Sakthivel, and Y. Shi. Multiobjective optimization technique for demand side management with load balancing approach in smart grid. *Neurocomputing*, 177:110–119, February 2016.

268. Afiqah Ngah Nasaruddin, Teruaki Ito, and Tee Boon Tuan. Digital twin approach to building information management. In *The Proceedings of Manufacturing Systems Division Conference 2018*, page 304. The Japan Society of Mechanical Engineers, 2018.

269. S. Naylor, M. Gillott, and T. Lau. A review of occupant-centric building control strategies to reduce building energy use. *Renewable and Sustainable Energy Reviews*, 96:1–10, 2018.

270. Elisa Negri, Luca Fumagalli, and Marco Macchi. A Review of the Roles of Digital Twin in CPS-based Production Systems. *Procedia Manufacturing*, 11:939–948, June 2017.

271. Andra-Ileana Neicu, Anamaria-Cătălina Radu, Gheorghe Zaman, Ivona Stoica, and Florian Răpan. Cloud Computing Usage in SMEs. An Empirical Study Based on SMEs Employees Perceptions. *Sustainability*, 12(12):4960, 2020.

272. Fátima Trindade Neves, Miguel de Castro Neto, and Manuela Aparicio. The impacts of open data initiatives on smart cities: A framework for evaluation and monitoring. *Cities*, 106:102860, 2020.

273. Mehrdad Niknam and Saeed Karshenas. A shared ontology approach to semantic representation of BIM data. *Automation in Construction*, 80:22–36, 2017.

274. nist. Project Website: https://www.nist.gov/.

275. Normcommissie 381184 'Informatie-integratie en Interopabiliteit'. NTA 8035:2020 Semantische gegevensmodellering in de gebouwde omgeving. Technical report, NEN, April 2020.

276. Christian Nothstein, Tim-Jonathan Huyeng, and Wendelin Sprenger. SCOPE Data Service, 2020.

277. Natalya F. Noy and Deborah L. McGuinness. Ontology development 101: A guide to creating your first ontology. Technical report, Stanford University, March 2001. Stanford Knowledge Systems Laboratory Technical Report KSL-01-05 and Stanford Medical Informatics Technical Report SMI-2001-0880.

278. J. O'Donnell, R. See, C. Rose, T. Maile, V. Bazjanac, and P. Haves. SimModel: A domain data model for whole building energy simulation. In *12th Conference of International Building Performance Simulation Association*, pages 382–389, Sydney, 2011.

279. Joseph O'Donovan, Declan O'Sullivan, and Kris McGlinn. A method for converting IFC geometric data into GeoSPARQL. In *Proceedings of the 7th Linked Data in Architecture and Construction Workshop Lisbon, CEUR Workshop Proceedings*, volume 2389, pages 7–20, Portugal, June 19-21, 2019.

280. Isaac Odun-Ayo, Chinonso Okereke, and H. E. Orovwode. Cloud and Application Programming Interface–Issues and Developments. In *Proceedings of The World Congress on Engineering 2018*, Lecture Notes in Engineering and Computer Science, pages 169–174, 2018.

281. OECD. *OECD Regions at a Glance 2016*. OECD Publishing, 2016.

282. University of Toronto. University of Toronto – Building Automation System, June 2020.

283. A. Ojo, E. Curry, and F. A. Zeleti. A Tale of Open Data Innovations in Five Smart Cities. In *2015 48th Hawaii International Conference on System Sciences*, pages 2326–2335, 2015.

284. Flauzac Olivier, Gonzalez Carlos, and Nolot Florent. SDN Based Architecture for Clustered WSN. In *Proceedings of the 9th International Conference on Innovative Mobile and Internet Services in Ubiquitous Computing (IMIS)*, pages 342–347, Santa Catarina, Brazil, 2015. IEEE.

285. Open Geospatial Consortium. OGC® Geography Markup Language (GML) – Extended schemas and encoding rules – v3.3.0. Implementation Standard 10-129r1, Open Geospatial Consortium, 2012.

286. Anthony M. Pagano and Matthew Liotine. Technology in Supply Chain Management and Logistics. In Mashrur Chowdhury, Amy Apon, and Kakan Dey, editors, *Technology in Technology in Supply Chain Management and Logistics: Current Practice and Future Applications*, chapter 2, pages 7–35. Elsevier, 2020.

287. E. Panagoulia and T. Rakha. Performance Analytics through Building Information Modelling (BIM) in Retrofitting Design. In *PLEA2020. 35th Passive and Low Energy Architecture Conference. Sustainable Architecture and Urban Design*, pages 1–8, A Coruna, Spain, 2010.

288. Alessandra De Paola, Marco Ortolani, Giuseppe Lo Re, Giuseppe Anastasi, and Sajal K. Das. Intelligent management systems for energy efficiency in buildings: A survey. *ACM Computing Surveys*, 47(1):1–38, July 2014.

289. P. Paolini, N. Di Blas, S. Copelli, and F. Mercalli. City4Age: Smart cities for health prevention. In *2016 IEEE International Smart Cities Conference (ISC2)*, pages 1–4, 2016.

290. Grigorios Papageorgiou, Nikolaos Alamanis, and Nikolaos Xafoulis. Methodology for Optimization of Road Works Schedule According to Local Climatic Data. *International Journal of Recent Technology and Engineering (IJRTE)*, 8(4):1470–1476, 2019.

291. Hergen Pargmann, Dörthe Euhausen, and Robin Faber. Intelligent big data processing for wind farm monitoring and analysis based on cloud-technologies and digital twins: A quantitative approach. In *2018 IEEE 3rd International Conference on Cloud Computing and Big Data Analysis (ICCCBDA)*, pages 233–237. IEEE, April 2018.

292. Kyung Gyu Park, Yoonkee Kim, Seon Mi Kim, Kwang Ho Kim, Wook Hyun Lee, and Hwa Choon Park. Building energy management system based on smart grid. In *INTELEC, International Telecommunications Energy Conference (Proceedings)*, 2011.

293. I. Patias and V. Georgiev. Cloud Federation Usage in Engineering and Construction Sector. In Vladimir Dimitrov and Vasil Georgiev, editors, *Proceedings of the Information Systems and Grid Technologies*, volume 2656, 2020.

294. P. Pauwels, E. Corry, and J. & O'Donnell. Representing SimModel in the web ontology language. *Computing in Civil and Building Engineering*, pages 2271–2278, 2014.

295. P. Pauwels and E. Petrova. *Information in Construction*. Eindhoven University of Technology, 2020.

296. P. Pauwels, W. Terkaj, T. Krijnen, and J. Beetz. Coping with lists in the ifcOWL ontology. In *22nd Workshop of the European Group of Intelligent Computing in Engineering*, pages 113–122, Eindhoven, Netherlands, 2015.

297. Pieter Pauwels, Thomas Krijnen, Walter Terkaj, and Jakob Beetz. Enhancing the ifcOWL ontology with an alternative representation for geometric data. *Automation in Construction*, 80:77–94, 2017.

298. Pieter Pauwels and Ana Roxin. SimpleBIM: From full ifcOWL graphs to simplified building graphs. In Symeon Christodoulou and Raimar Scherer, editors, *Proceedings of the 11th European Conference on Product and Process Modelling (ECPPM)*, pages 11–18, Limassol, Cyprus, 2016. CRC Press.

299. Pieter Pauwels and Walter Terkaj. EXPRESS to OWL for construction industry: Towards a recommendable and usable ifcOWL ontology. *Automation in Construction*, 63:100–133, March 2016.

300. Pieter Pauwels, Davy Van Deursen, Jos de Roo, Tim Van Ackere, Ronald de Meyer, Rik Van de Walle, and Jan Van Campenhout. Three-dimensional information exchange over the semantic web for the domain of architecture, engineering, and construction. *Artificial Intelligence for Engineering Design, Analysis and Manufacturing*, 25(4):317–332, 2011.

301. Pieter Pauwels, Sijie Zhang, and Yong-Cheol Lee. Semantic web technologies in AEC industry: A literature overview. *Automation in Construction*, 73:145–165, 2017.

302. Matthew Perry. PPT – OGC GeoSPARQL: Standardizing Spatial Query on the Semantic Web PowerPoint Presentation – ID:3402858, 2011.

303. Matthew Perry and John Herring. GeoSPARQL – A geographic query language for RDF data. techreport 11-052r4, Open Geospatial Consortium, 2012. Implementation Specification (IS).

304. Alexander Perzylo, Nikhil Somani, Markus Rickert, and Alois Knoll. An ontology for CAD data and geometric constraints as a link between product models and semantic robot task descriptions. In *2015 IEEE/RSJ International Conference on Intelligent Robots and Systems (IROS)*, pages 4197–4203, 2015.

305. Ioan Petri and Yacine Rezgui. *BIM for Energy Efficiency*. BRE Electronic Publications, 2019.

306. Raja Vara Prasad Y. and P. Rajalakshmi. Context aware building energy management system with heterogeneous wireless network architecture. In *Proceedings of 2013 6th Joint IFIP Wireless and Mobile Networking Conference, WMNC 2013*, 2013.

307. Freddy Priyatna, Oscar Corcho, and Juan Sequeda. Formalisation and Experiences of R2RML-Based SPARQL to SQL Query Translation Using Morph. In *Proceedings of the 23rd International Conference on World Wide Web*, WWW '14, pages 479–490, New York, NY, USA, 2014. Association for Computing Machinery.

308. C. Pungilă, T. F. Fortiş, and O. Aritoni. Benchmarking database systems for the requirements of sensor readings. *IETE Technical Review*, 26(5):342–349, 2009.

309. Qinglin Qi and Fei Tao. Digital Twin and Big Data Towards Smart Manufacturing and Industry 4.0: 360 Degree Comparison. *IEEE Access*, 6:3585–3593, 2018.

310. T. Qiu, N. Chen, K. Li, M. Atiquzzaman, and W. Zhao. How can heterogeneous internet of things build our future: A survey. *IEEE Communications Surveys & Tutorials*, 20(3):2011–2027, 2018.

311. Vivi Qiuchen Lu, Ajith Kumar Parlikad, Philip Woodall, Gishan Don Ranasinghe, and James Heaton. Developing a Dynamic Digital Twin at a Building Level: using Cambridge Campus as Case Study. In *International Conference on Smart Infrastructure and Construction 2019 (ICSIC)*, volume 2019, pages 67–75. ICE Publishing, January 2019.

312. Chenhao Qu, Rodrigo N. Calheiros, and Rajkumar Buyya. Auto-scaling web applications in clouds: A taxonomy and survey. *ACM Computing Surveys (CSUR)*, 51(4):1–33, 2018.

313. QUDT. QUDT CATALOG – Quantities, Units, Dimensions and Data Types Ontologies, 2020.

314. Kamel H. Rahouma, Farag M. Afify, and Hesham F. A. Hamed. Design of a New Automated Fault Detector based on Artificial Intelligence and Big Data Techniques. *Procedia Computer Science*, 163:460–471, 2019.

315. Mads Holten Rasmussen, Christian Aaskov Frausing, Christian Anker Hviid, and Jan Karlshøj. Demo: Integrating Building Information Modeling and Sensor Observations using semantic web. In Maxime Lefrançois, Raúl Garcia Castro, Amélie Gyrard, and Kerry Taylor, editors, *Proceedings of the 9th International Semantic Sensor Networks Workshop co-located with 17th International Semantic Web Conference (ISWC 2018)*, pages 48–55, Monterey, CA, United States, 2018.

316. Mads Holten Rasmussen, Maxime Lefrançois, Mathias Bonduel, Christian Anker Hviid, and Jan Karlshøj. OPM: An ontology for describing properties that evolve over time. In María Poveda-Villalón, Pieter Pauwels, and Ana Roxin, editors, *Proceedings of the 6th linked*

data in Architecture and Construction Workshop (LDAC), volume 2159 of *CEUR Workshop Proceedings*, pages 23–33, London, UK, 2018.

317. Mads Holten Rasmussen, Maxime Lefrançois, Pieter Pauwels, Christian Anker Hviid, and Jan Karlshøj. Managing interrelated project information in AEC Knowledge Graphs. *Automation in Construction*, 108:102956, 2019.

318. Mads Holten Rasmussen, Maxime Lefrançois, Georg Ferdinand Schneider, and Pieter Pauwels. BOT: the building topology ontology of the W3C linked building data group. *Semantic Web*, pages 1–19, 2019.

319. Tristan W. Reed, David A. McMeekin, and Femke Reitsma. Representing Spatial Relationships Within Smart Cities Using Ontologies. In Leila Ismail and Liren Zhang, editors, *Information Innovation Technology in Smart Cities*, pages 33–45. Springer, Singapore, 2018.

320. T. Reeves, S. Olbina, and R. R. Issa. Guidelines for using building information modeling for energy analysis of buildings. *Buildings*, 5(4):1361–1388, 2015.

321. Andrew P. Reimer and Elizabeth A. Madigan. Veracity in big data: How good is good enough. *Health Informatics Journal*, 25(4):1290–1298, 2019.

322. Y. Rezgui, S. Boddy, M. Wetherill, and G. Cooper. Past, present and future of information and knowledge sharing in the construction industry: Towards semantic service-based e-construction? *Computer-Aided Design*, 43(5):502–515, 2011.

323. Z. Riaz, E. A. Parn, D. J. Edwards, M. Arslan, C. Shen, and F. Pena-Mora. BIM and sensor-based data management system for construction safety monitoring. *Engineering, Design and Technology*, 15(6):738–753, 2017.

324. H. Rijgersberg, M. Wigham, and J. L. Top. How semantics can improve engineering processes: A case of units of measure and quantities. *Advanced Engineering Informatics*, 25(2):276–287, 2011.

325. Herbert S. Robinson, Patricia M. Carrillo, Chimay J. Anumba, and Ahmed M. Al-Ghassani. Perceptions and barriers in implementing knowledge management strategies in large construction organisations. In *Proceedings of the RICS COBRA Conference*, pages 451–460, 2001.

326. B. Rocha, E. Cavalcante, T. Batista, and J. Silva. A Linked Data-Based Semantic Information Model for Smart Cities. In *2019 IX Brazilian Symposium on Computing Systems Engineering (SBESC)*, pages 1–8, 2019.

327. Jonathan Rogers and Barry Kirwan. The Post-Occupancy Digital Twin : a Quantitative Report on Data Standardisation and Dynamic Building Performance Evaluation. *International Journal of Digital Innovation in the Built Environment*, 9(2):49, 2019.

328. Marco Rospocher, Chiara Ghidini, and Luciano Serafini. An Ontology for the Business Process Modelling Notation. In *Formal Ontology in Information Systems*, volume 267 of *Frontiers in Artificial Intelligence and Applications*, pages 133–146, 2014.

329. Douglas T. Ross and Jorge E. Rodriguez. Theoretical Foundations for the Computer-Aided Design System. *Simulation*, 2(3):R–3, 1964.

330. Derrick Rountree and Ileana Castrillo. Introduction to the Cloud. In Derrick Rountree and Ileana Castrillo, editors, *The Basics of Cloud Computing*, chapter 1, pages 1–17. Syngress, Boston, MA, USA, 2014.

331. C. Roussey, S. Bernard, G. André, and D. Boffety. Weather data publication on the LOD using SOSA/SSN ontology. *Semantic Web*, pages 1–11, 2019.

332. J. N. S. Rubí and P. R. de Lira Gondim. IoT-based platform for environment data sharing in smart cities. *International Journal of Communication Systems*, 34(2):e4515, 2021.

333. Marta Sabou, Stefan Biffl, Alfred Einfalt, Lukas Krammer, Wolfgang Kastner, and Fajar J. Ekaputra. Semantics for Cyber-Physical Systems: A cross-domain perspective. *Semantic Web*, 11(1):115–124, January 2020.

334. Anton Safiullin, Lyudmila Krasnyuk, and Zoya Kapelyuk. Integration of Industry 4.0 technologies for "smart cities" development. In *IOP Conference Series: Materials Science and Engineering*, volume 497. Institute of Physics Publishing, 2019.

335. S. Sagar, M. Lefrançois, I. Rebaï, M. Khemaja, S. Garlatti, J. Feki, and L. Médini. Modeling smart sensors on top of SOSA/SSN and WoT TD with the semantic smart sensor network (S3N) modular ontology. In *In ISWC 2018: 17th Internal semantic web Conference*, Volume 36: Emerging Topics in Semantic Technologies ISBN 978-1-61499-894-5, pages 163–177, Monterey, United States, 2018.

336. Tara Salman and Raj Jain. A Survey of Protocols and Standards for Internet of Things. *Advanced Computing and Communications*, 1(1), 2017.

337. Tunga Salthammer, Yinping Zhang, Jinhan Mo, Holger M. Koch, and Charles J. Weschler. Assessing Human Exposure to Organic Pollutants in the Indoor Environment. *Angewandte Chemie International Edition*, 57(38):12228–12263, September 2018.

338. Egwunatum I. Samuel, Esther Joseph-Akwara, and Akaigwe Richard. Assessment of energy utilization and leakages in buildings with building information model energy. *Frontiers of Architectural Research*, 6(1):29–41, 2017.

339. Abhijit Sarkar, Mark Fairchild, and Carl Salvaggio. Integrated daylight harvesting and occupancy detection using digital imaging. In *Sensors, Cameras, and Systems for Industrial/Scientific Applications IX*, volume 6816, page 68160F. International Society for Optics and Photonics, 2008.

340. Gabriela Nicoleta Sava, Stefanita Pluteanu, Vladimir Tanasiev, Roxana Patrascu, and Horia Necula. Integration of BIM solutions and IoT in smart houses. In *2018 IEEE International Conference on Environment and Electrical Engineering and 2018 IEEE Industrial and Commercial Power Systems Europe (EEEIC/I&CPS Europe)*, pages 1–4. IEEE, 2018.

341. Anil Sawhney, Mike Riley, and Javier Irizarry. *Construction 4.0: An innovation platform for the built environment*. Routledge, London, UK, first edition, February 2020.

342. R. J. Scherer and S. E. Schapke. A distributed multi-model-based Management Information System for simulation and decision-making on construction projects. *Advanced Engineering Informatics*, 25(4):582–599, 2011.

343. Raimar J. Scherer, Sven-Eric Schapke, and Helga Tauscher. *Mefisto: Management – Leadership – Information – Simulation in Construction*. Technische Universitaet Dresden, 2010.

344. G. F. Schneider, W. Terkaj, and P. Pauwels. Reusing domain ontologies in linked building data: the case of building automation and control. In *8th International Workshop on Formal Ontologies meet Industry*, volume 2050, 2017.

345. Georg Ferdinand Schneider, Mads Holten Rasmussen, Peter Bonsma, Jyrki Oraskari, and Pieter Pauwels. Linked building data for modular building information modelling of a smart home. In Jan Karlshøj and Raimar Scherer, editors, *Proceedings of the 12th European Conference on Product and Process Modelling*, pages 407–414. CRC Press, 2018.

346. Gerhard Schrotter and Christian Hürzeler. The Digital Twin of the City of Zurich for Urban Planning. *PFG – Journal of Photogrammetry, Remote Sensing and Geoinformation Science*, 88(1):99–112, February 2020.

347. Sule Selcuk. Predictive maintenance, its implementation and latest trends. *Proceedings of the Institution of Mechanical Engineers, Part B: Journal of Engineering Manufacture*, 231(9):1670–1679, 2017.

348. Madhumitha Senthilvel, Jyrki Oraskari, and Jakob Beetz. Common Data Environments for the Information Container for linked Document Delivery. In *Proceedings of the 8th Linked Data in Architecture and Construction workshop (LDAC 2020), CEUR Workshop Proceedings*, Vol. 2636, pages 132–145, Dublin, Ireland, 2020.

349. Aulon Shabani and Orion Zavalani. Predicting Building Energy Consumption using Engineering and Data Driven Approaches: A Review. *European Journal of Engineering and Technology Research*, 2(5):44–49, 2017.

350. Guodong Shao and Moneer Helu. Framework for a digital twin in manufacturing: Scope and requirements. *Manufacturing Letters*, 24:105–107, 2020.

351. Vadim Shapiro. *Solid Modeling*, volume 20, chapter 20, pages 473–518. O'Reilly, 2002.

352. Dennis R. Shelden, Pieter Pauwels, Pardis Pishdad-Bozorgi, and Shu Tang. *Data standards and data exchange for Construction 4.0*, chapter 12, pages 222–239. Taylor and Francis Ltd., 2020.

353. Li Shen, Laurie R. Margolies, Joseph H. Rothstein, Eugene Fluder, Russell McBride, and Weiva Sieh. Deep Learning to Improve Breast Cancer Detection on Screening Mammography. *Nature Scientific Reports*, 9, 2019.

354. Weiming Shen, Guy Newsham, and Burak Gunay. Leveraging existing occupancy-related data for optimal control of commercial office buildings: A review. *Advanced Engineering Informatics*, 33:230–242, 2017.

355. W. Shi, J. Cao, Q. Zhang, Y. Li, and L. Xu. Edge Computing: Vision and Challenges. *IEEE Internet of Things Journal*, 3(5):637–646, 2016.

356. Sara Shirowzhan, Willie Tan, and Samad M. E. Sepasgozar. Digital twin and CyberGIS for improving connectivity and measuring the impact of infrastructure construction planning in smart cities. *ISPRS International Journal of Geo-Information*, 9(4), 2020.

357. A. Sicilia, L. Madrazo, and J. Pleguezuelos. Integrating multiple data sources, domains and tools in urban energy models using semantic technologies. In *Proceedings of the 10th European Conference on Product and Process Modelling*, pages 837–844. CRC Press, 2014.

358. Alvaro Sicilia, German Nemirovski, and Andreas Nolle. Map-On: A web-based editor for visual ontology mapping. *Semantic Web*, 8(6):969–980, 2017.

359. Leslie F. Sikos. A novel ontology for 3D semantics: ontology-based 3D model indexing and content-based video retrieval applied to the medical domain. *International Journal of Metadata, Semantics and Ontologies*, 12(1):59–70, 2017.

360. D. Singh, C. Vishnu, and C. K. Mohan. Visual Big Data Analytics for Traffic Monitoring in Smart City. In *15th IEEE International Conference on Machine Learning and Applications (ICMLA)*, pages 886–891, 2016.

361. James Sinopoli. *Smart Building Systems for Architects, Owners and Builders*. Elsevier, 2010.

362. S. Skolthanarat, U. Lewlomphaisarl, and K. Tungpimolrut. Short-term load forecasting algorithm and optimization in smart grid operations and planning. In *2014 IEEE Conference on Technologies for Sustainability (SusTech)*, pages 165–171, 2014.

363. A. Slominski, V. Muthusamy, and R. Khalaf. Building a Multi-tenant Cloud Service from Legacy Code with Docker Containers. In *2015 IEEE International Conference on Cloud Engineering*, pages 394–396, 2015.

364. D. Snoonian. Smart buildings. *IEEE Spectrum*, 40(8):18–23, 2003.

365. Albert T.p. So and K. C. Wong. On the quantitative assessment of intelligent buildings. *Facilities*, 20:208–216, May 2002.

366. Ali Hassan Sodhro, Sandeep Pirbhulal, Zongwei Luo, and Victor Hugo C. de Albuquerque. Towards an optimal resource management for IoT based Green and sustainable smart cities. *Journal of Cleaner Production*, 220:1167–1179, 2019.

367. Ranjith K. Soman, Miguel Molina-Solana, and Jennifer K. Whyte. Linked-Data based Constraint-Checking (LDCC) to support look-ahead planning in construction. *Automation in Construction*, 120:103369, December 2020.

368. Manu Sporny, Dave Longley, GRegg Kellogg, Markus Lanthaler, Pierre-Antoine Champin, and Niklas Londström. JSON-LD 1.1. Technical report, W3C, 2014.

369. Jagjit Singh Srai, Ettore Settanni, Naoum Tsolakis, and Kaur Aulakh. Supply Chain Digital Twins: Opportunities and Challenges Beyond the Hype. In *Proceedings of the 23rd Cambridge International Manufacturing Symposium*, pages 26–27, 2019.

370. William Stallings. *Data and computer communications*. Pearson Education, Inc., Upper Saddle River, NJ, USA, eight edition, 2007.

371. Sander Stolk and Kris McGlinn. Validation of IfcOWL datasets using SHACL. In *Proceedings of the 8th Linked Data in Architecture and Construction Workshop*, CEUR Workshop Proceedings, volume 2636, pages 91–104, 2020.

372. E. Stroulia, M. El-Ramly, P. Sorenson, and R. Penner. Legacy systems migration in CELLEST. In *Proceedings of the 2000 International Conference on Software Engineering. ICSE 2000 the New Millennium*, page 790, 2000.

373. Eleni Stroulia, Mohammad El-Ramly, and Paul Sorenson. From legacy to web through interaction modeling. In *Proceedings of the 2002 International Conference on Software Maintenance*, pages 320–329. IEEE, 2002.

374. Ruben Taelman, Miel Vander Sande, and Ruben Verborgh. Bridges between GraphQL and RDF. In *W3C Workshop on Web Standardization for Graph Data. W3C*, 2019.

375. J. Tandy, L. van den Brink, and P. Barnaghi. Spatial data on the web best practices. Technical report, W3C, September 2017. W3C Working Group Note.

376. B. Tang, Z. Chen, G. Hefferman, S. Pei, T. Wei, H. He, and Q. Yang. Incorporating Intelligence in Fog Computing for Big Data Analysis in Smart Cities. *IEEE Transactions on Industrial Informatics*, 13(5):2140–2150, 2017.

377. Shu Tang, Dennis R. Shelden, Charles M. Eastman, Pardis Pishdad-Bozorgi, and Xinghua Gao. A review of building information modeling (BIM) and the internet of things (IoT) devices integration: Present status and future trends. *Automation in Construction*, 101:127–139, 2019.

378. Fei Tao, Jiangfeng Cheng, Qinglin Qi, Meng Zhang, He Zhang, and Fangyuan Sui. Digital twin-driven product design, manufacturing and service with big data. *The International Journal of Advanced Manufacturing Technology*, 94(9-12):3563–3576, February 2018.

379. Fei Tao, He Zhang, Ang Liu, and A. Y. C. Nee. Digital Twin in Industry: State-of-the-Art. *IEEE Transactions on Industrial Informatics*, 15(4):2405–2415, April 2019.

380. Fei Tao and Meng Zhang. Digital Twin Shop-Floor: A New Shop-Floor Paradigm Towards Smart Manufacturing. *IEEE Access*, 5:20418–20427, 2017.

381. Fei Tao, Meng Zhang, and A. Y. C. Nee. Background and Concept of Digital Twin. In *Digital Twin Driven Smart Manufacturing*, pages 3–28. Elsevier, 2019.

382. W. Terkaj and P. Pauwels. OWL ontology file for the IFC4 ADD1.exp EXPRESS schema, 2015.

383. Terkaj, Walter and Pauwels, Pieter. A method to generate a modular ifcOWL ontology. In *Proceedings of the Joint Ontology Workshops 2017 Episode 3: The Tyrolean Autumn of Ontology*, volume 2050, page 12, 2017.

384. Montbel Thibaud, Huihui Chi, Wei Zhou, and Selwyn Piramuthu. Internet of Things (IoT) in high-risk Environment, Health and Safety (EHS) industries: A comprehensive review. *Decision Support Systems*, 108:79–95, 2018.

385. David Thorpe and Ebrahim Parvaresh Karan. Method for calculating schedule delay considering weather conditions. In *Proceedings of the 24th annual conference of the Association of Researchers in Construction Management (ARCOM 2008)*, volume 2, pages 809–818. Association of Researchers in Construction Management (ARCOM), 2008.

386. Martin Tomko and Stephan Winter. Beyond digital twins – A commentary. *Environment and Planning B: Urban Analytics and City Science*, 46(2):395–399, February 2019.

387. Kiril Tonev, Simon Kappe, Preslava Krahtova, Hendro Wicaksono, and Jivka Ovtcharova. District-Scale Data Integration by Leveraging Semantic Web Technologies: A Case in Smart Cities. In Christophe Debruyne, Hervé Panetto, Georg Weichhart, Peter Bollen, Ioana Ciuciu, Maria-Esther Vidal, and Robert Meersman, editors, *On the Move to Meaningful*

Internet Systems. OTM 2017 Workshops, volume 10697 of *Lecture Notes in Computer Science*, pages 289–292, Cham, 2018. Springer International Publishing.

388. Mesut Toğaçar, Burhan Ergen, and Zafer Cömert. Application of breast cancer diagnosis based on a combination of convolutional neural networks, ridge regression and linear discriminant analysis using invasive breast cancer images processed with autoencoders. *Medical Hypotheses*, 135:109503, 2020.

389. Eric J. Tuegel, Anthony R. Ingraffea, Thomas G. Eason, and S. Michael Spottswood. Reengineering Aircraft Structural Life Prediction Using a Digital Twin. *International Journal of Aerospace Engineering*, 2011:1–14, 2011.

390. Shreshth Tuli, Nipam Basumatary, Sukhpal Singh Gill, Mohsen Kahani, Rajesh Chand Arya, Gurpreet Singh Wander, and Rajkumar Buyya. HealthFog: An ensemble deep learning based Smart Healthcare System for Automatic Diagnosis of Heart Diseases in integrated IoT and fog computing environments. *Future Generation Computer Systems*, 104:187–200, 2020.

391. Žiga Turk. Interoperability in construction–Mission impossible? *Developments in the Built Environment*, 4:100018, 2020.

392. F. M. Ugliotti, M. Dellosta, and A. Osello. BIM-based energy analysis using edilclima ec770 plug-in, case study archimede library EEB project. *Procedia engineering*, 161:3–8, 2016.

393. Pedro Valderas, Victoria Torres, and Vicente Pelechano. A microservice composition approach based on the choreography of BPMN fragments. *Information and Software Technology*, 127:106370, 2020.

394. Léon van Berlo. BIM bots – (Summer 2016), 2016.

395. Léon van Berlo, Peter Willems, and Pieter Pauwels. Creating Information Delivery Specifications using Linked Data. In Bimal Kumar, Farzad Rahimian, David Greenwood, and Timo Hartmann, editors, *Proceedings of the 36th CIB W78 2019 Conference*, pages 647–660, Northumbria, UK, 2019.

396. P. van den Brom, A. Meijer, and H. Visscher. Performance gaps in energy consumption: household groups and building characteristics. *Building Research & Information*, 46(1):54–70, 2018.

397. J. S. Van der Veen, B. Van der Waaij, and R. J. Meijer. Sensor data storage performance: SQL or NoSQL, physical or virtual. In *IEEE Fifth International Conference on Cloud Computing*, pages 431–438, 2012.

398. Sander van Nederveen, R. Beheshti, and P. Willems. Building Information Modelling in the Netherlands: A Status Report. In *Proceedings of the 18th CIB World Building Congress*, volume 361, pages 28–40, Salford, United Kingdom, 2010. CIB.

399. Jean-Philippe Vasseur and Adam Dunkels. *Interconnecting smart objects with IP: The next internet*. Morgan Kaufmann, 2010.

400. Ruben Verborgh, Miel Vander Sande, Olaf Hartig, Joachim Van Herwegen, Laurens De Vocht, Ben De Meester, Gerald Haesendonck, and Pieter Colpaert. Triple Pattern Fragments: A low-cost knowledge graph interface for the Web. *Journal of Web Semantics*, 37-38:184–206, 2016.

401. Peter C. Verhoef, Thijs Broekhuizen, Yakov Bart, Abhi Bhattacharya, John Qi Dong, Nicolai Fabian, and Michael Haenlein. Digital transformation: A multidisciplinary reflection and research agenda. *Journal of Business Research*, 122:889–901, 2021.

402. Rebekka Volk, Julian Stengel, and Frank Schultmann. Building Information Modeling (BIM) for existing buildings – Literature review and future needs. *Automation in construction*, 38:109–127, 2014.

403. Anna Wagner. *Linked product data : describing multi-functional and parametric building products using semantic web technologies*, volume 1/2020 of *Berichte des Instituts für Numerische Methoden und Informatik im Bauwesen*. Shaker Verlag, Düren, July 2020.

404. Anna Wagner, Mathias Bonduel, Pieter Pauwels, and Uwe Rüppel. Representing construction-related geometry in a semantic web context: A review of approaches. *Automation in Construction*, 115:103130, 2020.

405. Anna Wagner, Mathias Bonduel, Pieter Pauwels, and Rüppel Uwe. Relating geometry descriptions to its derivatives on the web. In *Proceedings of the European Conference on Computing in Construction (EC3 2019)*, pages 304–313, Chania, Greece, 2019.

406. Anna Wagner and Uwe Rüppel. BPO: The building product ontology for assembled products. In María Poveda-Villalón, Pieter Pauwels, Rui De Klerk, and Ana Roxin, editors, *Proceedings of the 7th Linked Data in Architecture and Construction Workshop (LDAC), CEUR Workshop Proceedings*, volume 2389, pages 106–119, Lisbon, Portugal, 2019.

407. Yun Wang, Sudha Ram, Faiz Currim, Ezequiel Dantas, and Luiz Alberto Sabóia. A big data approach for smart transportation management on bus network. In *IEEE 2nd International Smart Cities Conference: Improving the Citizens Quality of Life, ISC2 2016 – Proceedings*, page 7580839. Institute of Electrical and Electronics Engineers Inc., September 2016.

408. Greg Ward and Rob Shakespeare. *Rendering with Radiance: the art and science of lighting visualization*. Morgan Kaufmann Publishers, 1998.

409. Jeroen Werbrouck, Pieter Pauwels, Jakob Beetz, and Léon van Berlo. Towards a decentralised common data environment using linked building data and the solid ecosystem. In *36th CIB W78 2019 Conference*, pages 113–123, 2019.

410. Jeroen Werbrouck, Pieter Pauwels, Mathias Bonduel, Jakob Beetz, and Willem Bekers. Scan-to-graph: Semantic enrichment of existing building geometry. *Automation in Construction*, 119:103286, 2020.

411. Jeroen Werbrouck, Madhumitha Senthilvel, Jakob Beetz, and Pieter Pauwels. Querying heterogeneous linked building data with context-expanded graphql queries. In Maria Poveda-Villalón, Pieter Pauwels, Rui De Klerk, and Ana Roxin, editors, *Proceedings of the 7th Linked Data in Architecture and Construction Workshop*, CEUR Workshop Proceedings, pages 21–34. CEUR-WS.org, 2019.

412. H. Wicaksono, S. Rogalski, and E. Kusnady. Knowledge-based intelligent energy management using building automation system. In *2010 Conference Proceedings IPEC*, pages 1140–1145, 2010.

413. Hendro Wicaksono. Achieving global Sustainable Development Goals in Industry 4.0 Context through Collaborative Innovation. In *International Conference on Computing and Applied Informatics 2020*. OSF Preprints, December 2020.

414. Hendro Wicaksono. The roles of data management and analytics in industry 4.0 ecosystems, July 2020.

415. Hendro Wicaksono, Kiril Tonev, and Preslava Krahtova. Linked Data for Data Integration based on SWIMing Guideline: Use Cases in DAREED Project. In *4th International Workshop on Linked Data in Architecture and Construction (LDAC)*, pages 21–22, Madrid, Spain, 2016.

416. Willow Inc. Unlocking a new era of smart buildings through the digital twin, October 2019.

417. Tim Winter, Pascal Thubert, Anders Brandt, J. Hui, Richard Kelsey, Philip Levis, Kris Pister, Rene Struik, J. Vasseur, and R. Alexander. RPL: IPv6 routing protocol for low power and lossy networks, RFC 6550. Technical Report IETF RFC 6550, Internet Engineering Task Force (IETF), 2012.

418. J. K. W. Wong, H. Li, and S. W. Wang. Intelligent building research: a review. *Automation in Construction*, 14(1):143–159, 2005.

419. I. Wu and Chi-Chang Liu. A visual and persuasive energy conservation system based on BIM and IoT technology. *Sensors*, 20(1):139, 2020.

420. Junfeng Xie, F. Richard Yu, Tao Huang, Renchao Xie, Jiang Liu, Chenmeng Wang, and Yunjie Liu. A survey of machine learning techniques applied to software defined

networking (SDN): Research issues and challenges. *IEEE Communications Surveys & Tutorials*, 21(1):393–430, 2018.

421. Jinying Xu, Ke Chen, Anna Elizabeth Zetkulic, Fan Xue, Weisheng Lu, and Yuhan Niu. Pervasive sensing technologies for facility management: A critical review. *Facilities*, 2019.

422. J. K. Yates. *Productivity improvement for construction and engineering*. ASCE Press, 2014.

423. Raymond Yee. *Pro Web 2.0 mashups: remixing data and web services*. Apress, 2008.

424. Tan Yigitcanlar, Md. Kamruzzaman, Laurie Buys, Giuseppe Ioppolo, Jamile Sabatini-Marques, Eduardo Moreira da Costa, and JinHyo Joseph Yun. Understanding 'smart cities': Intertwining development drivers with desired outcomes in a multidimensional framework. *Cities*, 81:145–160, 2018.

425. Hyung-Jun Yim, Dongmin Seo, Hanmin Jung, Moon-Ki Baek, InA Kim, and Kyu-Chul Lee. Description and classification for facilitating interoperability of heterogeneous data/events/services in the Internet of Things. *Neurocomputing*, 256:13–22, 2017.

426. S. Yoon, Y. Yu, J. Wang, and P. Wang. Impacts of HVACR temperature sensor offsets on building energy performance and occupant thermal comfort. *Building Simulation*, 12(2):259–271, 2019.

427. Meng-Lin Yu and Meng-Han Tsai. ACS: Construction Data Auto-Correction System – Taiwan Public Construction Data Example. *Sustainability*, 13(1):362, 2021.

428. Baris Yuce, Monjur Mourshed, and Yacine Rezgui. A Smart Forecasting Approach to District Energy Management. *Energies*, 10(8), 2017.

429. Baris Yuce, Yacine Rezgui, and Monjur Mourshed. ANN-GA smart appliance scheduling for optimised energy management in the domestic sector. *Energy and Buildings*, 111:311–325, January 2016.

430. N. R. Yusupbekov, A. R. Marakhimov, H. Z. Igamberdiev, and Sh X. Umarov. An Adaptive Fuzzy-Logic Traffic Control System in Conditions of Saturated Transport Stream. *Scientific World Journal*, 2016, 2016.

431. Chad J. Zack, Conor Senecal, Yaron Kinar, Yaakov Metzger, Yoav Bar-Sinai, R. Jay Widmer, Ryan Lennon, Mandeep Singh, Malcolm R. Bell, Amir Lerman, and Rajiv Gulati. Leveraging Machine Learning Techniques to Forecast Patient Prognosis After Percutaneous Coronary Intervention. *JACC: Cardiovascular Interventions*, 12(14):1304–1311, 2019.

432. Iqra Zafar, Farooque Azam, Muhammad Waseem Anwar, Bilal Maqbool, Wasi Haider Butt, and Aiman Nazir. A novel framework to automatically generate executable web services from BPMN models. *IEEE Access*, 7:93653–93677, 2019.

433. Marijana Zekić-Sušac, Saša Mitrović, and Adela Has. Machine learning based system for managing energy efficiency of public sector as an approach towards smart cities. *International Journal of Information Management*, 58(102074), June 2020.

434. Chi Zhang, Jakob Beetz, and Bauke de Vries. BimSPARQL: Domain-specific functional SPARQL extensions for querying RDF building data. *Semantic Web Journal*, 9(6):829–855, 2018.

435. J. P. Zhang, Q. Liu, F. Q. Yu, Z. Z. Hu, and W. Z. Zhao. A framework of cloud-computing-based BIM service for building lifecycle. In *Computing in Civil and Building Engineering*, pages 1514–1521, Orlando, Florida, 2014.

436. Jun-Feng Zhao and Jian-Tao Zhou. Strategies and methods for cloud migration. *international Journal of Automation and Computing*, 11(2):143–152, 2014.

437. Yu Zheng, Sen Yang, and Huanchong Cheng. An application framework of digital twin and its case study. *Journal of Ambient Intelligence and Humanized Computing*, 10(3):1141–1153, 2019.

438. K. Zhou, C. Fu, and S. Yang. Big data driven smart energy management: From big data to big insights. *Renewable and Sustainable Energy Reviews*, 56:215–225, 2016.

439. Álvaro Sicilia, Gonçal Costa, Vincenzo Corrado, Alice Gorrino, and Fulvio Corno. A Semantic Decision Support System to optimize the energy use of public buildings. *CIB W78 conference*, pages 676–685, 2015.

440. Lin, Y.C. and Cheung, W.F., 2020. Developing WSN/BIM-based environmental monitoring management system for parking garages in smart cities. *Journal of Management in Engineering*, 36(3), p.04020012.

441. Service Delivery Protocols – DNS-SD, mDNS, uPnP and Simple Discovery Service Protocol: IOT Part 10. Engineers Garage. https://www.engineersgarage.com/service-delivery-protocols-dns-sd-mdns-upnp-and-simple-discovery-service-protocol-iot-part-10/ Last accessed on April 20, 2022.

442. Sreedhar Pelluru, et. al. AMQP 1.0 in Azure Service Bus and Event Hubs protocol guide. November 9, 2021. https://docs.microsoft.com/en-us/azure/service-bus-messaging/service-bus-amqp-protocol-guide Last accessed on April 20, 2022.

443. Allen Starke, Keerthiraj Nagaraj, Cody Ruben, Nader Aljohani, Sheng Zou, Arturo Bretas, Janise McNair, and Alina Zare. Cross-Layered Distributed Data-driven Framework For Enhanced Smart Grid Cyber-Physical Security. arXiv preprint arXiv:2111.05460. 2021 Nov 10. pp. 1-22.

Index